HANDBOOK OF ENVIRONMENTAL ENGINEERING

Volume 4
Water Resources and
Natural Control Processes

HANDBOOK OF
ENVIRONMENTAL ENGINEERING

Volume 1: Air and Noise Pollution Control

Volume 2: Solid Waste Processing and Resource Recovery

Volume 3: Biological Treatment Processes

Volume 4: Water Resources and Natural Control Processes

Volume 5: Physicochemical Technologies for Water and Wastewater Treatment

HANDBOOK OF ENVIRONMENTAL ENGINEERING

Volume 4

Water Resources and Natural Control Processes

Edited by
Lawrence K. Wang
*Lenox Institute for Research Inc.
Lenox, Massachusetts*

and

Norman C. Pereira
*Monsanto Company
St. Louis, Missouri*

The HUMANA Press · Clifton, New Jersey

Library of Congress Cataloging-in Publication Data

Water resources and natural control processes.

(Handbook of enviromental engineering ; v. 4)
includes bibliographies and index.
1. Water quality management--Handbooks, manuals,
etc. 2. Water, Underground--Management--Handbooks,
manuals, etc. 3. Waste disposal in the ground--
Handbooks, manuals, etc. I. Wang, Lawrence K.
II. Pereira, Norman C. III. Series.
TD170.H37 vol. 4 [TD365]   628 s   86-18632
ISBN 0-89603-059-8   [363.7'3947]

© 1986 The HUMANA Press Inc. · Crescent Manor · P.O. Box 2148 · Clifton, NJ 07015

All rights reserved. No part of this book may be reproduced, stored in a retrieval system, or transmitted in any form or by any means, electronic, mechanical, photocopying, microfilming, recording, or otherwise, without written permission from the Publisher.

Printed in the United States of America

Preface

The past few years have seen the emergence of a growing, widespread desire in this country, and indeed everywhere, that positive actions be taken to restore the quality of our environment, and to protect it from the degrading effects of all forms of pollution—air, noise, solid waste, and water. Since pollution is a direct or indirect consequence of waste, if there is no waste, there can be no pollution, and the seemingly idealistic demand for "zero discharge" can be construed as a demand for zero waste. However, as long as there is waste, we can only attempt to abate the consequent pollution by converting it to a less noxious form. In those instances in which a particular type of pollution has been recognized, three major questions usually arise: (1) How serious is the pollution? (2) Is the technology to abate it available? and (3) Do the costs of abatement justify the degree of abatement achieved? The principal intention of this series of books on environmental engineering is to help the reader formulate useful answers to the second and third of these questions, i.e., to outline the best currently available engineering solutions, and to examine their costs in the light of the real level of benefits afforded.

The traditional approach of applying tried-and-true solutions to specific pollution problems has been a major factor contributing to the success of environmental engineering, and in large measure has accounted for the establishment of a "methodology of pollution control." However, realizing the already great complexity of current environmental problems, and understanding that, as time goes on, these issues will become even more complex and interrelated, render it imperative that intelligent planning of pollution abatement systems be undertaken. Prerequisite to such planning is an understanding of the performance, potential, and limitations of the various methods of pollution abatement available for environmental engineering. In this series of books, we are reviewing at a practical tutorial level a broad spectrum of engineering systems (processes, operations, and methods) currently being utilized, or of potential utility, for such pollution abatement. We believe that the unification of the concepts and engineering methodology found in these books is a logical step in the evolution of environmental engineering.

The treatment of the various engineering systems presented will show how an engineering formulation of the subject flows naturally from the fundamental principles and theory of chemistry, physics, and mathematics. This emphasis on fundamental science is based on the recognition that engineering practice has of necessity in recent years become more firmly based on scientific principles, rather than depending so heavily on our empirical accumulation of facts, as was earlier the case. It was not intended, though, to neglect empiricism where such data lead quickly to the most economic design; certain engineering systems are not readily amenable to fundamental scientific analysis, and in these instances we have resorted to less science in favor of more art and empiricism.

Since an engineer must understand science within a context of applications, we first present the development of the scientific basis of a particular subject, followed by exposition of the pertinent design concepts and operations, and detailed explanations of their applications to environmental quality control or improvement. Throughout, methods of practical design calculation are illustrated by numerical examples. These examples clearly demonstrate how organized, analytical reasoning leads to the most direct and clear solutions. Wherever possible, pertinent cost data have been provided.

Our treatment of pollution-abatement engineering is offered in the belief that the trained engineer should more firmly understand fundamental principles, be more aware of the similarities and/or differences among many of the engineering systems, and exhibit greater flexibility and originality in the definition and innovative solution of environmental pollution problems. In short, the environmental engineer ought by conviction and practice be more readily adaptable to change and progress.

Coverage of the unusually broad field of environmental engineering has demanded an expertise that could only be provided through multiple authorship. Each author (or group of authors) was permitted to employ, within reasonable limits, the customary personal style in organizing and presenting a particular subject area, and consequently it has been difficult to treat all subject material in a homogeneous manner. Moreover, owing to limitations of space, some of the authors' favored topics could not be treated in great detail, and many less important topics had to be merely mentioned or commented on briefly. In addition, treatment of some well-established operations, such as distillation and solvent extraction, has been totally omitted. All of the authors have provided an excellent list of references at the end of each chapter for the benefit of the interested reader. Each of the chapters is meant to be self-contained and consequently some mild repetition among the various texts was unavoidable. In each case, all errors of omission or repetition are the responsibility of the editors and not the individual authors. With the current trend toward metrication, the question of using a consistent system of units has been a problem. Wherever possible the authors have used the British System (fps), along with the metric equivalent (mks, cgs, or SIU), or vice versa. The authors sincerely hope that this inconsistency of units usage does not prove to be disruptive to the reader.

The series has been organized in five volumes:
  I. Air and Noise Pollution Control
 II. Solid Waste Processing and Resource Recovery
III. Biological and Natural Control Processes
 IV. Solids Separation and Treatment
  V. Physicochemical Technologies for Water and Wastewater Treatment

As can be seen from the above titles, no consideration is given to pollution by type of industry, or to the abatement of specific pollutants. Rather, the above categorization has been based on the three basic forms in which pollutants and waste are manifested: gas, solid, and liquid. In addition, noise pollution control is included in Volume 1.

This Engineering Handbook is designed to serve as a basic text as well as a comprehensive reference book. We hope and expect it will prove of equal high value to advanced undergraduate or graduate students, to designers of pollution abatement systems, and to research workers. The editors welcome comments from readers. It is our hope that these volumes will not only provide information on the various pollution abatement technologies, but will also serve as a basis for advanced study or specialized investigation of the theory and practice of the individual engineering systems covered.

The editors are pleased to acknowledge the encouragement and support received from their colleagues at the Environmental and Energy Systems Department of Calspan Corporation during the conceptual stages of this endeavor. We wish to thank the contributing authors for their time and effort, and for having borne patiently our numerous queries and comments. Finally, we are grateful to our respective families for their patience and understanding during some rather trying times.

LAWRENCE K. WANG
Lenox, Massachusetts
NORMAN C. PEREIRA
St. Louis, Missouri

Contents

Preface v
Contributors xvii

CHAPTER 1

SURFACE WATER QUALITY ANALYSIS *1*

CLARK C. K. LIU

    I. Introduction 1
   II. Mathematical Simulation of the Surface Water System . 3
       A. Water Quality Model Formulation............... 3
       B. Transport and Transformation 6
       C. Mathematical Model and Computer Application ... 7
       D. Modeling Procedures 8
  III. Oxygen Consumption and Replenishment in Receiving Waters 11
       A. Biochemical Decay of Carbonaceous Waste Materials 11
       B. Nitrification................................ 15
       C. Photosynthesis and Respiration 18
       D. Sediment Oxygen Demand..................... 19
       E. Atmospheric Reaeration 21
  IV. Coliform Bacteria Die-Off 21
   V. Modeling Application in Surface Water Quality........ 24
       A. River Analysis 24
       B. Estuary Analysis............................. 40
       C. Lake Analysis............................... 44
      Nomenclature...................................... 49
      References 53

Appendix A	56
Appendix B	57
Appendix C	58

Chapter 2

WATER QUALITY CONTROL OF TIDAL RIVERS AND ESTUARIES 61

Mu Hao Sung Wang and Lawrence K. Wang

I. Introduction	61
II. Water Quality Parameters	62
A. Conservative Substances	62
B. Reactive Substances	62
C. Sequentially Reactive Constituents	62
D. Surface Transfer Coefficient, Reaeration Coefficient, and Deoxygenation Coefficient	63
E. Assimilation Ratio	63
F. Estuarine Number	64
G. Dispersion Coefficient	65
H. Advective Velocity	66
III. Basic Mathematical Models	66
IV. Working Models	70
A. Steady-State Equations for Waste Concentrations in Tidal Rivers and Estuaries Resulting from a Point Source of Pollution	70
B. Steady-State Equations for Waste Concentrations in Tidal Rivers and Estuaries Resulting from a Distributed Source of Pollution	74
C. Alternate Working Models and Systems Identification for Tidal Rivers	78
V. Practical Examples	84
Nomenclature	103
References	105
Appendix	106

Chapter 3

COOLING OF THERMAL DISCHARGES 107

William W. Shuster

I. Introduction	107

II.	Cooling Ponds	108
	A. Mechanism of Heat Dissipation	108
	B. Design of Cooling Lakes	109
III.	Cooling Towers	123
	A. Mechanism of Heat Dissipation	123
	B. Definitions	125
	C. Types of Towers	126
	D. Problems Associated with Cooling Tower Operations	134
	E. Costs	136
	References	138

Chapter 4

CONTROL OF RESERVOIRS AND LAKES *139*

Donald B. Aulenbach

I.	Introduction	139
II.	Special Features of Water	143
III.	Hydrology	146
IV.	Evaporation	157
V.	Transpiration	161
VI.	Evapotranspiration	161
VII.	Infiltration and Percolation	163
VIII.	Runoff	164
IX.	Groundwater	182
X.	Impact of Pollution on Lakes	193
XI.	Thermal Impacts on the Aquatic Environment	200
XII.	Toxics in Water Resources	207
XIII.	Goals of Water Pollution Control	210
	References	213

Chapter 5

DEEP-WELL DISPOSAL *215*

Charles W. Sever

I.	Introduction	215
II.	Basic Well Designs	217
III.	Evaluation of a Proposed Injection Well Site	222

	A. Confinement Conditions	223
	B. Potential Receptor Zones	224
	C. Subsurface Hydrodynamics	226
IV.	Potential Hazards—Ways to Prevent, Detect, and Correct Them	229
	A. Fluid Movement During Construction, Testing, and Operation of the System	230
	B. Failure of the Aquifer to Receive and Transmit the Injected Fluids	230
	C. Failure of the Confining Layer	231
	D. Failure of an Individual Well	233
	E. Failures Because of Human Error	233
V.	Economic Evaluation of a Proposed Injection Well System	234
VI.	Use of Injection Wells in Wastewater Management	235
	A. Reuse for Engineering Purposes	236
	B. Injection Wells as a Part of the Treatment System	236
	C. Storage of Municipal Wastewaters for Reuse	237
	D. Storage of Industrial Wastewaters	238
	E. Disposal of Toxic Wastewaters	238
	F. Disposal of Radioactive Wastes	239
	G. Disposal of Municipal and Industrial Sludges	241
VII.	Protection of Usable Aquifers	241
	A. Pathway 1: Migration of Fluids Through a Faulty Injection Well Casing	242
	B. Pathway 2: Migration of Fluids Upward Through the Annulus Between the Casing and the Well Bore	243
	C. Pathway 3: Migration of Fluids from an Injection Zone Through the Confining Strata	244
	D. Pathway 4: Vertical Migration of Fluids Through Improperly Abandoned and Improperly Completed Wells	245
	E. Pathway 5: Lateral Migration of Fluids from Within an Injection Zone into a Protected Portion of that Strata	249
	F. Pathway 6: Direct Injection of Fluids into or Above an Underground Source of Drinking Water	252
VIII.	Nomenclature	252
IX.	Practical Examples	254
	References	258

Chapter 6

CHEMICAL CONTROL OF PESTS AND VECTORS 261

Lenore S. Clesceri

    I. Introduction 261
    II. Pests and Poisons 261
   III. Control of Organisms Pathogenic to Humans 263
   IV. Chemical Control of Vector Organisms 264
        A. Vertebrates 264
        B. Arthropods. 265
    V. Chemical Control of Organisms Destructive or
       Pathogenic to Plants 268
   VI. Chemical Control of Nuisance Organisms 270
        A. Terrestrial Plants 270
        B. Aquatic Plants 271
        C. Arthropods 272
        D. Vertebrates 273
  VII. Pollution from Chemical Poisons 274
        A. Zone of Influence 274
        B. Biological Magnification 276
        C. Toxic Effects on Human Health 277
 VIII. Alternatives to Chemical Poisoning 278
        A. Prevention Programs 278
        B. Conclusion 281
        References 281

Chapter 7

MANAGEMENT OF RADIOACTIVE WASTES 283

Donald B. Aulenbach and Robert M. Ryan

    I. Introduction 283
        A. Historical 283
        B. Effects of Radioactivity on Matter 285
        C. Effects of Radioactivity on Humans 287
        D. Energy Relations 290
        E. Magnitude of the Problem 294
    II. Sources of Radioactive Wastes 298
        A. Nuclear Fuel Cycle 298

xiv CONTENTS

	B. Research, Development, and Commercial Applications	315
	C. Medical	315
III.	Transport Mechanisms	324
	A. Air	324
	B. Water	326
	C. Concentration	328
IV.	Waste Management	332
	A. Principles of Treatment	332
	B. Plan for Waste Management	335
	C. Methods of Treatment	336
	D. Ultimate Disposal	341
V.	Surveillance	352
	A. Monitoring	352
VI.	Protection	356
	A. Principles	356
	B. Population Protection	357
VII.	Costs	359
	A. Various Treatment and Disposal Methods	359
	B. Relation to Benefits	361
VIII.	Anticipated Future Problems	363
	A. Long-Lived Isotopes	363
	B. High-Temperature Gas-Cooled Reactors	364
	C. Liquid-Metal Fast-Breeder Reactors	366
	D. Krypton-85	368
	E. Fusion	368
IX.	Summary	369
	References	370

Chapter 8

DRYING AND EVAPORATION PROCESSES *373*

George P. Sakellaropoulos

I.	Introduction	373
II.	Natural Dewatering	374
	A. Drying Sand Beds	375
	B. Drying Lagoons	385
III.	Thermal Drying	386
	A. Tray Dryers	388
	B. Rotary Dryers	389
	C. Flash Dryers	390

D.	Spray Dryers	393
E.	Toroidal Dryers	394
F.	Multiple Hearth Furnaces	395
IV.	Evaporation	397
V.	Theory of Drying	400
A.	Water Evaporation and Thermal Drying	400
B.	Theory of Gravity Dewatering	423
VI.	Theory of Evaporation	432
A.	Heat Transfer	432
B.	Heat and Material Balance	434
C.	Multiple-Effect Evaporators	436
	Nomenclature	440
	References	443

Chapter 9

DREDGING OPERATIONS AND WASTE DISPOSAL 447

Lawrence K. Wang

I.	Introduction	447
II.	Type of Dredges	448
A.	Clamshell Dredge	448
B.	Dipper Dredge	449
C.	Pipeline Dredge	449
D.	Hopper Dredge	450
III.	Sources of Pollution from Dredging Operations	451
IV.	Dredge Disposal and Environmental Enhancement Alternatives	457
A.	Open Water Disposal	457
B.	Creating New Land in Diked Disposal Areas	459
C.	Wildlife Enhancement in Diked Disposal Areas	461
D.	Additional Disposal Alternatives	464
E.	Environmental Enhancement by Modifications of Dredging Schedules and Equipment	465
V.	Dredge Transport Alternatives	465
A.	Long Distance Piping	466
B.	Mechanical Transport Methods	467
VI.	Dredge Containment Alternatives	467
A.	Retaining Dike with Stone, Sand, or Carbon Filter	468
B.	Retaining Dike with Sluices	470
C.	Retaining Dike with Divider	472
VII.	Dredge Treatment Alternatives	472

	A. Processes Available for Sediment–Water Separation .	473
	B. Processes Available for Treatment of Spoil Water. .	474
VIII.	Quality Criteria of Treated Sediment and Water.	483
IX.	Economic Aspects of Dredging Treatment	485
X.	Conclusions and Summary	486
XI.	Practical Examples	488
	References	490
	Index	493

Contributors

DONALD B. AULENBACH • *Department of Chemical and Environmental Engineering, Rensselaer Polytechnic Institute, Troy, New York*

LENORE S. CLESCERI • *Department of Biology, Rensselaer Polytechnic Institute, Troy, New York*

CLARK C. K. LIU • *Nanyang Technological Institute, Singapore*

ROBERT M. RYAN • *Division of Radiation and Nuclear Safety, Rensselaer Polytechnic Institute, Troy, New York*

GEORGE P. SAKELLAROPOULOS • *Department of Chemical Engineering, University of Thessaloniki, Thessaloniki, Greece*

CHARLES W. SEVER • *US Environmental Protection Agency, Dallas, Texas*

WILLIAM W. SHUSTER • *Department of Chemical and Environmental Engineering, Rensselaer Polytechnic Institute, Troy, New York*

LAWRENCE K. WANG • *Lenox Institute for Research Inc., Lenox, Massachusetts*

MU HAO SUNG WANG • *Department of Environmental Conservation, State of New York, Albany, New York*

1
Surface Water Quality Analysis

Clark C. K. Liu
Nanyang Technological Institute, Singapore

I. INTRODUCTION

The objective of the 1972 Federal Water Pollution Control Act Amendments is to restore and maintain the chemical, physical, and biological integrity of the Nation's waters [45]. In order to achieve this objective, the Act requires each State to establish specific water quality standards for natural waters within its boundary. Furthermore, the Act establishes nationwide minimum effluent limits for all dischargers. In situations in which these minimum treatment levels are not sufficient to prevent water quality contravention, the Act provides for the imposition of additional treatment.

The US Environmental Protection Agency (EPA) issued guidelines [39] that require all surface waters be:
(a) Free from substances attributable to municipal, industrial, or other discharges, or agricultural deposits that will settle to form objectionable sludge deposits.
(b) Free from floating debris, scum, and other floating material attributable to municipal, industrial, or other discharges, or agricultural practices producing amounts sufficient to be unsightly or deleterious.
(c) Free from materials attributable to municipal, industrial, or other discharges, or agricultural practices producing color,

odor or other conditions in such degree as to create a nuisance.
(d) Free from substances attributable to municipal, industrial, or other discharges or agricultural practices in concentrations or combinations that are toxic or harmful to human, animal, plant, or aquatic life.

Surface waters are often classified according to potential beneficial usages. In the guidelines, EPA suggested water quality criteria for fresh surface waters based on two use classifications. Appendix A summarizes the quality criteria associated with each of these classifications and Appendix B gives examples of additional quality parameters pertinent to both.

A number of states classify their fresh surface waters into more classes. In New York State, for instance, detailed water quality criteria have been established for six distinct classes of fresh waters. Within most of these classes, the dissolved oxygen standard varies, depending on the type of fish found in the waters (see Appendix C).

Every natural water body has its own capacity to assimilate wastes; this is referred to as the water's self-purfication potential. Pollution is both a relative and a subjective term. Indiscriminate waste discharges cannot and should not be tolerated, while a complete prohibition of all waste discharges, without taking into account the natural self-purification of a water body, is economically infeasible and environmentally unjustified. Consequently, effluent limitations must be established based on a detailed waste assimilative capacity analysis of the receiving water. Because the waste assimilative capacity of a water body is influenced by a variety of hydrodynamic, biochemical, and meteorological factors, its characteristics value is difficult to establish. As a result, the use of mathematical models in surface water quality analysis has gained popularity in recent years.

A mathematical model is a set of equations that describes the dynamic response of an actual system under different conditions. In other words, without looking at every component of the actual system in detail, modeling analysis provides a needed link in the cause–effect relationship. Although "modeling" gained its eminence in practical applications only after the arrival of the computer era, it had been a major cornerstone in the development of modern science [29].

In surface water quality analysis, a mathematical model for a natural water body, i.e., a river, estuary, or lake, is often used to simulate the processes of hydrodynamic transport and biological, chemical, and physical transformations. Model structure may be simple or complex, depending on data availability and the desired level of accuracy.

Data used in the development of a model are, ideally, derived from several intensive field surveys designed to define the particular kinetics sought and conducted under different field conditions. In general, conditions of a water body and its waste loadings during the intensive surveys do not match exactly with those that have been selected for allowable loading determinations. Therefore, one of the most important characteristics of a water quality model is its predictability; a model must be able to provide the desired water quality information under a varying waste loading and hydrodynamic conditions.

The following sections will introduce the theoretical basis of model formulation, estimations of model parameters, and model applications in various types of surface water quality analysis.

Although the modeling applications illustrated in this Chapter are mainly for the investigation of organic wastes and corresponding dissolved oxygen (DO) depression in freshwater systems, these modeling techniques should be equally useful in the analysis of transport and mixing of various toxic substances in aquatic environment when sufficient knowledge on their reaction kinetics becomes available.

II. MATHEMATICAL SIMULATION OF THE SURFACE WATER SYSTEM

A. Water Quality Model Formulation

A general water quality model with transport and transformation components can be readily derived based on the mass conservation principle and Fick's diffusion law.

Considering a small volume V in a free flowing water system bounded by surface S (Fig. 1), the principle of mass conservation dictates that the time rate of change of a total mass within control the volume must be equal to the rate of total net mass flow through this volume, plus any internal sources and sinks. It is expressed mathematically as follows:

$$-\frac{\partial M}{\partial t} = \int_s \bar{F}\bar{n}ds \qquad (1)$$

where
M = Total mass of the transported substance

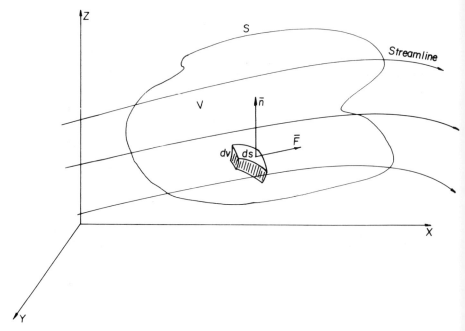

Fig. 1. Control volume in a flow field.

$\bar{F}$ = Flux vector or mass flow rate
$\bar{n}$ = Unit vector normal to the surface element ds
t = Time

Let the concentration of a pollutant substance be $C(x, y, z, t)$ or $M = \int_V C dV$ the left hand side of Eq. (1) becomes,

$$-\frac{\partial M}{\partial t} = \int_V \frac{\partial C}{\partial t} dV \qquad (2)$$

By the divergence theorem, the surface integral on the right hand side of Eq. (1) can be replaced by a volume integral

$$\int_S \bar{F} \bar{n} ds = \int_V \nabla \bar{F} dV \qquad (3)$$

Where $\nabla \bar{F}$ denotes a divergence operation of vector $\bar{F}$

With these changes, Eq. (1) is simplified as

$$\int_V \left(\frac{\partial C}{\partial t} + \nabla \bar{F} \right) dV = 0 \qquad (4)$$

Since V is an arbitrary volume in the flow field, the integrand must vanish at every point, giving the equation of mass conservation:

$$\frac{\partial C}{\partial t} = -\nabla \bar{F} \qquad (5)$$

Equation (5) is derived relative to a conservative substance. If a nonconservative substance is encountered, an additional term would be required such that:

$$\frac{\partial C}{\partial t} = -\nabla \bar{F} + \Sigma S \qquad (6)$$

Where S represents the rate of internal mass production and reduction. The terms of conservative substance and nonconservative substance are defined in the next subsection.

In a natural water system the rate of total mass flow, $\bar{F}$ is determined by advection as well as dispersion.

Advection is a temporal change of pollutants following the bulk motion of the water. For pollutant constituents with concentrations $C(x, y, z, t)$, advection is expressed by

$$\bar{F}_A = \bar{U}C$$

where $\bar{U}$ is the velocity vector, and $\bar{F}_A$ is the flux caused by the bulk motion of the water.

Essentially, all natural water flows are turbulent. Random temporal fluctuation of velocity in a flow field causes turbulent diffusion. The detailed physical basis and statistical properties of turbulent diffusion are yet to be developed. Fortunately, in surface water quality analysis, the classical theory of turbulent diffusion was found satisfactory [24]. By a semiempirical approach, this theory introduced a coefficient of eddy diffusion that expresses the flux arising from turbulent diffusion analogous to that of diffusivity defined in Fick's Law of molecular diffusion:

$$\bar{F}_T = A\nabla C \qquad (7)$$

where

$\bar{F}_T$ = Flux caused by turbulent diffusion
$\bar{A}$ = Eddy diffusion coefficient

Equation (7) was formulated by assuming a flow field with stationary and homogeneous turbulence. In other words, the velocity fluctuates only with respect to time, but is uniformly distributed across the entire flow field. In most natural water systems, however, velocity and turbulent intensity change markedly in both vertical and horizontal directions. Thus, the influence of mean velocity gradients on the movement and spread of pollutants must be superimposed onto the temporal phenomenon of turbulent diffusion. This new form of dispersion was first discussed by Taylor [33], and was sometimes called turbulent dispersion or simply dispersion (7). In simulation of an estuarine system, tidal mixing may also be included in the dispersion term.

On the basis of the above discussions, the rate of total mass flow in any control volume inside a flow field is described mathematically by:

$$\bar{F} = \bar{F}_A + \bar{F}_T = \bar{U}C - \bar{E}\nabla C \qquad (8)$$

where $\bar{F}$ = Total mass flux
$\bar{E}$ = Dispersion coefficient

By introducing the total mass flux representation of Eq. (8) into Eq. (6), the mass conservation principle is represented by an expanded form

$$\frac{\partial C}{\partial t} = -\nabla(\bar{U}C) + \nabla\bar{E}\nabla C + \Sigma S \qquad (9)$$

Equation (9) is a second-order partial differential equation, constituting the general form of a three-dimensional time-variable water quality model. It may be expressed in terms of cartesian coordinates as

$$\frac{\partial C}{\partial t} = -\frac{\partial(U_x C)}{\partial x} - \frac{\partial(U_y C)}{\partial y} - \frac{\partial(U_z C)}{\partial z} + \frac{\partial}{\partial x}\left(\bar{E}_x \frac{\partial C}{\partial x}\right) + \qquad (10)$$

$$\frac{\partial}{\partial y}\left(\bar{E}_y \frac{\partial C}{\partial y}\right) + \frac{\partial}{\partial z}\left(\bar{E}_z \frac{\partial C}{\partial z}\right) + \Sigma S$$

In practice, water quality modeling is often conducted in one or two dimensional formulations, requiring system simplification by verticle and/or horizontal averaging. The term longitudinal dispersion, as associated with a one- or two-dimensional model, consists not only of the turbulent dispersion phenomenon as discussed above, but also a derivative involving the spatial mean values of the three-dimensional eddy diffusivity.

B. Transport and Transformation

Equation (9) indicates that the concentration of pollutant constituents at any location in a water body varies according to the nature of hydrodynamic transport in terms of advection and turbulent dispersion, and the nature of biochemical transformation in terms of constituent production and decay.

Limited hydrodynamic information is required in the analysis of a simple water system. For example, the traditional approach to stream analysis developed by Streeter and Phelps [32] investigated the combined effects of organic waste decay and atmospheric reaeration on the dissolved oxygen (DO) content. Hydrodynamic information required for any stream segment would be simply a constant velocity and a constant water depth, which can be obtained from previous hydrologic investigations of the stream, or by conducting conjunctive field measurements. Thus, the governing equation is simplified to a first-order ordinary differential equation. Its derivation will be presented in a later subsection.

In the analysis of more complex river systems or estuaries, however, processes of hydrodynamic transport are often unsteady. Variables such as flow velocity and water surface elevation or tidal elevation can only be obtained by a simultaneous solution of a hydrodynamic model consisting of the continuity and momentum equations for the water system. The continuity equation is derived by the application of the mass conservation principle to the water in the flow field. The equation of momentum is the application of Newton's law to fluid motion; it describes the temporal and spatial status of hydrodynamic variables as a result of external forces, i.e., gravitational force and shear force at boundaries. Solution of hydrodynamic equations to compute the velocity field will not be discussed in great detail in this chapter. Some aspects of hydrodynamic simulation are illustrated later in a subsection.

Pollutant constituents in a receiving water may be conservative or nonconservative. Nonconservative substances are those that decay as a result of bacterial decomposition or that are otherwise removed because of chemical reaction. Nonconservative substances commonly considered in water quality analyses are organic materials, coliform bacteria, other biochemical oxygen demanding materials (BOD), and so on. Conservative substances include salts, chlorides and total phosphorus. They are "conservative" in the sense that they do not easily decay in the water over time.

The modeling of conservative substances is relatively simple because their distribution is mainly influenced by hydrodynamic transport. For nonconservative substances, however, impacts of both transport and transformation must be investigated.

Several major processes of biochemical transformation will be presented in other sections.

C. Mathematical Model and Computer Application

As water quality modeling analysis extends to more complex water bodies, the mathematical structure of these models becomes increasingly sophisticated. A general system equation, Eq. (9), takes the form of a second-order partial differential equation. This type of equation, because of its widespread application to engineering problems, is not limited simply to water quality analysis [48]. Analytical solutions available to these equations, e.g., the Laplace and Fourier transforms, or separation of variables, can be applied only to a very special class of problems. In engineering practice, approximation methods are used that give acceptable solutions by different limiting processes. In water quality modeling, one of the most popular approximation methods is the finite difference method.

Depending on specific problems in water quality modeling analysis, the use of the finite difference method on the governing equation may lead to a set of algebraic equations with hundreds of unknowns. Their solutions can be feasible only with the aid of a high-speed digital computer. That is why the term "computer model" has been used indistinguishably with the term "mathematical model."

In numerical calculations by means of a high-speed digital computer, a step-by-step, or flow charting, procedure that translates a particular engineering problem into a machine "program" must be utilized. For a full appreciation of the techniques employed, the reader may refer to the quoted literature [3].

D. Modeling Procedures

Although the conceptual framework of water quality modeling is well-founded scientifically, application of a model to specific water quality analyses will likely be unsuccessful, or of limited value, unless certain procedures are followed. Unfortunately, many past modeling efforts were attempted without the availability of a good and adequate data base. These kinds of modeling efforts have unfortunately been used in modeling critiques to discredit the role of modeling in water quality management as a whole. It is fair to say that the failure of modeling practitioners to achieve a reasonable validation of a model before its application is at least partially responsible for these critiques.

A successful modeling program in water quality analysis must include at least the following steps (Fig. 2):

1. Mathematical Formulation

Depending on the nature of the receiving water and the objective of the water quality analysis, a mathematical structure is selected for model formulation. This mathematical structure must be sufficiently detailed to

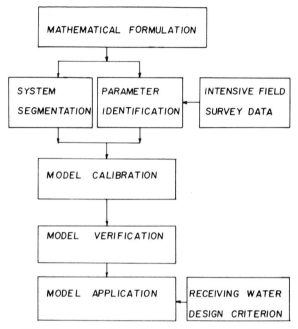

Fig. 2. Modeling procedures in surface water quality analysis.

describe the dynamic responses relative to the desired water quality analysis, and yet simple enough to allow successful calibration and verification within the limits of available resources and field data.

2. System Segmentation

For numerical computation it is necessary to segment a water system into many small volumes, or geometric grids. Solutions are obtained through simultaneous application of mass conservation equations within these grids. An intelligent system segmentation provides the basis for more efficient numerical computation on a high-speed digital computer.

For a complex water system requiring a sequential hydrodynamic and water quality simulation, a complete understanding of the hydrodynamic characteristics demands more detailed geometric grids than would be necessary for normal water quality simulation alone. Thus, it is often advisable to provide separate, more detailed hydrodynamic grids, the solutions of which may then be incorporated into the water quality model [35].

3. Parameter Identification

In a water quality model, biochemical transformation processes are represented by certain simplified kinetics. Reaction rate coefficients as well as the hydrodynamic factors that are included in these kinetic equations should be developed by intelligent interpretation of the field-derived data. Because of the complexity of a particular water system and the variability of its environmental conditions, developing a range of coefficient values may prove to be more appropriate than attempting to define a single value.

4. Model Calibration

Once the estimated range of values of both transport and transformation parameters are defined, they are plugged into the model, which is then solved to obtain predicted values of the water quality parameter(s) of concern, e.g., dissolved oxygen content. Model calibration is completed when these computed values are reasonably close to the observed ones.

5. Model Verification

After successful calibration, the model is then subjected to test with field data from an independent intensive water quality survey conducted on the same water body. This is necessary to verify the model's predictive capability.

6. Model Projection

Those effluent limitations that are imposed on wastewater treatment facilities in order to meet the applicable quality standards in the receiving waters are established by projecting the model's response to critical conditions of stream flow and water temperature. For a free-flowing stream, the MA7CD/10 (or 7Q/10) flow condition, which is the statistically derived minimum average streamflow for seven consecutive days occurring once in ten years, is often adopted. Water temperature during such critical flow situations are normally assumed unless, of course, more precise information is available.

Because of practical limitations, one cannot normally expect to conduct intensive field work precisely during the period wherein the defined critical conditions manifest themselves; however, if the field work is to be at all useful for modeling purposes, such conditions must at least be approached. At any rate, streamflows that are somewhat higher and temperatures that are somewhat lower than those expected under critical conditions are normally encountered in the field. In addition, the waste loadings that are encountered may not be representative of "design" conditions.

III. OXYGEN CONSUMPTION AND REPLENISHMENT IN RECEIVING WATERS

Except for a number of anaerobic microorganisms, dissolved oxygen (DO) is required for the respiration of all aquatic life forms. Its content depicts a water's ability to maintain and propagate a balanced ecological system. With human or other organic wastes being discharged into a water way, one of the most important signs of water quality degradation is the depression of its oxygen content. Therefore, dissolved oxygen content has classically been the most popular index in water quality management. Most of the municipal and industrial wastewater treatment facilities have been built mainly to reduce oxygen demanding organic materials.

In addition to oxygen demanding organic materials (BOD), DO in a receiving water also depends on a variety of other environmental factors. In this section, these factors are discussed with an emphasis on evaluating rate coefficients of DO-related mechanisms. These rate coefficients can be directly used as parameters in water quality modeling analysis.

A. Biochemical Decay of Carbonaceous Waste Materials

Biochemical oxygen demand (BOD) in wastewater consists of both carbonaceous and nitrogenous materials. The BOD of carbonaceous materials (CBOD) is normally exerted first because of a "lag" in the growth of the nitrifying bacteria necessary for oxidation of the nitrogen form (NBOD) [12]. In the Ohio River Study by Streeter and Phelps [32] and many subsequent works, only CBOD was considered.

CBOD exertion is commonly assumed to be a first-order decay process

$$\frac{dL_c}{dt} = K_1 L_c \qquad (11)$$

where L_c = Carbonaceous BOD remaining at time t (mg/L or lb/d).
K_1 = Laboratory CBOD deoxygenation coefficient (L/d).

Example 1

An effluent sample from a municipal wastewater treatment plant was analyzed in the laboratory and showed long term BOD exertions as below. Determine the CBOD deoxygenation coefficient, K_1.

Day	Oxygen consumption, mg/L	BOD remaining, mg/L
1	8	102
3	28	82
5	36	74
7	41	69
10	48	62
14	61	49
17	79	31
21	100	10
28	110	0

Laboratory data of BOD series are plotted on log-normal paper (Fig. 3). Two straight lines are fitted, representing the first and second stages of BOD decay. As mentioned previously, CBOD decay takes place first. So, the deoxygenation coefficient, K_1, is the slope of the first straight line, or

$$K_1 = \frac{1}{4.8} \log \left(\frac{6}{4.5}\right) = 0.026 \text{ d}^{-1} \text{ (base 10)} = 0.06 \text{ d}^{-1} \text{ (base } e)$$

It is noted that the value of K_1 base e is 2.3 times that of base 10.

In the natural environment, hydraulic and other physicochemical conditions are somewhat different from those in the laboratory; therefore, CBOD deoxygenation in a stream normally exhibits different, usually higher, rates from those observed in the laboratory. This field rate, or K_d, can be determined from the data collected from sequential sampling sites during intensive water quality surveys.

An empirical equation was developed [35] that provides a relationship between K_d and K_1

$$K_d = K_1 + n\left(\frac{U}{h}\right)$$

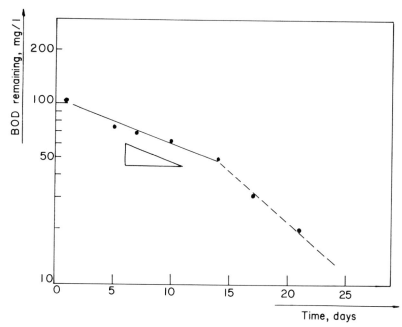

Fig. 3. Results of laboratory long term BOD test. The solid line indicates the carbonaceous stage of BOD decay.

where K_d = Stream deoxygenation rate coefficient, d^{-1}, base e
K_1 = Laboratory deoxygenation rate coefficient, d^{-1}, base e
U = Stream velocity, ft/s
h = Stream depth, ft
n = Coefficient of stream bed activity

Coefficient n is taken as a step function of stream slope, as shown in Table 1.

In natural waters, CBOD may be removed by actions of sedimentation and biological absorption, in addition to deoxygenation. In such a situation, a new rate coefficient, K_r, is defined such that

$$K_r = K_3 + K_d \qquad (13)$$

where

K_r = Stream removal rate coefficient of CBOD, d^{-1} (base e)
K_3 = Rate coefficients representing other factors of CBOD reduction, d^{-1} (base e)

TABLE 1
Change in Coefficient of Bed Activity
by Stream Slope

Stream slope, ft/mile	n
2.5	0.1
5.0	0.15
10.0	0.25
25.0	0.40
50.0	0.60

Example 2

An intensive water quality survey was conducted in a small polluted river. Figure 4 shows the profile of BOD_5 in the river below a municipal wastewater treatment plant outfall. Average flow velocity in this river reach is estimated at 2.6 miles/d. Determine stream CBOD decay rate coefficients.

Figure 4 shows that immediately downstream from the plant outfall there was a sharp reduction of CBOD, this represents the combined effects of deoxygenation, sedimentation, and biological absorption. Further downstream, a significantly smaller reduction rate was observed, representing mainly CBOD deoxygenation. So, K_r and K_d are the slopes of these two straight lines, respectively.

$$K_r = \text{slope of straight line A} \times \text{average flow velocity} = \frac{\ln 2.0}{2.9} \times 2.6$$

$$= 0.62 \text{ d}^{-1} \text{ (base } e\text{)}$$

$$K_d = \text{slope of straight line B} \times \text{average flow velocity} = \frac{\ln 1.2}{2.7} \times 2.6$$

$$= 0.18 \text{ d}^{-1} \text{ (base } e\text{)}$$

Rate of CBOD decay is temperature dependent. Values used at different water temperatures can be determined with the following relationship:

$$K_d^T = K_d^{20} \theta_c^{(T-20)} \qquad (14)$$

where K_d^T = CBOD deoxygenation rate at temperature $T°C$

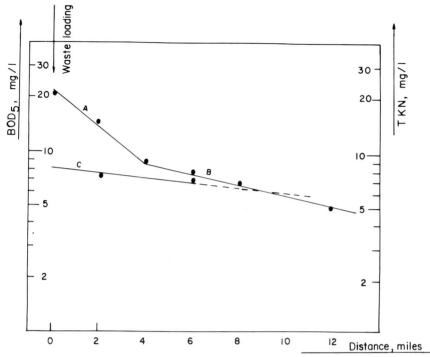

Fig. 4. Distribution of BOD_5 and TKN in a receiving stream from a single waste input. Line A indicates the reduction of BOD_5 in a stream from sedimentation, bio-absorption, and deoxygenation. Line B indicates the reduction of BOD_5 in a stream from deoxygenation alone. Line C indicates the reduction of TKN in a stream from nitrification.

K_d^{20} = CBOD deoxygenation rate at temperature 20°C
θ_c = Temperature correction factor. Approximately, θ_c = 1.047 [36]

B. Nitrification

Organic nitrogen and ammonia in wastewater, in the presence of nitrifying bacteria, undergo a series of chemical reactions in a receiving water. During these processes, dissolved oxygen is consumed. In the early studies of Streeter [31] he recognized the existence of nitrogenous BOD, but because of the "lag," it was considered insignificant in the downstream reaches close to the sources of wastes, where the greatest

DO depletion occurred. It was later reported that in a relatively shallow rapidly flowing stream, early nitrification may take place. This is especially true for treated waste effluent [8]. Therefore, in waste assimilative capacity analyses for a receiving water with the treated effluent inputs, both CBOD and NBOD should be included [22].

The process of nitrification is a sequential oxidation of ammonia to nitrite and finally to nitrate, involving nitrosomonas and nitrobacter, or,

$$2NH_3 + 3O_2 \xrightarrow{\text{Nitrosomonas}} 2NO_2^- + 2H_2O + 2H^+$$

$$2NO_2^- + O_2 \xrightarrow{\text{Nitrobacter}} 2NO_3^-$$

These stoichiometric relationships of oxygen to NO_2—N and to NO_3—N indicate that each mg/L of NH_3—N oxidized requires 4.57 mg/L of oxygen.

In the natural water environment, organic nitrogen is broken down to ammonia and then reduced to the form of nitrite and nitrate. The sum of organic and ammonia nitrogen is also called total Kjeldahl nitrogen, or TKN. In practice, nitrogenous biochemical oxygen demand (NBOD) is derived from multiplying the TKN by 4.57.

Example 3

Results of the laboratory analysis of the effluent sample used in Example 1 has an initial total Kjeldahl nitrogen (TKN) of 12 mg/L. Determine both CBOD and NBOD.
Long term BOD data were plotted on a normal scale paper (Fig. 5).
$$NBOD = 4.57 TKN = 54.84 \text{ mg/L}$$
As shown in Fig. 5, NBOD is the second stage of BOD. The first-stage BOD, or CBOD is the difference between ultimate BOD (BOD_{28}) and NBOD.
$$CBOD = 110 - 54.84 = 55.16 \text{ mg/L}$$

Nitrification is simulated as a first-order, single-stage process

$$\frac{dL_n}{dtc} = K_n L_n \qquad (15)$$

where L_n = Nitrogenous BOD remaining at time t, mg/L
K_n = Nitrogenous deoxygenation coefficient, d^{-1} (base e)
The rate of nitrification in a receiving water is extremely difficult to determine, because each component of total nitrogen is subject to envi-

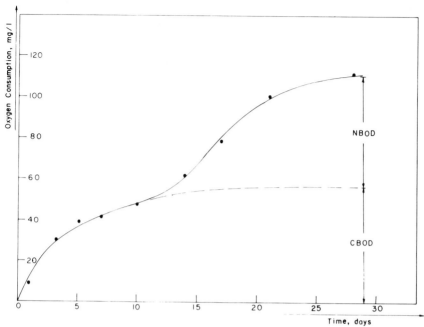

Fig. 5. Determination of the amounts of carbonaceous and nitrogenous BOD in an effluent sample.

ronmental processes other than nitrification. Ammonia and nitrate, for instance, may be utilized by aquatic plants and thus be removed from the water; organic nitrogen may also be produced because of algal decay. Therefore, K_n values determined from the in-stream TKN profile must be used with caution in the modeling analysis.

Temperature has a strong effect on nitrification rates. Within the temperature range of 10–30°C the temperature effects can be simulated by the following expression:

$$K_n^T = K_n^{20} \, \theta_n^{(T-20)} \tag{16}$$

where K_n^{20} = Nitrification rate coefficient at temperature, 20°C
θ_n = Temperature correction factor, approximately, θ_n = 1.08 [36]

Example 4

Using the profile of total Kjeldahl nitrogen (TKN) in a polluted river (Fig. 4), determine the nitrogenous deoxygenation coefficient, K_n.

$$K_n = \text{Slope of line C} \times \text{average velocity}$$
$$= \frac{\ln(6.9/6.0)}{2} \times 2.6 = 0.18 \text{ d}^{-1} \text{ (base } e\text{)}$$
$$= 0.18 \text{ d}^{-1} \text{ (base } e\text{)}$$

C. Photosynthesis and Respiration

In a natural water containing phytoplankton biomass and benthic plants, the processes of photosynthesis and respiration constitute the major source and sink of oxygen. Photosynthesis is the biological synthesis of organic compounds by chlorophyll-bearing plants in the presence of solar energy. A byproduct of this process is oxygen. On the other hand, oxygen is continually consumed by living organisms during the process of respiration.

The photosynthesis rate of aquatic plants changes with light intensity, and thus shows a pattern of diurnal variation. Respiration rate, on the other hand, is relatively constant. The US Environmental Protection Agency (EPA) [39] suggested that the optimum rate of photosynthesis and respiration rates in water are related to the chlorophyll-a concentration, in terms of the following equations:

$$P = 0.25 \text{ Chla} \qquad (17)$$

where P = Optimum rate of phytosynthesis, mg/L-d
Chla = Chlorophyll-a concentration, mg/L
and

$$r = m\text{Chla} \qquad (18)$$

where r = Respiration rate, mg/L-d
m = Constant ranging from 0.05 to 0.05, with 0.025 a common value.

With Eqs. (17) and (18), dirunal DO variations in a receiving stream can be simulated using a method developed by DiToro and O'Connor [44]. Equations (17) and (18) should be used only for preliminary estimations. The combined effect of photysynthesis and respiration (α) is expressed by

$$\alpha = P - r$$

Other environmental factors, such as nutrient level, turbidity, and the availability of sunlight, may also affect photosynthesis and respiration actions. Also, in natural water, rooted plants may contribute major biological activity.

In lake and estuary analysis the processes of photosynthesis and respiration have to be investigated in detail since they affect the biological productivity of aquatic lives, a major concern in water quality management. But, in modeling analyses of a stream, where the main concern is DO depression because of organic wastes, inclusion of the processes of photosynthesis and respiration is often not necessary, except during the model calibration.

As indicated previously, model calibration consists of the comparison of model output with observed field data. Since the processes of photosynthesis and respiration are time-dependent, they often produce significant diurnal DO fluctuations, even in a steady-state stream with constant waste loadings. Interpretation of diurnal DO data in the calibration of a steady-state stream model often demands some subjective engineering judgments. To resolve this weakness, a new method has been developed analytically by Liu [18] that separates time-varying effects from observed dissolved oxygen data before they are used in model calibration. As a result, both the stream water quality model and the dissolved oxygen data used in its calibration and verification are completely in the steady-state mode. This leads to a reasonable model formulation in the sense that evaluated model parameters represent the true hydrodynamic and biochemical behaviors of a stream.

According to this method, a stream's theoretical dissolved oxygen content in the absence of photosynthesis and respiration can be determined by the following relationship:

$$D_w = D_r - C_b(D_r + D_p) \tag{19}$$

where D_r and D_p = Maximum and minimum DO deficit in an observed DO diurnal curve, respectively

D_w = Theoretical value of the stream DO when photosynthetic and respiration actions are excluded.

C_b = Biological rate constant ranging from 0.1 to 0.3.

Figure 6 demonstrates the application of Eq. (19) in a biologically active stream. Values of D_w are computed and connected by a solid line, which serves as a reasonable data base for the calibration of a steady-state stream model.

D. Sediment Oxygen Demand

The muds found in the bottom of rivers, lakes, and estuaries often constitute a significant oxygen sink, absorbing large amounts of oxygen. These

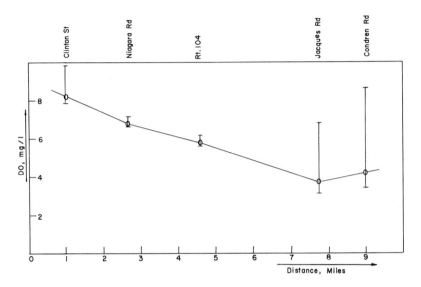

Fig. 6. Separation of photosynthesis and respiration effects from observed diurnal DO data. The circles indicate the stream DO after exclusion of photosynthesis and respiration action.

muds usually result from the decomposition of organic materials that accumulate over a period of time.

Ordinarily, sediment oxygen demand (SOD) in a water body is determined by collecting mud samples with a dredge. Samples are then transported to a laboratory for oxygen uptake measurements. However, this method has been criticized on the grounds that the reconstruction of natural layering of solids in a laboratory flask, a necessary condition for obtaining realistic results, is extremely difficult. As a compromise, a benthic respirometer capable of measuring the oxygen uptake rates of bottom mud *in situ* has been developed [19]. A benthic respirometer can trap and confine a volume of overlying water in contact with the bottom while observing any reduction in dissolved oxygen content resulting from uptake by the bottom mud.

In addition to the texture and composition of bottom muds, the rate of sediment oxygen demand is influenced by the overlying dissolved oxygen concentration. The rate of sediment oxygen demand in a river may range from 0.05 to 10.0 g/m²/d [36].

Sediment uptake rate in natural water also changes with water temperature. The relationship can be expressed as

$$K_S^T = K_S^{20} \theta_s^{(T - 20)} \qquad (20)$$

where K_S^T = Sediment oxygen uptake rate at temperature $T°C$, g/m²-d
K_S^{20} = Sediment oxygen uptake rate at temperature 20°C, g/m²-d
θ_S = Temperature correction factor for sediment oxygen demand rate.

The temperature correction factor, θ_s, has been determined in several previous studies. Its reported values range between 1.04 and 1.13 [35].

E. Atmospheric Reaeration

Oxygen in the atmosphere will enter a water body where a dissolved oxygen deficit exists. This natural process of oxygen transfer is called atmospheric reaeration. A water with a high potential of atmospheric reaeration can receive a large amount of organic wastes and still meet its minimum dissolved oxygen standard. Because of this phenomenon, the relative magnitudes of the rate of atmospheric reaeration and the rate of BOD decay has been called the self-purification factor of a receiving water [10].

Numerous studies have been conducted for accurate determinations of atmospheric reaeration rate. A field measurement technique was developed by Tsivoglou [37] that determines the stream reaeration rate based on observed concentration variations of gaseous tracers. This technique uses radioactive krypton as the gaseous tracer, tritium as the dispersion and dilution tracer, and Rhodamine-WT dye to determine when to sample for the radioactive tracers. A modified method not requiring radioactive tracer was later developed by Rathbun et al. [28]. When field determination is not feasible, the atmospheric reaeration rate may be estimated in terms of a receiving water's hydraulic characteristics.

Detailed discussions of the theoretical basis of atmospheric reaeration and its rate coefficient determination are presented in another chapter on Surface and Spray Aeration in this volume.

IV. COLIFORM BACTERIA DIE-OFF

Protection of human health against threats of pathogenic bacteria has traditionally been the most important task in water pollution control. Now even long after the elimination of typhoid fever epidemics, the danger of transmission of virus diseases through contaminated water still exists. In many instances, therefore, effluent disinfection is still deemed necessary.

There are a large variety of pathogenic bacteria, and the number of any particular species present in a polluted water are few. Consequently, the amount of coliform bacteria present is usually taken as an index of bacteria pollution. Densities of total and fecal coliforms are determined by analyzing water samples in a laboratory by either a multiple-tube fermentation procedure or by membrane filtration methods [1]. The multiple-tube procedure determines coliform concentration based on a statistical probability number of coliform organisms present in a given volume of water, expressed as the most probable number (MPN) per unit volume of water. The membrane filtration method, on the other hand, gives a direct organism count.

In 1974, the US Environmental Protection Agency formed a task force to develop the necessary background information on wastewater disinfection requirements and the use of chlorine. The task force had reviewed all aspects of wastewater disinfection with regard to public health and water quality requirements, toxic effects, and availability of alternate processes. The task force report [40] provided the following main findings:

1. Disinfection of sewage effluents does provide an effective means of reducing to a safe level the hazards of infectious disease in receiving waters. However, under certain circumstances and at some locations, the benefits of disinfection for the protection of public health are minimal and may not be needed. Furthermore, the reaction byproducts of certain disinfectants have been identified with potential health hazards; these properties must be considered when disinfection is practiced.
2. The toxic effects of total residual chlorine on freshwater organisms have been further confirmed at very low concentrations. Dechlorination greatly reduces or eliminates the toxicity caused by residual chlorine, its effect on reducing chlorinated organics is not known.
3. There are satisfactory alternate disinfection processes using ozone, bromine chloride, or ultraviolet light, that could be substituted in place of chlorination.

As a result of this task force report, the US Environmental Protection Agency amended the secondary treatment requirements by deleting fecal coliform limitations from the nationwide minimum level of wastewater treatment [39]. Instead, assimilative capacity analysis must be conducted to establish effluent limitations for coliform contents on a case-by-case basis.

In surface water quality analysis, the amount of coliform bacteria is often estimated on the basis of dilution and die-off. Besides, increases in coliform bacteria counts in wastewater effluents after they have mixed in the receiving waters have been observed [17]. However, the exact nature and extent of the bacterial "regrowth" in the receiving water is still unknown [16].

In modeling analysis, coliform die-off in a receiving water may be represented by a first-order decay process [41].

$$\frac{dB}{dt} = -K_b B \tag{21}$$

where B = Coliform residual after any time t, MPN/100 mL
K_b = Coliform die-off rate, d^{-1} (see Table 3)

Equation (21) denotes an exponential decay; in other words, the amount of coliform diminishes along a straight line on a log-normal paper. However, actual data frequently show nonlinear log-scale decays. As an alternative, a two-stage equation was proposed by Frost and Streeter [41] that assumes coliform organisms consist of two distinguishable types with different decay rates or,

$$B = B_o \exp(-K_B t) + B'_b \exp(-K'_B t) \tag{22}$$

where B_o, B'_o = Concentration of each of two hypothetical organism types, MPN/100 mL
K_b, K'_o = Decay rates for the two organism types, d^{-1}

Values of fractions B_o and B'_o with the corresponding decay rates were determined by Phelps based on Ohio River data (Table 4).

TABLE 3
Values for Coliform Specific Die-off Rates Used in Several Modeling Studies

System	K_b@20°C, d^{-1}	Reference
Various streams	0.010–3.504	Baca and Arnett (1976) [2]
Lake Ontario	0.480–1.992	US Army Corps of Engineers (1974) [37]
Lake Washington	4.80	Chen and Orlob (1975) [5]
Boise River, Idaho	0.480	Chen and Wells (1975) [6]
Long Island Estuaries, New York	0.480–7.992	Tetra Tech (1976) [34]

TABLE 4
Values of $B_o, B_o', K_b,$ and K_b'
From the Ohio River (Phelps, 1944) [27]

Parameter	Warm weather	Cold weather
B_o, % total concentration	99.51	97
K_b, d^{-1}, base e	1.075	1.165
B_o', % total concentration	0.49	3.0
K_b', d^{-1} base e	0.1338	0.0599

Example 5

A secondary municipal wastewater treatment plant has a design flow of 0.7 ft^3/s. Its effluent is discharged into a stream that is tributary to a large lake. Critical low streamflow is 3.2 ft^3/s, having a traveling time to the lake of 4 d. Lake water has been classified to be a water supply source and the total coliform content of the lake water must be less than 2400 MPN/100 mL. Coliform concentration of the existing effluent is 10^6 MPN/100 mL. Die-off rate was estimated to be 0.9 d^{-1} (base e) at 20°C. It is further assumed that no bacterial aftergrowth would take place in the receiving water. Determine the requirement of effluent disinfection.

Coliform concentration at mixing point =

$$\frac{10^6 \times 0.7}{0.7 + 3.2} = 179{,}500 \text{ MPN/100 mL}$$

from Eq. (21), residual concentration of the coliform at the mouth is,

$$B = B_o \exp(-K_B t) = 179{,}500 \ (e^{-0.9 \times 4})$$
$$= 179{,}500 \times 0.027 = 4850 \text{ MPN/100 mL}$$

It would be larger than the permitted amount, or 2400 MPN/100 mL; thus, effluent disinfection is required.

V. MODELING APPLICATION IN SURFACE WATER QUALITY ANALYSIS

A. River Analysis

The oxygen sag equation developed by Streeter and Phelps [32] in their Ohio River Water Pollution Study provides a basic relationship between oxygen depression and replenishment in a river or stream subjected to organic (point source) waste loadings. It has been the most popular tool in water quality analysis.

Analytically, the Streeter-Phelps equation can be derived directly from Eq. (10), which is the general formulation of water quality models. In the derivation, the arbitrary substance C in Eq. (10) is replaced by dissolved oxygen deficit D, which is the difference of saturation concentration of DO and the actual concentration. In addition, the following assumptions must be made:

1. The dispersion effect is minimal and thus advection is the only recognized transport mechanism, i.e., $E_x = E_y = E_z = 0$.
2. Stream flow is one dimensional, i.e., terms relative to the y and z coordinates can be dropped. In other words, the bulk motion of the water can be expressed simply by a longitudinal velocity U.
3. Stream system and waste loadings are in steady state, i.e., $\partial D/\partial t = 0$
4. Biochemical oxidation of organic wastes and atmospheric reaeration are the only recognized sink and source of stream dissolved oxygen

$$\Sigma S = K_d L + K_2 D$$

where K_d is the stream deoxygenation rate coefficient; L is the total biochemical oxygen demand in a receiving stream; and K_2 is the stream reaeration rate coefficient.

With these assumptions, Eq. (10) becomes

$$U\frac{dD}{dX} = K_r L - K_2 D \quad (22)$$

where D = DO deficit; U = longitudinal velocity of a one-dimensional stream flow; and X = downstream distance.

Direct integration of equation (22) leads to

$$D = \frac{K_r L_o}{K_2 - K_r}\left[\exp\left(\frac{-K_r X}{U}\right) - \exp\left(\frac{-K_2 X}{U}\right)\right] + D_o \exp\left(\frac{-K_2 X}{U}\right) \quad (23)$$

where D_o = DO deficit at the beginning point of the stream reach under investigation, mg/L
D = DO deficit at stream distance X, mg/L
L_o = BOD at the beginning point of the stream reach under investigation, mg/L

Because the stream flow is in steady state, the term X/U also denotes the time of travel t of a waste slug moving down the stream. So, DO deficit is often presented in terms of the time of travel,

$$D = \frac{K_d L_o}{K_2 - K_r}[\exp(-K_r t) - \exp(-K_2 t)] + D_o \exp(-K_2 t) \quad (24)$$

Organic waste materials discharged into a stream undergo a process of biochemical decay that utilizes the stream's dissolved oxygen. If the loading is sufficiently large, this leads to a pronounced dissolved oxygen deficit that, in turn, would induce the process of atmospheric reaeration. Finally, an equilibrium situation is reached such that the rates of oxygen consumption and replenishment balance each other. At this point, the dissolved oxygen deficit is at its maximum, and is called sag or critical DO. After that, more oxygen enters the stream because of reaeration than oxygen is taken out because of consumption, and the dissolved oxygen starts to recover. Equation (24) describes this kind of DO variation in a receiving stream and, therefore, is often called the oxygen sag equation.

The time to the critical DO point, or t_c is determined by differentiating Eq. (24) and setting the first derivative to zero, so that

$$t_c = \frac{1}{K_2 - K_r} \ln\left\{\frac{K_2}{K_r}\left[1 - \left(\frac{K_2 - K_r}{K_d}\frac{D_o}{L_o}\right)\right]\right\} \quad (25)$$

By assuming $K_r = K_d$ and replacing the term K_2/K_r with the stream self-purification coefficient "f" defined earlier, Eq. (25) can be simplified into,

$$t_c = \frac{f}{K_2(f-1)} \ln\left\{f\left[1 - (f-1)\frac{D_o}{L_o}\right]\right\} \quad (26)$$

Equation (26) indicates that the time required to reach the critical DO in a stream depends on the self-purification nature of the receiving stream, but not on the rate of organic waste decay.

At the DO sag point, oxygen consumption because of organic waste decay equals replenishment from atmospheric reaeration, so

$$K_d L_c = K_2 D_c \quad (27)$$

Biochemical demand of wastes (BOD) at any downstream point, L, and at the sag point, L_c, can be determined based on the first-order kinetics of BOD decay, or

$$L = L_o \exp(-K_d t) \quad (28)$$
$$L_c = L_o \exp(-K_d t_c) \quad (28a)$$

The stream's dissolved oxygen content at the sag point, or D_c can be derived by combining Eqs. (27) and (28a)

$$D_c = \frac{K_d L_c}{K_2} \exp(-K_d t_c) \qquad (29)$$

It indicates that, contrary to the time to sag, the value of critical dissolved oxygen depression is also a function of the organic wastes decay rate.

Example 6

A municipal wastewater treatment facility provides a secondary treatment. Its effluent flow, which averages 1.0 mgd on a daily average basis, exhibits an ultimate BOD strength of 135 mg/L and a DO of 5.0 mg/L. The receiving stream has a critical low flow, or MA7CD/10 of 5.0 ft³/s with a dissolved oxygen content of 8.5 mg/L and an ultimate BOD of 3.0 mg/L. Rate coefficients of BOD decay and reaeration were estimated to be 0.15 and 0.30 d^{-1}, respectively, both to the base e and at a temperature of 20°C. Determine the critical stream DO at the sag point and the time necessary to reach it. The water quality standards of this receiving stream require its DO to be greater than 3.0 mg/L at any time. Is advanced treatment required at this facility?

The stream water temperature during the critical flow period, for the purpose of this preliminary analysis, is also assumed to be 20°C. It should be noted that, in actual practice, higher water temperatures are normally assumed for critical flow analysis. In any event, for 20°C the dissolved oxygen saturation in the stream is 9.17 mg/L. Thus, the DO deficits associated with the effluent (D_e) and the stream water (D_s) are:

$$D_e = 9.17 - 5.0 = 4.17 \text{ mg/L}$$
$$D_s = 9.17 - 8.5 = 0.67 \text{ mg/L}$$

Since instant mixing in the receiving stream is normally asumed, the initial in-stream DO deficit (D_o) and BOD (L_o) can be determined by considering the mass balance at the mixing point.

$$D_o = \frac{1.55 \times 4.17 + 5.0 \times 0.67}{1.55 + 5.0} = 1.50 \text{ mg/L}$$

$$L_o = \frac{1.55 \times 135 + 5.0 \times 3.0}{1.55 + 5.0} = 34.24 \text{ mg/L}$$

Note that the effluent flow is normally given in terms of million gallons per day (MGD) which can be converted to ft³/s by multiplying by a factor of 1.55.

The time necessary to reach critical DO is determined by utilizing Eq. (25)

$$t_c = \frac{1}{K_2 - K_d} \ln\left\{\frac{K_2}{K_d}\left[1 - \frac{(K_2 - K_d)D_o}{K_d}\frac{}{L_o}\right]\right\}$$

$$= \frac{1}{0.3 - 0.15} \ln\left[\left(\frac{0.3}{0.15}\right)\left(1 - \frac{0.15 \times 1.50}{0.15 \times 34.24}\right)\right]$$

$$= 4.4$$

From Eq. (29), the critical DO deficit is determined to be:

$$D_c = \frac{K_d}{K_2} L_c \exp(-K_d t_c)$$

$$= \frac{0.15}{0.30} \times 34.24 \times \exp(-0.15 \times 4.40)$$

$$= 8.84 \text{ mg/L}$$

or critical DO is

$$\text{DO} = 9.17 - 8.84 = 0.33 \text{ mg/L} < 3.0 \text{ mg/L}$$

Thus, advanced wastewater treatment would be required.

The 1972 Federal Water Pollution Control Act Amendments [45] established minimum effluent limits, referred to as Best Practical Treatment (BPT), for all dischargers. For municipal treatment facilities, EPA has defined BPT as 85% removal of carbonaceous BOD and suspended solids. In this particular example, incidental to the CBOD removal, approximately 50% of the NBOD is also removed; thus, the treatment facility already exceeds the minimum treatment requirement. In any event, the foregoing preliminary analysis indicates that even more stringent effluent limits would be necessary if the stream DO standard is to be maintained.

Wastewater treatment facilities become very costly as higher levels of advanced treatment are imposed. As a result, there is an increasing demand for greater accuracy in stream analysis.

A number of more comprehensive stream models have been devleoped to meet these needs with many of them essentially being modified versions of the Streeter-Phelps formulation. These modified models consider a number of additional factors affecting dissolved oxygen balance in a receiving stream, while the streamflow remains one dimensional and steady state. Among new factors introduced are nitrification, sediment oxygen demand, and photosynthesis–respiration. Mechanisms associated with these factors have already been discussed in Section III.

Mathematical formulation of a typical water quality model based on a modified Streeter-Phelps equation is shown below [4]:

1. Distribution of carbonaceous oxygen demanding materials (CBOD) in a receiving stream

$$L_C = L_{cp} \exp(-K_r t) + L_{cd} \frac{1}{K_r}[1 - \exp(-K_r t)] \quad (30)$$

where L_C = in-stream CBOD, mg/L
L_{cp} = point source CBOD, mg/L
L_{cd} = distributed, or nonpoint source CBOD, mg/L
K_r = CBOD removal rate, d^{-1}, base e

2. Distribution of nitrogenous oxygen demanding materials (NBOD) in a receiving stream

$$L_N = L_{np} \exp(-K_n t) + L_{nd} \frac{1}{K_n}[1 - \exp(1 - K_n t)] \quad (31)$$

where L_N = in-stream NBOD, mg/L
L_{np} = point source NBOD, mg/L
L_{nd} = distributed, or nonpoint, source NBOD, mg/L
K_n = nitrification rate, d^{-1}, base e

3. Distribution of dissolved oxygen deficit (D) in a receiving stream.

$$D = D_1 + D_2 + D_3 + D_4 + D_5 + D_6 + D_7 \quad (32)$$

where

$$D_1 = D_o \exp(-K_2 t) \quad (32a)$$

$$D_2 = \frac{K_d L_{cp}}{K_2 - K_r}[\exp(-K_r t) - \exp(-K_2 t)] \quad (32b)$$

$$D_3 = \frac{K_n L_{np}}{K_2 - K_n}[\exp(-K_n t) - \exp(-K_2 t)] \quad (32c)$$

$$D_4 = \frac{K_d L_{cd}}{K_2 K_r}[1 - \exp(-K_2 t)] - \frac{K_d L_{cd}}{(K_2 - K_r)K_r}[\exp(-K_r t) - \exp(-K_2 t)] \quad (32d)$$

$$D_5 = \frac{L_{nd}}{K_2}[1 - \exp(-K_2 t)] - \frac{L_{nd}}{K_2 - K_n}[\exp(-K_n t) - \exp(-K_2 t)] \quad (32e)$$

$$D_6 = -\frac{\alpha}{K_2}[1 - \exp(-K_2 t)] \quad (32f)$$

$$D_7 = \frac{K_s}{K_2}[1 - \exp(-K_2 t)] \quad (32g)$$

where K_2 is the rate of atmospheric reaeration, α is the rate of net oxygen production from photosynthesis and respiration, and K_s is the rate of sediment oxygen demand. The elements of the deficit equation are as follows:

D_1 = point source DO deficit, initial value of DO deficit
D_2 = deficit due to point source of CBOD
D_3 = deficit due to point source of NBOD
D_4 = distributed source of deficit due to CBOD input with no significant addition to river flow
D_5 = distributed source of deficit due to NBOD input with no significant addition to river flow
D_6 = deficit due to distributed net algal oxygen production
D_7 = distributed deficit due to sdeiment oxygen demand effect

Point source in the Eqs. (32a), (32b), and (32c) refers to all input that occurs at the upstream end of the section to which the equation applies. This may include effluent point loads, minor tributary loads, and all input from the downstream ends of the sections directly upstream from the section in question. In other words, the total point source represents the boundary condition at the upstream end of the section.

Equations (30), (31), and (32) constitute the analytical basis of the stream water quality model SNSIM, developed by the US Environmental Protection Agency, Region II [44]. In actual solution, a river and its tributaries are segmented into numerous sections with constant hydrodynamic and biochemical characteristics. System equations are then applied to sections sequentially to determine their CBOD, NBOD, and DO deficit in relation to various point and nonpoint waste loadings.

Example 7

An intensive stream water quality survey was conducted in Eighteen Mile Creek, a tributary to Lake Ontario (Fig. 7). Nine sampling stations were selected for diurnal DO measurements. Composite samples were collected at six stations and were sent to a laboratory for the determination of long-term biochemical oxygen

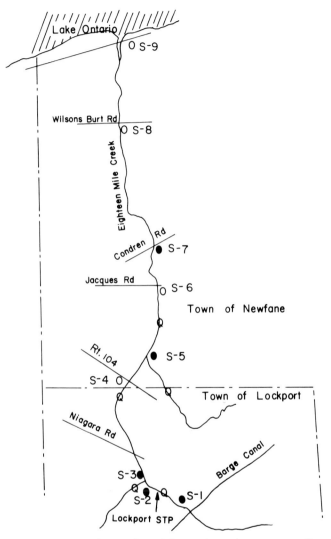

Fig. 7. Sampling stations selected for an intensive water quality survey in the Eighteen Mile Creek, New York. The circles indicate the diurnal DO stations. The dots indicate stations for both diurnal DO and composite sampling. Q indicates the streamflow measurement station.

demand (BOD) and nitrogen series, among others. Table 5 lists hydrodynamic data measured in the field. Rate coefficients shown in Table 5 were estimated, based on survey data, using methods presented in Section III.

Survey data show significant diurnal DO fluctuations in Eighteen Mile Creek, especially downstream from the Rt. 104 bridge. Theoretical values of creek DO when photosynthesis and respiration action were excluded were derived by Eq. (7). Results are shown on Fig. 6.

The SNSIM model [44] was implemented on a computer with the parameter values measured during the intensive survey (Table 6). Comparisons of the computed and observed DO in Eighteen Mile Creek (Fig. 8) show good agreement and suggest a reasonable simulation of the Creek's hydrodynamic and biochemical behavior.

The above example presents the details of a successful river model calibration. Next, the model must be operated for a completely new set of data to establish its validity.

After the calibration and verification in a particular river or stream, the water quality model is ready for the predictive analysis by introducing a set of design waste loadings and receiving water conditions. During this process, some of the model parameters need further modification. The atmospheric reaeration rate, for instance, depends on the hydraulic characteristics of the receiving stream; normally, different values must be used in the projection runs from those for calibration and verification runs because of variations in temperature and flow regime. Nevertheless, these modifications must be made on the basis of established relationships.

Example 8

The effluent from a small village sewage treatment plant must meet the receiving stream water quality standards, which require a DO above 3.0 mg/L at all times. The stream under design low flow conditions has a discharge of 0.25 ft^3/s. Village raw waste has a design flow of 0.3 mgd, which contains a CBOD concentration of 300 mg/L and an NBOD concentration of 180 mg/L. The decay rates, as determined by model calibration and verification, are $K_r = K_d = 0.40$ d^{-1} (base e), and $K_n = 0.23$ d^{-1} (base e), both at 20°C. It also was found that reaeration rates can be determined adequately for this stream by an energy dissipation formula developed by Tsivoglou [4], which has the form

TABLE 5
Results of Parameter Identification Based on a Water Quality Survey at Eighteen Mile Creek, New York

Section	Section length, mile	Discharge, ft^3/s	Velocity, ft/s	Depth, ft	Water temperature, °C	K_d, d^{-1} (base e)	K_r, d^{-1} (base e)	K_n, d^{-1} (base e)	K_2, d^{-1} (base e)
1	0.4	37.5	0.80	1.95	20.0	1.09	1.09	1.07	2.24
2	4.4	37.5	0.93	1.83	22.0	1.09	1.09	1.07	2.16
3	4.4	54.8	0.20	1.50	22.0	1.09	1.09	1.07	0.76

TABLE 6
Waste Characteristics; Eighteen Mile Creek Survey

Waste source	Discharge, ft³/s	CBOD, mg/L	NBOD, mg/L	DO, mg/L
Background condition	37.5	7.65	4.16	8.0
Lockport STP	13.0	1.50	14.90	5.2
The Gulf	4.3	4.95	24.7	3.4

$$K_2 = C_e \frac{H}{T_f}$$

where H = difference in water surface elevation between the beginning and end of the stream section, ft
T_f = time of water travel in the section, d
C_e = escape coefficient, ft^{-1}

Design conditions in the receiving water are indicated below:

TABLE 6 (continued)

Section	Length, mile	Velocity, ft/s	H, ft	Average depth, ft	Escape coefficient, ft^{-1}
1	0.5	0.1	15	0.3	0.07
2	1.0	0.15	30	0.3	0.07
3	0.5	0.15	20	0.3	0.07

Determine effluent limitations for oxygen demanding waste. First, reaeration rates of the three stream sections are determined under design conditions

Section 1

$$K_2 = \frac{\Delta H}{CT_f} = 0.07 \times \frac{15}{(0.5 \times 5280/0.1 \times 60^2 \times 24)} = 0.86 \text{ d}^{-1} \text{ (base } e\text{)}$$

Section 2

$$K_2 = \frac{\Delta H}{CT_f} = 0.07 \frac{30}{(1.0 \times 5280/0.15 \times 60^2 \times 24)} = 1.44 \text{ d}^{-1} \text{ (base } e\text{)}$$

Section 3

$$K_2 = \frac{\Delta H}{CT_f} = 0.07 \frac{20}{(0.5 \times 5280/0.15 \times 60^2 \times 24)} = 1.72 \text{ d}^{-1} \text{ (base } e\text{)}$$

The stream water temperature under design conditions is 24°C,

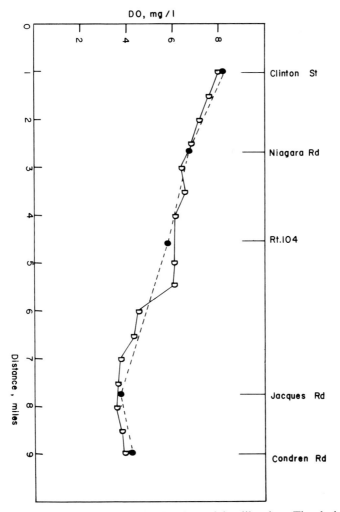

Fig. 8. Results of the Eighteen Mile Creek model calibration. The dashed line indicates observed DO and the solid line indicates computed DO.

therefore, decay rates determined in model calibration and verification must be modified, based on Eqs. (13) and (15),

$$K_r = K_d = 0.40 \times (1.40)^{24-20} = 0.481 \text{ d}^{-1} \text{ (base } e\text{)}$$
$$K_n = 0.23 \times (1.08)^{24-20} = 0.313 \text{ d}^{-1} \text{ (base } e\text{)}$$

Three wastewater treatment levels with increasing BOD removal efficiencies were used in the determination of an acceptable treatment level. Figure 9 gives the DO profiles as the result of these

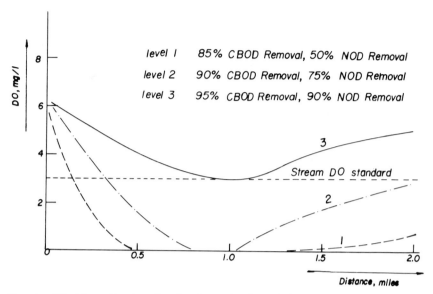

Fig. 9. Model application for the determination of the acceptable wastewater treatment level.

waste loadings; it indicates that a plant effluent with 95% CBOD removal and 90% NBOD removal will meet the stream DO standard. These correspond to effluent loadings of CBOD = 15 mg/L and NBOD = 18 mg/L.

In recent years, two-dimensional water quality analyses have been conducted in a number of rivers where the one-dimensional analyses are inadequate because of the complexities of these rivers' geometric configuration and hydrodynamic patterns. Complete solution of a two-dimensional river model often demands, in addition to the water quality simulation, a hydrodynamic simulation, in order to provide necessary information regarding the flow field.

A two-dimensional water quality simulation of a river system can be accomplished by assuming constant pollutant distribution across the river channel or along its depth. In the case of a vertical averaging model, concentration of a nonconservative substance may be described mathematically as follows:

$$W\frac{\partial C}{\partial t} = -W\frac{\partial (U_x C)}{\partial x} - W\frac{\partial (U_z C)}{\partial z} + \frac{\partial}{\partial x}\left(E_x W \frac{\partial C}{\partial x}\right) + \quad (33)$$

$$\frac{\partial}{\partial z}\left(E_z W \frac{\partial}{\partial z}\right) + W\Sigma S$$

where W is the width of the river.

The other parameters have the same definitions as those of Eq. (10).

Applying the finite difference method, Eq. (33) is transformed into a set of simultaneous algebraic equations. The exact form of these equations would depend on the water body under investigation and the particular system segmentation scheme adopted. Subsequently, these equations can be solved on a computer by employing one of the matrix inversion techniques, such as the Gauss-Seidel iteration method.

Waste assimilative capacity analysis in the Lower Genesee River conducted by Hydroscience Inc. [15] is a typical example of two-dimensional modeling analysis. The geometric configuration and hydrodynamic pattern of the Lower Genesee River is shown in Fig. 10. System segmentations required for numerical computation by the computer model SPAM (Table 7) are indicated on Fig. 11. Figure 11 also gives the

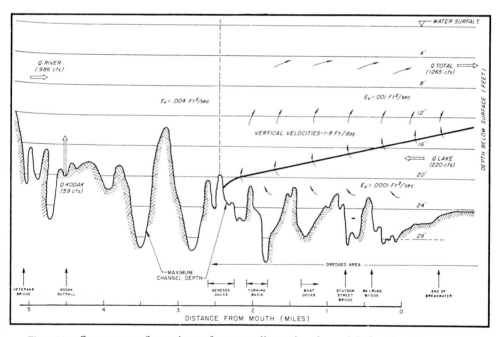

Fig. 10. System configuration of a two-dimensional model for the Lower Genesee River, New York (Courtesy Hydroscience, Inc.)

TABLE 7
Summary of Several Popular River Water Quality Models

Name	Basic assumptions	Objective	Previous applications	Author	Solution techniques
DOSAG-I	One dimensional steady-state, first-order decay kinetics	Determining minimum treatment requirement	San Antonia River, Texas	Water Resource Engineers, Texas Water Development Board	Analytical Solution
EXPLORE-I	One- or two-dimensional time variable, simulating both hydrodynamic and water quality responses	River basin planning and water resources studies	Willamette River Basin, OR	Battelle-NW Laboratories	Finite difference method
PIONEER	One-dimensional steady state, first-order decay kinetics. Simulating water quality and biological processes	Water quality management planning	Chemung River, NY	Battelle-NW Laboratories	Analytical Solution

(continued)

QUAL-II	One-dimensional steady state, first-order decay kinetics	Water quality management planning	Upper Mississippi River	Water Resources Engineers	Finite difference method
RECEIV-II	Two-dimensional time variable, first-order decay, and simplified nutrient cycle	Evaluating receiving water quality from storm water input	Used by US Corps of Engineers and US EPA Region III and IV	Metcalf & Eddy, Univ. of Florida Water Resources Engineers	Finite difference method
RIVER	One-dimensional steady state, first-order decay kinetics	Waste assimilative capacity analysis	Fall Creek, NY, W. Branch Delaware River	Hydroscience, Inc.	Analytical Solution
SPAM	Two-dimensional steady state, first-order decay kinetics	Waste assimilative capacity analysis	Lower Genesee River, NY	Hydroscience, Inc.	Finite difference method
SNSIM	One-dimensional steady state, first-order decay kinetics	Evaluation of various treatment schemes. Waste assimilative capacity analysis	Used by US EPA and NYS Dept. of Environmental Conservation	US EPA Region II	Analytical Solution

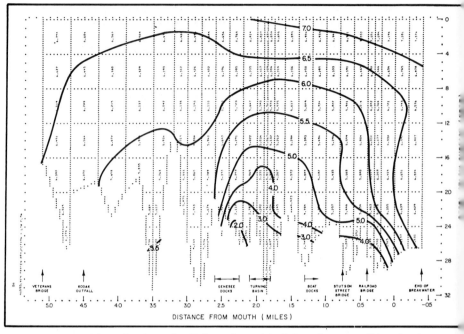

Fig. 11. Model projections for DO under critical low-flow condition in the Lower Genesee River, New York. It also shows geometric grids used in the system segmentation. (Courtesy Hydroscience, Inc.)

output of model projections for dissolved oxygen concentration under critical low flow conditions.

Numerous water quality models have been developed for river analysis based on different system segmentations and numerical methods. Table 7 summarizes some of the more popular ones.

B. Estuary Analysis

An estuary is the lower portion of a river where river flow is affected by tidal current. A more comprehensive definition is contained in Section 104(n)(4) of the US Water Pollution Control Act Amendment of 1972 [40]:

> For the purpose of the Subsection, the term "estuarine zones" means an environmental system consisting of an estuary and those transitional areas which are consistently influenced or affected by water from an estuary such as, but not limited to, salt marshes, coastal and intertidal areas, bays, harbors, lagoons, inshore waters, and channels, and the term "estuary" means all or part of the

mouth of a river or stream or other body of water having unimpaired natural connection with open sea and within which the sea water is measurably diluted with fresh water derived from land drainage.

Water quality analysis of an estuary system is rather different from and often more difficult than that of a river. Difficulties arise as a result of the following:

(a) The dispersive effect is more significant in an estuary system and must be incorporated in model formulation.
(b) In estuary analysis, steady-state formulation is often inadequate to simulate tidal height variation and tidal current fluctuation, two important transport mechanisms. On the other hand, application of a time variable model in estuary analysis is often complicated because of the small time-scale required.
(c) The Coriolis effect and wind forces may play important roles in transport processes of an estuary system and must be included in many detailed hydrodynamic simulations.

The estuary model can be formulated either on a real-time or tidal-averaged basis. A real-time estuarine model simulates actual hydrodynamics within each tidal cycle. The time interval used in numerical simulation are normally 30 min or less. A tidal-averaged model, on the other hand, does not treat variations caused by tidal oscillations explicitly; rather, effects of tidal mixing are manifested through inclusion in a bulk dispersion coefficient. Generally, real-time simulation leads to more accurate results, but investigation of steady-state concentration distribution in terms of a tidal-averaged model may be a useful planning strategy, especially when the fresh water inflow of an estuary is low and constant.

Although the geometric configuration of an estuary system is highly irregular, most past modeling efforts were conducted on the basis of a one-dimensional or quasi two-dimensional formulation. This was mainly because of the relative simplicity of one-dimensional models and to the limitations of data available for model calibration and verification.

One-dimensional estuary representation assumes that an estuary system is well-mixed vertically and horizontally, thus system segmentation is performed only along its main longitudinal axis. The quasi two-dimensional model is also called a ''link-node'' model [5]. In this formulation, an estuary system is divided into many nodes that are defined by constant physical characteristics. Nodes are interconnected by channels or links (Fig. 12). Water in the system is constrained to flow from one node to another through the defined channel. The quasi two-dimensional model is capable of describing substance distribution longitudinally as well as horizontally, but its mathematical formulation remains essentially one-dimensional.

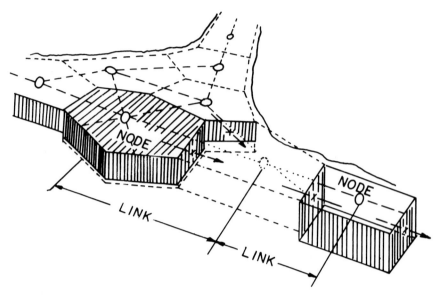

Fig. 12. Quasi two-dimensional geometric representation for an estuary system. (Source: US EPA-600/3-78-105.)

A one-dimensional estuary model can be constructed based on Eq. (10) by introducing a term of constant cross-sectional area, and by dropping unnecessary terms, it can be written as

$$\frac{\partial (AC)}{\partial t} = \frac{\partial}{\partial x}\left(AE\frac{\partial C}{\partial x}\right) - \frac{\partial (QC)}{\partial x} + \Sigma S \qquad (34)$$

where C = substance concentration, mg/L
 A = cross-sectional area of the estuary system, ft^2
 Q = net advective flow of the estuary system, ft^3-s
 ΣS = sources and sinks of substance C in the system, mg/L/s
 $\frac{\partial}{\partial t}$ = instantaneous time rate of change
 $\frac{\partial}{\partial x}$ = rate of change in longitudinal direction
 E = estuarine dispersion coefficient, ft^2/s

The estuary dispersion coefficient E consists of the effects of turbulent diffusion, velocity and density gradients, and tidal mixing. Accurate determination of this coefficient is a major task of any estuary modeling effort. In a tidal-averaged estuary model with relatively small advective transport, estimation of the estuary dispersion coefficient can be made on the basis of a "salt balance"; or the downstream advection of salt by the mean flow is in balance with upstream transport by dispersion [25]. That is,

$$US = E\frac{dS}{dX} \tag{35}$$

where U = tidal velocity, mile/d
X = distance upstream from the ocean boundary, mile
S = salinity at distance X, ppt (parts per thousand)

Direct integration of Eq. (35) leads to

$$E = \frac{U/x}{\ln S_o/S} \tag{36}$$

where S_o = salinity at ocean boundary, ppt.

Example 9

Salinity at the ocean boundary of an estuary is 35 ppt which decreases to 25 ppt upstream at a distance of 2.0 miles. Tidal velocity has been measured to be about 3 miles/d. Estimate the estuary dispersion coefficient.
From Eq. (35)

$$E = \frac{U/x}{\ln S_o/S} = \frac{3/2}{\ln 35/25} = 4.5 \text{ mile}^2/\text{d, or } 1450 \text{ ft}^2/\text{s}$$

When real-time simulation is required in an estuary analysis, the hydrodynamic terms in Eq. (34) must be determined by simultaneous solution of the continuity and momentum equations.

A one-dimensional real-time estuary model has been developed by Harleman et al. at the Massachusetts Institute of Technology, and is called the MIT Transient Water Quality Network Model [14]. It was later successfully applied to the study of St. Lawrence River and Estuary System. Transformation processes of BOD-DO and the nutrient cycle can be investigated by this model by replacing the source and sink term, or ΣS in Eq. (34) with proper system kinetics formulations.

The governing equations for the hydrodynamic simulation in the MIT model take the following forms:
Continuity equation

$$W\frac{\partial h}{\partial t} + \frac{\partial Q}{\partial x} + q = 0 \tag{36}$$

Momentum equation

$$\frac{\partial Q}{\partial t} + U\frac{\partial Q}{\partial x} + Q\frac{\partial U}{\partial x} + g\frac{\partial h}{\partial x}h + g\frac{Ad_c}{\rho}\frac{\partial \rho}{\partial x} + g\frac{Q|Q|}{C_h^2 R_h} = 0 \quad (37)$$

where W = channel top width, ft
h = depth from water surface to an arbitrary horizontal datum, ft
Q = cross-sectional discharge, ft^3/s
q = lateral inflow per unit length in ft^3/s/ft
U = average cross-sectional longitudinal velocity, ft/s
g = gravitational acceleration, ft/s^2
R_h = hydraulic radius, ft
A = the cross-sectional area where there is longitudinal flow in the channel, ft^2
C_h = Chezy coefficient
ρ = fluid density
d_c = depth to the centroid of the channel cross-section, ft

These equations, in addition to the water quality equation, or Eq. (34) can be solved numerically on a computer for an estuary system by imposing proper initial and boundary conditions [14].

Table 8 provides a summary of a number of popular estuary models.

C. Lake Analysis

In lake water quality analysis, one of the major tasks is the investigation of its trophic level or the intensity of eutrophication caused by nutrient loadings. Without human interference, eutrophication might take place slowly in a lake as the normal aging process. But accelerated eutrophication caused by human activity is undesirable; it causes prolific weed growth, nuisance algal blooms, deteriorating fisheries and other recreational usages and, therefore, must be controlled.

Two nutrients most often cited as causes of accelerated eutrophication are nitrogen and phosphorus. Nitrogen and phosphorus in various forms may enter a lake as a result of municipal and industrial wastewater discharges or from nonpoint sources of pollution such as urban runoff and soil erosion. A number of mathematical models have been developed that evaluate nutrient loadings from a drainage basin under different management alternatives. Since this chapter limits its scope to the analysis of water quality impacts of waste loading in receiving waters, watershed modeling techniques are not included. Readers interested in this subject may refer to the quoted literature [42].

Shallow lakes are more susceptible to eutrophication caused by excessive nutrient loadings than are deep lakes. Vollenweider [47] reviewed trophic characteristics of a number of lakes throughout the world and, as a result, established a preliminary relationship that specifies admissible nutrient loadings per unit surface area of a lake according to its mean water depth (Table 9). It is noted that metric units instead of British units are used in the lake analysis according to the current practice.

Since both nitrogen and phosphorus are essential for biological activity in a lake, a deficiency of either one would reduce a lake's biological productivity, and thus, the pace of human-induced eutrophication. Therefore, a prime consideration in eutrophication control is the identification of the limiting nutrient. Limiting nutrient is defined as the nutrient whose quantity change would be most critical to the trophic level of a lake.

Table 9 indicates that the ratio of permissible loadings of total nitrogen and total phosphorus (N/P) is 15. In other words, for a lake subjected to excessive nutrient loadings with an N/P ratio smaller than 15, the control of total nitrogen loading would be more effective; for N/P larger than 15, on the other hand, phosphorus is the limiting nutrient and should be the prime control target.

Example 10

A lake has a surface area of 7.0 km^2 and a mean water depth of 10.0 m. It receives nutrients from upland runoff estimated 455 kg/d of total nitrogen and 28.6 kg/d of total phosphorus. A municipal wastewater treatment plant also discharges its effluent directly into this lake, which contains an average total nitrogen and phosphorus of 500 kg/d and 240 kg/d, respectively. What are the permissible and dangerous nutrient loadings for this lake based on the preliminary relationship recommended by Vollenweider? Which one is the limiting nutrient?

From Table 9, permissible loadings for this lake are 1.5 g of N/m^2/yr and 0.10 g of P/m^2/yr, dangerous loadings are 3.0 g of N/m^2/yr 0.20 g of P/m^2/yr.

TABLE 8
Summary of Several Popular Estuary Water Quality Models

Name	Basic assumptions	Objective	Previous applications	Author	Solution technique
Dynamic Estuary Model (DEM)	Real-time link-node formulation with hydraulic and water quality submodels. Wind and Coriolis forces are ignored. Vertical stratification is absent. It is linked to the Tidal Temperature Model (TTM) for heat budget	Analyzing estuary water quality relative to DO, eutrophication and bacteria pollution	Pearl Harbor Delaware Estuary Potomac Estuary	Water Resources Engineers	Finite difference method
Hydrodynamic-Numerical Model	Real-time two-dimensional formulation, provides the mean	Simulates near-shore current and exchange processes	Prudhoe Bay Coastal area of the Beaufort Sea	University of Hamburg, Germany	Finite difference method

(*continued*)

Model	Description	Application	Case Study	Developer	Method
	flow velocity and turbulent flux in terms of the Monte Carlo scheme				
Estuarine Water Quality Model (ES001)	Tidal-averaged steady state one-dimensional, first-order decay	Evaluation of a number of varying estuary and wste load conditions	New York Habor Hudson–Champlain	Hydroscience, Inc.	Finite difference method
MIT Hydrodynamic Model	Real-time one dimensional, first-order decay, and complete nitrogen cycle	Analysis of eutrophication and DO depression caused by distributed and point waste source	St. Lawrence River and Estuary	MIT Parsons Laboratory	Finite element method
Tidal Temperative Model (TTM)	Real-time quasi two-dimensional formulation, wind and Coriolis forces ignored	Simulate the heat budget and dispersion characteristics	Columbia River	EPA Pacific Northwest Laboratory	Finite difference method

TABLE 9
Permissible Loading Levels for Total Nitrogen and Total Phosphorus (Biochemically Active) in Lakes (g/m²/yr)

Mean depth up to, m	Permissible loading, up to		Dangerous loading, in excess of	
	N, g	P, g	N, g	P, g
5	1.0	0.07	2.0	0.13
10	1.5	0.10	3.0	0.20
50	4.0	0.25	8.0	0.50
100	6.0	0.40	12.0	0.80
150	7.5	0.50	15.0	1.00
200	9.0	0.60	18.0	1.20

Nutrient loadings from both upland runoff and treated municipal wastewater effluent are:

$$N(\text{loading}) = \frac{(455{,}000 + 500{,}000) \times 365}{(7 \times 1000)^2} = 7.10 \text{ g/m}^2/\text{yr}$$

$$P(\text{loading}) = \frac{(28600 + 24{,}000) \times 365}{(7 \times 1000)^2} = 0.39 \text{ g/m}^2/\text{yr}$$

$$N/P = \frac{7.0}{0.39} = 18.21$$

Since N/P is larger than 15.0, phosphorus would be the substance that limits the lake productivity. The results also indicate that the total phosphorus loading is beyond the dangerous limit for this lake and requires immediate management attention.

The empirical approach to lake analysis presented above is rather preliminary and only valid under certain ideal conditions. In this type of empirical analysis, one must assume that a lake is a well-mixed and homogeneous water system. Also eliminated from consideration are complex environmental interactions in the entire community of organisms. In order to provide a sound basis in the decision-making process for the management of large surface water bodies such as the Great Lakes, comprehensive water quality–ecological modeling is often required. These models simulate hydrodynamic transport, lake circulation, heat transfer, biological transformation, and chemical reaction taking place in the lake, and provide an integrated interpretation of physical, chemical, and biological data observed in the field [30].

Lake ecological models have been formulated and applied in a number of recent lake water quality analyses (Table 10). In general, these models require the superpositions of transport and circulation characteristics to various water quality and ecological activities in the water system. Although transport dynamics of lake systems have been simulated successfully [43], significant improvements are required to make the water quality–ecological model a useful analytical tool for the management of fresh water lakes.

NOMENCLATURE

A	cross-sectional area of a surface water system
$\bar{A}$	eddy diffusion coefficient
B	coliform concentration
C	concentration of pollutant substance
C_b	biological rate constant
C_e	escape coefficient in Tsivoglou reaeration formula
C_h	Chezy coefficient
Chl_a	Chlorophyll-a concentration
d_c	depth to the controid of a channel cross-section
D	dissolved oxygen deficit in stream
D_c	maximum dissolved oxygen (DO) deficit in a stream from waste input
D_e	dissolved oxygen deficit from effluent waste loading
D_o	dissolved oxygen deficit at the beginning of a stream reach
D_p	minimum dissolved oxygen deficit in an observed diurnal DO curve
D_r	maximum dissolved oxygen deficit in an observed diurnal DO curve
D_s	dissolved oxygen deficit in a stream from upstream pollutant
D_w	theoretical stream dissolved oxygen deficit when photosynthesis and respiration actions were excluded
f	stream self-purification factor
F	flux vector or mass flow rate
F_A	flux from the bulk motion of the water
F_T	flux due to turbulent diffusion
h	channel depth
K_b	coliform die-off rate
K_d	in-stream CBOD deoxygenation rate
K_n	NBOD deoxygenation rate

TABLE 10
Summary of Several Popular Lake Water Quality Models

Name	Basic assumptions	Objective	Previous applications	Author	Solution technique
LAKE-1 (Lake-1 Ecologic Model)	One-dimensional time variable formulation including temperature, nitrogen, phosphorus in a two-segment fresh water lake system	Simulates the ecological process of photosynthesis, growth, decay and respiration	Lake Ontario Lake Huron Lake Erie	Manhattan College	Finite difference method
MS. CLEANER (Multi-Segment Comprehensive Lake Ecosystem Analyzer)	A biologically realistic aquatic ecosystem model that simulates physicochemical and biotic characteristics in	Evaluation of environmental problems in lakes and reservoirs	Lake George, NY DeGray Reservoir, Arkansas & others	Rensselaer Polytechnic Institute	

(*continued*)

Model	Description	Application	Developer	Method
	multiple segments by model decompositions			
Ecological-Hydrodynamic Model	Three-dimensional, time variable, including three basic modules—hydrodynamic, interface, and water quality. Wind and Coriolis forces, bottom friction are considered.	Evaluation of lake eutrophication and DO depression from excessive nutrient input	Tetra Tech, Inc.	Finite difference method
LAKE-3 (Lake-3 Ecologic Model)	Three dimensional real-time formulation including eight variables of nutrient, phytoplankton, and zooplankton	Provide detailed phytoplankton-nutrient interaction for lake eutrophication investigation	Lake Ontario and Rochester Bay	Manhattan College	Finite difference method

K_r	in-stream CBOD removal rate
K_s	sediment oxygen demand rate
K_1	laboratory CBOD deoxygenation rate
K_2	atmospheric reaeration rate
K_3	rate of in-stream CBOD reduction from factors other than deoxygenation
L	total biochemical oxygen demand (BOD) in a receiving water
L_c	carbonaceous biochemical oxygen demand (CBOD) remaining at time t
L_C	in-stream CBOD
L_{cp}	point source of CBOD
L_{cd}	distributed or nonpoint source of CBOD
L_n	nitrogenous biochemical oxygen demand (NBOD) remaining at time t
L_N	in-stream NBOD
L_{np}	point source of NBOD
L_{nd}	distributed or nonpoint source of NBOD
L_o	total BOD at the beginning point of a stream reach under investigation
M	total mass of a transported substance
n	stream bed activity coefficient
$\vec{n}$	unit vector normal to a surface element
P	optimum rate of photosynthesis
q	lateral inflow
R	respiration rate of phytoplankton biomass in surface waters
R_h	hydraulic radius of an open channel
S	surface of the control volume in a free flowing water system
S_s	slope of the profile of minimum observed DO along a stream
t_c	time to critical DO in a stream from waste input
T_f	time of water travel
U	longitudinal velocity of a one-dimensional stream flow
$\vec{U}$	velocity vector, with component U_x, U_y, and U_x
V	control volume in a free-flowing water system
W	channel top width
∇	divergence operator
α	net daily algal oxygen production rate
θ_c	temperature correction factor for CBOD deoxygenation rate
θ_n	temperature correction factor for nitrification rate
θ_s	temperature correction factor for sediment oxygen demand rate
ρ	fluid density
ΣS	rate of all internal mass production and reduction

REFERENCES

1. American Public Health Association, *Standard Methods for the Examination of Waste and Wastewater*, 14th Ed. Washington, DC, 1976, pp. 919–941.
2. R. G. Baca and R. C. Arnett, *A Limnological Model for Eutrophic Lakes and Impondments*, Battelle Pacific Northwest Laboratories, 1976.
3. R. Beckett and J. Hurt, *Numerical Calculations and Algorithms*, McGraw-Hill, New York, 1967.
4. R. E. S. Braster, S. C. Chapra, and G. A. Nossa, *A Computer Program for the Steady State Water Quality Simulation of a Stream Network*, US EPA Region II, New York, 1975.
5. C. W. Chen and G. T. Orlob, *Ecological Simulation for Aquatic Environment, System Analysis and Simulation in Ecology*, Vol. III, Academic Press, New York, 1975.
6. C. W. Chen and T. Wells, "Boise River Water Quality–Ecological Model for Urban Planning Study," Tetra Tech technical report prepared for US Army Engineering District, Walla Walla, Washington, 1975.
7. G. T. Csanady, *Turbulent Diffusion in the Environment*, Reidel, Boston, 1973.
8. W. W. Eckenfelder, Jr., *Water Quality Engineering for Practicing Engineers*, Barnes & Noble, New York, 1970.
9. N. Edeberg and B. V. Hofsten, *Water Res.* **7,** 1285 (1973).
10. G. W. Fair, J. C. Geyer, and J. C. Morris, *Water Supply and Wastewater Disposal*, Wiley, New York, 1959.
11. W. H. Frost, and H. W. Gould, *Public Health Bulletin 143*, US Public Health Service, Washington, DC 1975.
12. J. J. Gannon, *J. San. Engr. Div. ASCE* **92,** No. SA1., 135 (1966).
13. D. R. F. Harleman, "Real-Time Models for Salinity and Water Quality Analysis in Estuary," in *Estuaries, Geophysics, and the Environment*, National Academy of Sciences, Washington, DC, 1977.
14. D. R. F. Harleman, et al., "User's Manual" for the MIT Transient Water Quality Network Model, US EPA Report, Corvallis Environmental Research Laboratory, Corvallis, Oregon, EPA-600/3-77-010, 1977.
15. Hydroscience, Inc., *Water Quality Analysis of the Lower Genesee River*, Westwood, New Jersey, 1976.
16. E. C. Kinney, et al., *Water Poll. Control Fed.* **50,** No. 10, 2307 (1978).
17. F. W. Kittrell and S. A. Furfari, *Water Poll. Control Fed.* **35,** No. 11, 1361 (1963).
18. C. C. K. Liu, Filtering of Dissolved Oxygen Data in Stream Water Quality Analysis, *Water Resources Bulletin*, **18,** No. 1, 15 (1982).
19. A. M. Lucas and N. A. Thomas, Sediment Oxygen Demand in Lake Erie's Central Basin, in "Project Hypo.", EPA TS-05-71-208-24 US EPA Region V, 1972, pp. 45–70.
20. A. J. McDonnel and S. D. Hall, *J. Water Poll. Control Fed.* **41,** No. 8, Part 2 (1969).
21. Metcalf & Eddy, Inc., *Wastewaer Engineering*, McGraw-Hill, New York, 1972.
22. R. C. Mt. Pleasant and W. Schlickenrieder, Implication of Nitrogenous BOD in Treatment Plant Design, paper presented at 1971 ASCE Phoenix National Meeting, 1971.
23. New York State Environmental Conservation Law, Classifications and Standards

Governing the Quality and Purity of Waters of New York State, Part 700, 701, and 702, Title 6, 1974.
24. G. Neumann and W. J. Pierson, Jr., *Principles of Physical Oceanography*, Prentice-Hall, Englewood Cliffs, New Jersey, 1966.
25. D. J. O'Connor, *Estuarine Analysis*, Manhattan College Summer Institute in Water Pollution Control, New York, 1977.
26. M. M. Pamatmat, *Int. Rev. Ges. Hydrobiol. Hydrogr.*, **56,** (1971).
27. E. B. Phelps, *Stream Sanitation*, Wiley, New York, 1944.
28. R. E. Rathbun, D. J. Schultz, and D. W. Stephens, Preliminary Experiments with a Modified Tracer Technique for Measuring Stream Reaeration Coefficients, US Geological Survey Open-File Report No. 75-256, Bay St. Louis, MI, 1975.
29. J. Roger, "Scientific Thinking as a Destructive Process," *Cornell Review*, **2,** 22 Ithaca, New York, 1977.
30. D. Scavia, and A. Robertson, *Perspectives on Lake Ecosystem Modeling*, Ann Arbor Science, Ann Arbor, MI, 1979.
31. H. W. Streeter, *Sewage Works J.* **7,** No. 2 (1935).
32. H. W. Streeter and E. B. Phelps, *A Study of the Pollution and Natural Purification of the Ohio River*, US Public Health Bulletin, No. 146, 1925.
33. G. I. Taylor, *Proc. Roy. Soc. London* **A219,** 186 (1953).
34. Tetra Tech, Inc., "Estuary Water Quality Models, Long Island, New York, User's Guide," Technical Report Prepared for Nassau Suffolk Regional Planning Board, 1976.
35. Tetra Tech, Inc., "Rates, Constants, and Kinetics Formulation in Surface Water Quality Modeling," EPA-600/3-78-105, EPA Environmental Research Laboratory, Athens, Georgia, 1978.
36. R. V. Thomann, *System Analysis and Water Quality Management*, Environmental Research and Application, New York, 1971.
37. E. C. Tsivoglou, et al. *J. Water Poll. Control Fed.* **40,** No. 2, 285 (1967).
38. US Army Corps of Engineers, *Water Quality for River-Reservoir Systems*, Hydrologic Engineering Center Technical Report, Davis, California, 1974.
39. US Environmental Protection Agency, *Guidelines for Developing or Revising Water Quality Standards Under the Federal Water Pollution Control Act Amendments of 1972*, Water Planning Division, Washington, DC, 1973.
40. US Environmental Protection Agency, *Task Force Report on Disinfection of Wastewater*, Office of Research and Development, Washington, DC, 1975.
41. US Environmental Protection Agency, "An Amendment to Secondary Treatment Information Regulation (40 CFR 133)," *Fed. Reg.* **41,** No. 144, 1976.
42. US Environmental Protection Agency, *Urban Stormwater Managment and Technology*, Office of Research and Development, Report EPA-600/8-77-014, Cincinnati, Ohio, 1977.
43. US Environmental Protection Agency, Results of a Joint USA/USSR Hydrodynamic and Transport Modeling Project, EPA-600/3-79-013, Environmental Research Laboratory, Duluth, MN, 1979.
44. US Environmental Protection Agency, *Environmental Modeling Catalogue*, Management Information and Data Systems Division, EPA-68-01-4723, Washington, DC, 1979.
45. US Federal Water Pollution Control Act Amendment (PL 92-500), US Government Printing Office, Washington, DC, 1972.
46. C. J. Velz, *Applied Stream Sanitation*, Wiley-Interscience, New York, 1970.

47. R. A. Vollenweider, *Scientific Fundamentals of the Eutrophication of Lakes and Flowing Waters, with Particular Reference to Nitrogen and Phosphorus as Factors in Eutrophication*, Organization for Economic Cooperation and Development, Paris, France, 1968.
48. H. F. Weinberger, *A First Course in Partial Differential Equations*, Xerox College Publishing, Lexington, MA, 1965.

APPENDIX A
Water Quality Criteria Suggested by US EPA, Part 1 (36)[a]

Use class	Microbiological	Dissolved oxygen	Temperature	Hydrogen ion	Dissolved solids[b]	Taste and odor producing substances	Dissolved gas	Color and turbidity producing substances
Class A primary contact recreation (swimming, water skiing, etc.)	Shall not exceed a geometric mean of 200 fecal coliform per 100 mL	Not less than 5 mg/L. Class B levels also apply	90°F, max. Class B levels also apply	Hydrogen ion concentrations expressed as pH shall be maintained between 6.5–8.3	Shall not exceed 500 mg/L or one third above that characteristic of natural condition (whichever is less)	None in amounts that will interfere with water contact use.	Class B levels apply	Secchi disk visible at minimum depth of 1 m
Class B Desirable species of aquatic life and secondary contact recreation (boating, fishing, etc.)	Shall not exceed a geometric mean of 10,000 total coliform of 2000 per 100 mL (fecal coliform counts are preferred)	Not less than 5 mg/L (except for 4 mg/L for short periods of time within a 24-h period.) Not less than 6 mg/L in trout waters. Not less than 5 mg/L in marine waters	*Cold water* (trout) 5°F rise. Max. of 68°F *Warm Water* (bass etc.) 5°F rise in streams. 3°F degree in impoundments. Mas. 90°F *Marine water* 1½°F rise	Hydrogen ion concentrations expressed as pH shall be maintained between 6.0 and 9.0	Shall not exceed one-third above that characteristic of natural conditions	Shall contain no substances that will render any undesirable tastes to fish flesh or in any other way make fish inedible	*Cold water* Total dissolved gas pressure not to exceed 100% of existing atmospheric pressure	Cold waters, 10 JU warm waters, 50 JU marine waters Secchi disk visible a minimum depth of 1 m

[a] The water quality criteria for Classes A and B are compatible with uses for PWS, agricultural, industrial, and navigation.
[b] Not applicable to marine water. The criteria for this parameter cannot be uniformly applied to all surface waters. Criteria commensurate with natural regional differences should be applied regionally.

APPENDIX B
Water Quality Criteria Suggested by US EPA, Part 2

Radioactivity

Gross beta	1000 pCi/L
Radium-226	3 pCi/L
Strontium-90	10 pCi/L

Phosphorus

Phosphorus as P shall not exceed 100 μg/L in any stream nor exceed 50 μg/L in any reservoir or lake, or in any stream at the point where it enters any reservoir or lake.

Suspended Collodial or Settleable Solids

None from a waste water source that will permit objectionable deposition or be deleterious for the designated uses.

Oil and Floating Substances

No residue attributable to waste water nor visible film oil or globules of grease.

Mixing Zones

The total area and/or volume of a receiving stream assigned to mixing zones be limited to that which will: (1) not interfere with biological communities or populations of important species to a degree that is damaging to the ecosystem; (2) not diminish other beneficial uses disproportionately.

APPENDIX C
Classifications and Standards for Fresh Surface Waters in New York State

Classification	Best usage	Conditions of best usage	Dissolved oxygen standards					
				Trout waters		:c	Nontrout waters	:c
			Trout waters spawning	Minimum daily average	Minimum		Minimum daily average	Minimum
Class AA	Water supply for drinking or food processing	Waters will meet Health Department standards	7 mg/L	6 mg/L	5 mg/L		5 mg/L	4 mg/L
Class A	Water supply for drinking or food processing	Waters will meet Health Department standards for drinking water with approved treatment	7 mg/L	6 mg/L	5 mg/L		5 mg/L	4 mg/L
Class B	Contact recreation and other uses except water supply and food processing	—	7 mg/L	6 mg/L	5 mg/L		5 mg/L	4 mg/L
Class C	Fishing and other uses except water supply, food processing, and contact recreation	—	7 mg/L	6 mg/L	5 mg/L		5 mg/L	4 mg/L
Class D	Secondary contact recreation. Waters are not suitable for propagation of fish	Waters must be suitable for fish survival	—	—	—		—	3 mg/L
Class N	Enjoyment of water in its natural condition for whatever compatible purposes	No waste discharges whatsoever permitted without approved filtration through 200' of unconsolidated earth	Natural	Natural	Natural		Natural	Natural

SURFACE WATER QUALITY ANALYSIS

Coliform standard						Radioactivity standards		
Monthly median value	80% of sample	Monthly geometric mean	pH	Total dissolved solids	Phenolic compounds	Gross beta	Radium 226	Strontium 90
Less than 50/100 mL coliforms	Less than 240/100 mL coliforms	—	6.5–8.5	As low as practicable. Less then 500 mg/L	Less than 0.001 mg/L (phenol)	Less than 1000 pCi/L (in absence of Se^{90} and alpha emitters)	Less than 3 pCi/L	Less than 10 pCi/L
Less than 5000/100 mL coliforms	Less than 20,000/100 mL coliforms	Less than 200/100 mL fecal coliforms	6.5–8.5	As low as practicable. Less than 500 mg/L	Less than 0.005 mg/L (phenol)	Less than 1000 pCi/L (in absence of Sr^{90} and alpha emitters)	Less than 3 pCi/L	Less than 10 pCi/L
Less than 2400/100 mL coliforms	Less than 5000/100 mL coliforms	Less than 200/100 mL fecal coliforms	6.5–8.5	None detrimental to aquatic life. Waters currently less than 500 mg/L shall remain below this limit	—	—	—	—
—	M	Less than 10,000/100 mL coliforms and 2,000/100 mL fecal coliforms	6.5–8.5	None detrimental to aquatic life. Waters currently less than 500 mg/L shall remain below this limit	—	—	—	—
—	—	—	6.0–9/.5	—	—	—	—	—
Natural	Natural	Natural	Natural	Natural	—	Natural	Natural	Natural

2
Water Quality Control of Tidal Rivers and Estuaries

Mu Hao Sung Wang
Department of Environmental Conservation, State of New York, Albany, New York

Lawrence K. Wang
Lenox Institute for Research Inc., Lenox, Massachusetts

I. INTRODUCTION

Estuaries are those water bodies in which a significant hydrodynamic transport mechanism is caused by astronomical tides and other similar mechanisms. For the purposes of water quality analysis, the "estuary" is that portion of the river that is under the influence of tidal action in which the dispersion factor is always significant. An official, more comprehensive definition is contained in Section 104 (n) (4) of the US Water Pollution Control Act Amendment of 1972 [1], and documented in a general chapter on "Surface Water Quality Analysis."

Estuaries normally consist of two sections that are characterized by the relative magnitude of advective flow to tidal mixing or dispersion. In purely estuarine systems the upper reaches are referred to as tidal river reaches and characterized by a significant advective transport component. The downstream portion is generally dominated by tidal mixing and freshwater advective flow and is less important in transporting physical and chemical constituents. The length of the estuary may be much greater than the length of the salt water intrusion.

This report considers methods for evaluating water quality responses in estuaries and tidal rivers. Mathematical simulation and analysis of water quality are introduced. Practical examples useful to the environmental engineers in the initial development steps of the 208 water quality management plan [2] are provided by the US Environmental Protection

Agency (EPA), Cincinnati, Ohio, and are presented in this report for demonstrating the analytical methods for both point an nonpoint sources of pollution.

II. WATER QUALITY PARAMETERS

A. Conservative Substances

Conservative constituents are those that are not subject to reactive change and remain dissolved or suspended in the receiving water. Typical conservative constituents are total dissolved solids, total nitrogen, total phosphorus, total suspended solids, and other materials that decay at such slow rates that they may be regarded as conservative.

B. Reactive Substances

Nonconservative substances are subject to change within the receiving water as a result of physical, chemical, or biological reactions, and can also be termed reactive constituents. Typical nonconservative substances include BOD, coliform bacteria, and nutrients. Although total nitrogen and phosphorus are treated as conservative on an annual average basis, they are considered reactive during the summer low flow period because of the algal uptake of the nutrients and their subsequent removal by sedimentation.

The reactions of reactive constituents in tidal rivers and estuaries are analyzed based on the following assumptions:
1. The reaction is first-order with a decay coefficient, K (see Table 1).
2. Steady-state conditions exist.
3. Constant coefficients exist, i.e., flow, cross-sectional area, reaction kinetics, and dispersion characteristics are all constant along the length of the estuary under investigation.

C. Sequentially Reactive Constituents

A sequentially reactive constituent, such as dissolved oxygen deficit (D), is affected or changed by the reacting constituents, such as ultimate oxygen demand (UOD).

TABLE 1
First-Order Range of Reaction Coefficients for Tidal Rivers and Estuaries[a]

Substance	Reaction coefficients, d^{-1}	K, base e, 20°C
Coliform bacteria	2–4	(K_b)
BOD_5	0.2–0.5	(K_d or K_1)
Nutrients	0.1–0.25	(K_n or K_p)

[a]Source: ref. [3].

D. Surface Transfer Coefficient, Reaeration Coefficient, and Deoxygenation Coefficient

The surface transfer coefficient (K_L, ft/d) is related to the volumetric reaeration coefficient (K_2, d^{-1}) by the depth:

$$K_2 = K_L/h \qquad (1)$$

where h is the average depth at mean tide in ft.

Table 2 indicates the range of transfer and reaeration coefficients for tidal rivers and estuaries. Both K_L and K_2 are functions of the velocity (average tidal current) and depth of flow.

The deoxygenation coefficient (K_d) ranges from 0.2 to 0.5, as indicated in Table 1, assuming that the estuary is no shallower than about 5 ft.

E. Assimilation Ratio

The assimilation ratio (R) characterizes water quality in tidal rivers and estuaries, and is a function of the reaeration coefficient (K_2) and the deoxygenation coefficient (K_d), as indicated in Table 3. From Tables 2 and 3, it can be seen that the deep main channel estuaries (10–30 ft in depth) have assimilation ratios ranging from 0.2 to 0.8, while the shallower tidal tributaries (5–10 ft in depth) are in the range 0.8–3.0. At the lower limits of the assimilation ratio ranges are found the more restricted tidal bodies with lower velocity, higher temperatures, and plant effluents from secondary treatment or less. The upper limits include the free-flowing, higher velocity estuary, with more moderate water temperature and effluents from advanced treatment.

TABLE 2
Range of Transfer and Reaeration Coefficients Estimated for
Tidal Rivers and Estuaries (K_L in ft/d, K_2 in d^{-1})

Mean tidal depth, ft	Average tidal velocity, ft/s					
	1	1	1–2	1–2	2	2
	K_L	K_2	K_L	K_2	K_L	K_2
10	4	0.5	5.5	0.6	7	0.8
10–20	3	0.2	4.5	0.3	6	0.4
20–30	2.5	0.1	3.5	0.14	5	0.2
30	2	0.06	2.5	0.08	4	0.12

F. Estuarine Number

The estuarine number (N) is defined as

$$N = KE/U^2 \tag{2}$$

where E = the dispersion coefficient ranging from 1 to 20 mi^2/d; and U = advective velocity ranging from 0.1 to 10 mi/d. The dispersion coefficient and advective velocity are discussed in subsequent subsections. The estuarine numbers developed from the practical ranges of E and U at $K = 3$ d^{-1} for BOD$_5$ are summarized in Table 4. For waste substances with higher reaction rates, the N value increases proportionally according to Eq. (2).

A summary of Tables 3 and 4 with approximate physical descriptions of the types of tidal rivers and estuaries is presented in Table 5. Tables 3–5 are all abstracted from ref. [3].

TABLE 3
Assimilation Ratio of Tidal Rivers and Estuaries

Reaeration coefficient, K_2, d^{-1}	Assimilation ratio			
	$K_d = 0.2$	$K_d = 0.3$	$K_d = 0.4$	$K_d = 0.5$
0.08	0.40	0.27	0.20	0.16
0.15	0.75	0.50	0.38	0.30
0.30	1.50	1.00	0.75	0.60
0.60	3.00	2.00	1.50	1.20

TABLE 4
Range of Estuarine Number for Tidal Rivers[a]

Tidal dispersion, mi^2/d	Estuarine number at various advective velocities[b]			
	0.5	1.0	2.0	4.0
2	2.4	0.6	0.15	0.04
5	6.0	1.5	0.38	0.10
10	12.0	3.0	0.75	0.19
20	24.0	6.0	1.50	0.75

[a] At BOD_5 reaction rate = 0.3 d^{-1}.
[b] Unit of advective velocity = mi/d.

G. Estaurine Dispersion Coefficient

The primary difference between estuaries and the one-dimensional river flow situation is the dispersive mass transport resulting from the tidal mixing occasioned by tidal flow reversals. The longitudinal dispersion coefficient can be determined by the salinity gradient in an estuary sys-

TABLE 5
Classification of Tidal Rivers and Estuaries
(K = 0.3/d)

Description	Assimilation ratio, R		Estuary number, N	
	Average value	Range	Average value	Range
Large, deep, main channel in vicinity of mouth	0.3	0.1–0.5	15	5–30
Moderate navigation channel, upstream from mouth, saline, large tidal tributaries	0.5	0.2–1.0	5	2–10
Minimum navigation upstream, smaller saline or nonsaline tidal tributaries	1.0	0.5–2.0	2	0.5–5
Tidal tributaries, shallow and nonsaline	2.0	1.0–3.0	1	0.2–2

tem. An estimate of the dispersion coefficient may be obtained by plotting the salinity versus distance on semilog graphical paper, fitting a straight line to the data, and obtaining E. Other field and mathematical methods for determination of E values are possible [4–8].

As discussed earlier, the practical range of the dispersion coefficient is from 1 to 20 mi^2/d. The upper limit describes the highly saline, high-tidal-velocity stretches in the vicinity of the estuarine mouth, while the lower limit applies to the upstream, nonsaline, low tidal sections of the estuary.

H. Advective Velocity

The advective velocity (U) associated with the freshwater flow is determined by dividing the freshwater flow (Q) by the average cross-sectional area (A).

III. BASIC MATHEMATICAL MODELS

Since a summary of several popular estuary water quality models has been presented in another chapter entitled, "Surface Water Quality Control," this chapter will not be an exhaustive review. The objective here is to introduce and describe the basic dispersion process in an estuarine system.

The movement of water in an estuary is more complex than in nontidal streams principally because of tidal mixing and the time-scale of estuarine response to continuous and intermittent loadings. In general, the water movement in an estuary can be divided into two major categories: bulk motion (e.g., freshwater flow, tidal flow) and turbulent motion (e.g., eddies). Soluble point-source and non-point source pollutants that are discharged to an estuary are transported along the flow direction by the bulk water motion, and are spread out, both along and perpendicular to the direction of flow, by the diffusive effects of the turbulent motion. By assuming that the mass flux (M/L^2/T) in a given direction is proportional to the concentration gradient in that direction, the total mass flux can be expressed by Eq. (3).

$$M_T = M_l + M_v + M_t \tag{3}$$

where

M_1 = longitudinal mass flux, M/L²/T
 = $(U_x C - \bar{E}_x \partial C/\partial x)$ (3a)
M_v = vertical mass flux, M/L²/T
 = $(U_y C - \bar{E}_y \partial C/\partial y)$ (3b)
M_t = transverse mass flux, M/L²/T
 = $(U_z C - \bar{E}_z \partial C/\partial z)$ (3c)
C = soluble substance concentration, M/L³
$\bar{E}_x$ = turbulent diffusion coefficient in longitudinal direction x, L²/T
$\bar{E}_y$ = turbulent diffusion coefficient in vertical direction y, L²/T
$\bar{E}_z$ = turbulent diffusion coefficient in transverse direction z, L²/T
U_x = longitudinal bulk water velocity, L/T
U_y = vertical bulk water velocity, L/T
U_z = transverse bulk water velocity, L/T

Equation (4) is mass balance equation [9, 10] derived from Eq. (3).

$$\frac{\partial C}{\partial t} = -\frac{\partial M_1}{\partial x} - \frac{\partial M_v}{\partial y} - \frac{\partial M_t}{\partial z} + S \qquad (4)$$

where t is the time, T; and the term $S(x, y, z, t)$ is added to account for external sources and sinks of substance C in the surface water system, M/L³/T. The above equation is a second-order partial differential equation, constituting the general form of a three-dimensional time variable water quality model. Subject to appropriate boundary conditions and a knowledge of velocities and turbulent dispersion coefficients in longitudinal, vertical, and transverse directions, Eq. (4) can be solved to determine the spatial and temporal distribution of the concentration C. The computational effort required to solve the equation for realistic estuarine conditions, however, is enormous and costly. The common procedure adopted to simplify the problem is to average Eq. (4) over one or more of the space dimensions [10]:

1. Averaging vertically yields a two-dimensional plan model suitable for the investigation of wide vertically mixed estuaries and bays.
2. Averaging transversely yields a two-dimensional elevation model suitable for the study of narrow stratified estuaries.
3. Averaging both vertically and transversely yields a one-dimensional model suitable for estuaries that are well mixed both vertically and transversely.

The last one-dimensional model is commonly used for its comparative computational simplicity. Equation (4) can be reduced to the following basic equation for the one-dimensional estuary, after some regrouping of the advective and diffusive terms:

$$\partial C/\partial t = (E/A)\partial(A\partial C/\partial x)/\partial x - (Q/A)(\partial C/\partial x) \pm S \qquad (5)$$

where A = cross-sectional area of the estuary system, L^2
Q = net advective flow of the estuary system, L^3/T
E = estuarine dispersion coefficient, L^2/T

Equation (5) defines the time rate of change of a pollutant in an estuary, with freshwater flow. There is no tidal velocity term since the analysis is restricted to slack water at high or low tide. The variable C in Eq. (5) may apply in general to any soluble substances including dissolved oxygen concentration. In the following sections the term "C" will apply specifically to the concentration of conservative substance, such as salinity or chlorides; the term "L" to any substance that decays in accordance with a first-order reaction and in particular to the concentration of the oxygen demanding material.

The majority of estuaries in which there are pollutional problems are probably of variable cross-sectional area. The area functions can be the liner, the power, and the exponential forms.

In order to simplify the water quality analysis problem further, the estuarine area may be assumed constant for considerable distances. Typical examples of estuaries whose areas may be assumed constant are Hudson River and East River in New York Metropolitan area, shown in Fig. 1. Assuming the sink term to be first-order decay, Eq. (5) becomes:

$$\partial C/\partial t = E(\partial^2 L/\partial x^2) - U(\partial L/\partial x) - KL \qquad (6)$$

where K is the first-order reaction rate, T^{-1} (base e).

Section IV presents working models derived from the general one-dimensional estuary model, Eq. (6), assuming the cross-sectional area of channel is constant. Detailed derivation of the working models can be found elsewhere [9, 10]. Also presented are alternate working models [11–16] and system identification techniques for tidal rivers [8, 14, 15].

It should be noted that all working models presented in the next section are suitable to tidal rivers and estuaries only. The upstream nontidal portion of the receiving water system may be analyzed for its water quality by some mathematical models in a general chapter, "Surface Water

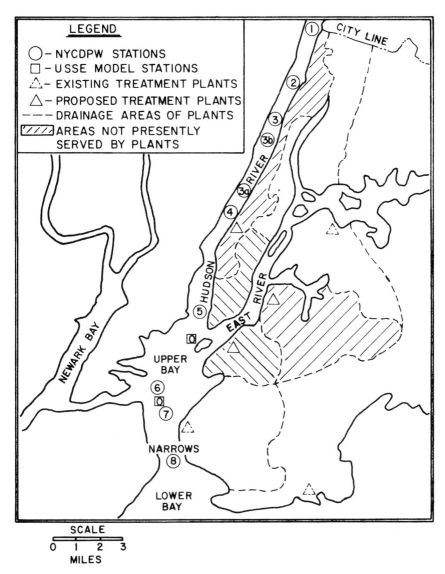

Fig. 1. Hudson River and East River in New York Metropolitan Area.

Quality Analysis," and by the EPA models summarized in Tables 6A and 6B. All water quality parameters are clearly defined in the Nomenclature section.

TABLE 6A
Pollutant Concentrations in Nontidal Rivers Resulting from a Point Source of Pollution

Conservative substances	$C = C_o + W/Q$
Reactive substances	$L = L_o \exp(-K_r X/U) + (W/Q) \exp(-K_r X/U)$
Sequentially reactive constituents	$D = D_o \exp(-K_2 X/U)$ $+ L_o [K_d/(K_2 - K_r)] [\exp(-K_r X/U) - \exp(-K_2 X/U)]$ $+ (W/Q) [K_d/(K_2 - K_r)] [\exp(-K_r X/U) - \exp(-K_2 X/U)]$

TABLE 6B
Pollutant Concentrations in Nontidal Rivers Resulting from a Nonpoint Source of Pollution

Conservative substances	$C = C_o + wX/Q$
Reactive substances	$L = L_o \exp(-K_r X/U) + [w/(AK_r)] [1 - \exp(-K_r X/U)]$
Sequentially reactive constituents	$D = D_o \exp(-K_2 X/U)$ $+ L_o [K_d/(K_2 - K_r)] [\exp(-K_r X/U) - \exp(-K_2 X/U)]$ $+ [w/(AK_r)] [K_d/(K_2 - K_r)] [K_r K_2^{-1} \exp(-K_2 X/U) - \exp(-K_r X/U) + (K_2 - K_r)/K_2]$

IV. WORKING MODELS

A. Steady-State Equations for Waste Concentrations in Tidal Rivers and Estuaries Resulting from a Point Source of Pollution

The following working models are derived from the general estuary model in Section III and provided by the US EPA, Cincinnati, Ohio. Figure 2 illustrates the stream and estuary conditions. Figure 3 shows the distribution of waste concentrations in tidal rivers and estuaries resulting from a point source of pollution.

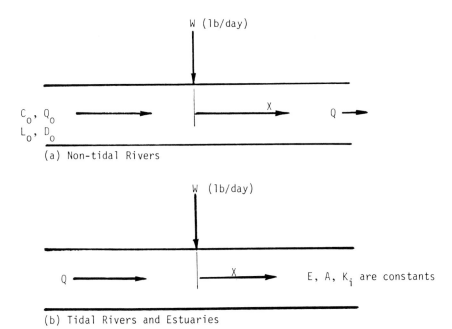

Fig. 2. Discharge of point source of pollutants into a receiving water system.

1. At Point Source of Discharge ($X = 0$)

a. Conservative Substances. The maximum concentration of a conservative substance at the point source discharge location ($X = 0$) is simply the mass rate of waste discharge divided by the freshwater flow:

$$C_o = W/Q \tag{7}$$

where C_o = the maximum concentration of a conservative substance at the point source discharge location, M/L^3; W = mass of point source pollution discharge rate, M/T; Q = freshwater flow, L^3/T; and X = downstream distance from the point source of pollution, L.

b. Reactive Substances (System 1). The concentration of reactive substances at the point source discharge (L_o, in M/L^3) can be calculated by Eq. (8):

$$L_o = W/(Qm_1) \tag{8}$$

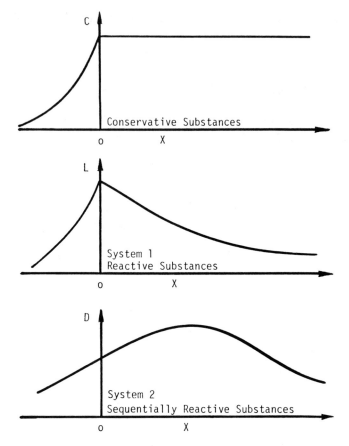

Fig. 3. Distribution of various types of waste substances in tidal rivers and estuaries resulting from a point source of pollution.

where K_1 = decay rate, system 1 (e.g., BOD), T^{-1};
E = dispersion coefficient, L^2/T; and
$m_1 = [1 + 4K_1E/U^2]^{0.5}$ (9)

c. *Sequentially Reactive Constituents (System 2).* The dissolved oxygen deficit at the point source discharge (D_o, in M/L^3) can be calculated by Eq. (10):

$$D_o = [L_oK_{12}/(K_2 - K_1)] [1 - m_1/m_2] \quad (10)$$

where K_2 = reaction rate, system 2 (e.g., reaeration rate), T^{-1}; K_{12} =

reaction rate between systems 1 and 2 (e.g., deoxygenation rate K_d), T^{-1}; and

$$m_2 = [1 + 4K_2E/U^2]^{0.5} \quad (11)$$

2. Upstream Reach ($X \leqq 0$)

a. *Conservative Substances.* The concentration of conservative substances (C, in M/L^3) can be determined by

$$C = C_o \exp(UX/E) \quad (12)$$

where X = downstream distance from the point source of pollution, L.

b. *Reactive Substances (System 1).* The concentration of reactive substances (L, in M/L^3) can be determined by

$$L = L_o \exp(g_1 X) \quad (13)$$

where

$$g_1 = (U/2E)(1 + m_1) \quad (14)$$

c. *Sequentially Reactive Constituents (System 2).* Dissolved oxygen deficit at any point upstream of the point source of pollution (D, in M/L^3)

$$D = [L_o K_{12}/(K_2 - K_1)] [\exp(g_1 X) - m_1 m_2^{-1} \exp(g_2 X)] \quad (15)$$

where

$$g_2 = (U/2E)(1 + m_2) \quad (16)$$

3. Downstream Reach ($X \geqq 0$)

a. *Conservative Substances*

$$C = C_o = W/Q \quad (17)$$

b. *Reactive Substances (System 1)*

$$L = L_o \exp(j_1 X) \quad (18)$$

where

$$j_1 = (U/2E)(1 - m_1) \quad (19)$$

c. *Sequentially Reactive Constituents (System 2)* The dissolved oxygen deficit (D) at downstream distance X can be determined by Eq. (20):

$$D = [L_oK_{12}/(K_2 - K_1)] [\exp(j_1X) - m_1m_2^{-1}\exp(j_2X)] \quad (20)$$

where

$$j_2 = (U/2E)(1 - m_2) \quad (21)$$

and the critical distance (X_c, in L) where the critical dissolved oxygen deficit (D_c, in M/L^3) occurs can be determined by Eq. (22):

$$X_c = \ln[m_1(1 - m_2)m_2^{-1}(1 - m_1)^{-1}]/[(U/2E)(m_2 - m_1)] \quad (22)$$

The critical dissolved oxygen deficit can then be calculated by:

$$D_c = [L_oK_{12}/(K_2 - K_1)] [\exp(j_1X_c) - m_1m_2^{-1}\exp(j_2X_c)] \quad (23)$$

B. Steady-State Equations for Waste Concentrations in Tidal Rivers and Estuaries Resulting from a Distributed Source of Pollution

The following working models are also derived from the general estuary model in Section III. They are provided by the US EPA, Cincinnati, Ohio. Figure 4 illustrates the tidal river and estuary conditions. Figure 5

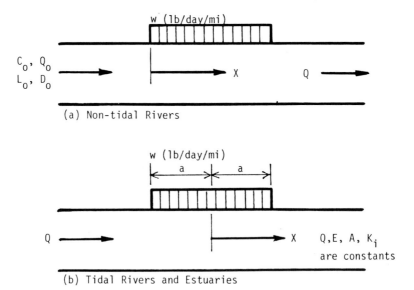

Fig. 4. Discharge of nonpoint source of pollutants into a receiving water system.

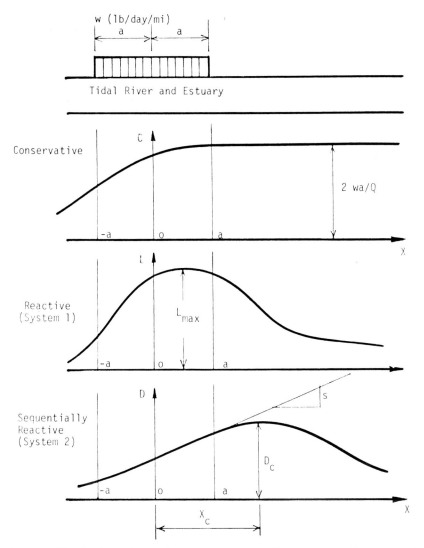

Fig. 5. Distribution of estuarine waste concentrations resulting from nonpoint source of pollution.

shows the distribution of estuarine waste concentrations for conservative substances, reactive substances, and sequentially reactive constituents. Table 7 further lists equations for calculating X_c and D_c for distributed sequentially reactive wastes. It is assumed that the distributed source (or

TABLE 7
Location of Maximum Concentration for Distributed Sequentially Reacting Waste[a]

$s = [dD/dx]$ at $x = a$		$s = \dfrac{e^{2aj_1} - 1}{m_1} - \dfrac{e^{2aj_2} - 1}{m_2}$	(65)

Value of s	Location of D_c	Implicit Equation for x_c and for D_c when $s = 0$	
I Positive	$x_c > a$	$\dfrac{e^{j_1(x_c + a)} - e^{j_1(x_c - a)}}{m_1} = \dfrac{e^{j_2(x_c + a)} - e^{j_2(x_c - a)}}{m_2}$	(66)
II Negative	$x_c < a$	$\dfrac{e^{j_1(x_c + a)} - e^{g_1(x_c - a)}}{m_1} = \dfrac{e^{j_2(x_c + a)} - e^{g_2(x_c - a)}}{m_2}$	(67)
III Zero	$x_c = a$	$D_c = \dfrac{W}{Q} \cdot \dfrac{K_{12}}{K_2 - K_1} \cdot \left[\dfrac{e^{2aj_1} - 1}{m_1 j_1} - \dfrac{e^{2aj_2} - 1}{m_2 j_2} \right]$	(68)

[a] Source: US EPA.

nonpoint source) of pollution (w, in lb/d/mi) is discharged into the receiving water according to Fig. 4, and has a total length of $2a$. The following subsections present the steady state equations for three types of waste concentrations (C, L, and D) in three zones ($X \leq a$, $-a \leq X \leq +a$, and $X \geq a$) assuming that the flow, dispersion coefficients, cross-sectional area, and reaction rates are all constants and X is the downstream distance starting from the center of the distributed source of pollution.

1. In the Reach Covered by the Distributed Source of Pollution ($-a \leq X \leq +a$)

a. Conservative Substances

$$C = wa/Q + X/a + (E/aU)\{1 - \exp[UE^{-1}(X - a)^{-1}]\} \quad (24)$$

b. Reactive Substances (System 1)

$$L = \dfrac{w}{Qm_1} \left[\dfrac{\exp[j_1(X + a)] - 1}{j_1} - \dfrac{\exp[g_1(X - a)] - 1}{g_1} \right] \quad (25)$$

c. Sequentially Reactive Constituents (System 2)

$$D = \frac{w}{Q}\frac{K_{12}}{K_2 - K_1}\left\{\left[\frac{\exp[j_1(X+a)] - 1}{m_1 j_1} - \frac{\exp[g_1(X-a)] - 1}{m_1 g_1}\right] - \left[\frac{\exp[j_2(X+a)] - 1}{m_2 j_2} - \frac{\exp[g_2(X-a) - 1}{m_2 g_2}\right]\right\} \quad (26)$$

2. **Upstream Reach ($X \leq -a$)**

   a. *Conservative Substances*

   $$C = [wE/(QU)]\{\exp[U/E(X+a)] - \exp[U/E(X-a)]\} \quad (27)$$

   b. *Reactive Substances (System 1)*

   $$L = [w/(Qm_1 g_1)]\{\exp[g_1(X+a)] - \exp[g_1(X-a)]\} \quad (28)$$

   c. *Sequentially Reactive Constituents (System 2)*

   $$D = \frac{w}{Q}\frac{K_{12}}{K_2 - K_1}\left\{\frac{\exp[g_1(X+a)] - \exp[g_1(X-a)]}{m_1 g_1} - \frac{\exp[g_2(X+a)] - \exp[g_2(X-a)]}{m_2 g_2}\right\} \quad (29)$$

3. **Downstream Reach ($X \geq +a$)**

   a. *Conservative Substances*

   $$C = 2wa/Q \quad (30)$$

   b. *Reactive Substances (System 1)*

   $$L = [w/(Qm j_1)]\{\exp[j_1(X+a)] - \exp[j_1(X-a)]\} \quad (31)$$

   c. *Sequentially Reactive constituents (System 2)*

   $$D = \frac{w}{Q}\frac{K_{12}}{K_2 - K_1}\left\{\frac{\exp[j_1(X+a)] - \exp[j_1(X-a)]}{m_1 j_1} - \frac{\exp[j_2(X+a)] - \exp[j_2(X-a)]}{m_2 j_2}\right\} \quad (32)$$

Four equations in Table 7 are presented for estimation of the location of maximum concentration (X_c) and critical dissolved oxygen deficit (D_c) of distributed sequentially reactive wastes. The symbol s in the table represents the slope of the dissolved oxygen deficit curve.

It should be noted that all water quality analyses are expected to be carried out using the steady-state Eqs. (7)–(32) for calculating waste concentrations in estuaries with constant geometry, hydrology, and kinetics and having continuous pollutant inputs. For study areas having significantly varying goemetry, flows, or kinetic rates (e.g., reaeration coefficients), appropriate analyses may be conducted to determine the sensitivity of the result to the varying parameters [11, 12].

C. Alternate Working Models and Systems Identification for Tidal Rivers

A procedure for systematic identification of characteristic parameters in water quality models by a moment method [8, 14, 15] is presented in detail. More specifically, the parameters identified in a steady-state tidal stream model include the reaeration coefficient (K_2), the coefficient of BOD settling rate and other variables (K_3), the longitudinal dispersion coefficient (E), the rate of BOD addition to the overlying water from the bottom deposits as well as the local run-off (K_s), and the oxygen production rate by photosynthesis (α). This section initially introduces the working models for tidal rivers, then presents the Moment Method. A mathematical model describing the effect of salinity on reaeration coefficients can be found in Appendix [18].

1. BOD and DO Deficit Models

In a tidal river, the following two simplified mathematical models for BOD and DO deficit can be derived from Eq. (3) based on the assumptions of: the tidal river is under steady-state conditions; C can be BOD concentration or DO concentration; DO deficit is equal to the saturation DO minus the DO in river water; and the tidal river system is well mixed vertically and horizontally, thus system segmentation is performed only along its main longitudinal axis.

$$E\frac{d^2L}{dX^2} - U\frac{dL}{dX} - (K_d + K_3)L + K_s = 0 \qquad (33)$$

$$E\frac{d^2D}{dX^2} - U\frac{dD}{dX} + K_dL - K_2D - \alpha = 0 \qquad (34)$$

in which, K_s is the rate of BOD addition to the overlying water from the bottom deposits and local run-off, α is the oxygen production rate by photosynthesis, E is the longitudinal dispersion coefficient, U is the net mean velocity of water flow, and X is the distance from the discharge point of pollutant. Integrating Eqs. (33) and (34), and substituting the initial and boundary conditions into them, one can obtain the following working mathematical models:

$$L = \left[L_o - \frac{K_s}{K_d + K_3}\right] \exp(J_1 X) + \frac{K_s}{K_d + K_3} \quad (35)$$

$$D = \frac{K_d}{K_2 - K_d - K_3}\left[L_o - \frac{K_s}{K_d + K_3}\right][\exp(J_1 X) - \exp(J_2 X)]$$

$$+ \frac{K_d}{K_2}\left[\frac{K_s}{K_d + K_3} - \frac{\alpha}{K_d}\right][1 - \exp(J_2 X)] + D_o \exp(J_2 X) \quad (36)$$

where

$$J_1 = \frac{U}{2E} - \left[\frac{U^2}{4E^2} + \frac{K_d + K_3}{E}\right]^{1/2} \quad (37)$$

$$J_2 = \frac{U}{2E} - \left[\frac{U^2}{4E^2} + \frac{K_2}{E}\right]^{1/2} \quad (38)$$

L_o = the initial concentration of remaining ultimate BOD, and D_o = the initial concentration of dissolved oxygen deficit.

2. Moment Method

The moment method [8] is introduced in this section by defining

$$P_k = \frac{K_s}{K_d + K_3} \quad (39)$$

$$a = \frac{K_d}{K_2 - K_d - K_3}\left[L_o - \frac{K_s}{K_d + K_3}\right] \quad (40)$$

and

$$b = \frac{K_d}{K_2}\left[\frac{K_s}{K_d + K_3} - \frac{\alpha}{K_d}\right] \quad (41)$$

Equations (35) and (36) become

$$L = L_o \exp(J_1 X) + P_k[1 - \exp(J_1 X)] \quad (42)$$

$$D = a[\exp(J_1X) - \exp(J_2X)] + b[1 - \exp(J_2X)] + D_o\exp(J_2X) \quad (43)$$

in which a, b, and P_k are constants. Both L and D are the functions of X. Equations (42) and (43) can be multiplied by X^β ($\beta = 0,1,2\ldots$), and the products are defined as the βth order moment about y axis. If there are n sets of observed data of L and X, by the theory of the moment method, the sum of moments on the right side of Eqs. (42) and (43) must be equal to the sum of the left side, that is

$$\sum_1^n X^\beta L = \sum_1^n X^\beta \{L_o\exp(J_1X) + P_k[1 - \exp(J_1X)]\} \quad (44)$$

$$\sum_1^n X^\beta D = \sum_1^n X^\beta \{a[\exp(J_1X) - \exp(J_2X)]\} + b[1 - \exp(J_2X)] + D_o\exp(J_2X) \quad (45)$$

Since L_o, P_k, a, b, and D_o are all constants, Eqs. (44) and (45) can be rearranged as

$$\sum_1^n X^\beta L = L_o\sum_1^n X^\beta \exp(J_1X) + P_k\sum_1^n X^\beta [1 - \exp(J_1X)] \quad (46)$$

$$\sum_1^n X^\beta D = a\sum_1^n X^\beta [\exp(J_1X) - \exp(J_2X)] + b\sum_1^n X^\beta [1 - \exp(J_2X)] + D_o\sum_1^n X^\beta \exp(J_2X) \quad (47)$$

Now there are only two unknown parameters (P_k and J_1) in Eq. (46). If one takes the zero order ($\beta = 0$) and the first-order ($\beta = 1$) moments from Eq. (44), the following two equations are obtained:

$$\beta = 0, \quad \sum_1^n L = L_o\sum_1^n \exp(J_1X) + P_k\sum_1^n [1 - \exp(J_1X)] \quad (48)$$

$$\beta = 1, \quad \sum_1^n LX = L_o\sum_1^n X\exp(J_1X) + P_k\sum_1^n X[1 - \exp(J_1X)] \quad (49)$$

The P_k and J_1 can then be solved by the numerical methods from Eqs. (48) and (49). The advantage of taking the lower order moments (the zero and the first) for estimating parameters is that the calculation is very simple, and the error is less than that of the higher orders.

Similarly, it will be able to solve J_2, a and b by taking the third-order moments from Eq. (47):

$$\beta = 0, \sum_1^n D = a \left[\sum_1^n \exp(J_1 X) - \sum_1^n \exp(J_2 X) \right]$$
$$+ b \left[n - \sum_1^n \exp(J_2 X) \right] + D_0 \sum_1^n \exp(J_2 X) \quad (50)$$

$$\beta = 1, \sum_1^n DX = a \left[\sum_1^n X\exp(J_1 X) - \sum_1^n X\exp(J_2 X) \right]$$
$$+ b \left[\sum_1^n X - \sum_1^n X\exp(J_2 X) \right] + D_0 \sum_1^n \exp(J_2 X) \quad (51)$$

$$\beta = 2, \sum_1^n DX^2 = a \left[\sum_1^n X^2\exp(J_1 X) - \sum_1^n X^2\exp(J_2 X) \right]$$
$$+ b \left[\sum_1^n X^2 - \sum_1^n X^2\exp(J_2 X) \right] + D_0 \sum_1^n X^2\exp(J_2 X) \quad (52)$$

3. *Procedures for System Identification*   Equations (48) to (52) are nonlinear, but can be solved by the numerical methods or a graphical method. The procedures of calculation are listed below:
   (a) Rearrange Eqs. (48) and (49), and obtain Eqs. (53) and (54), respectively.

$$P_k = \frac{\sum_1^n L - L_0 \sum_1^n \exp(J_1 X)}{n - \sum_1^n \exp(J_1 X)} \quad (53)$$

$$P_k = \frac{\sum_1^n LX - L_0 \sum_1^n \exp(J_1 X)}{\sum_1^n X - \sum_1^n X\exp(J_1 X)} \quad (54)$$

   (b) Substitute the observed data of X, L, L_o and n into Eqs. (53) and 54).

(c) Assume a suitable J_1 value and substitute it into Eqs. (53) and (54) in order to calculate the P_k values. Repeat this procedure for different assumed values of J_1.
(d) Plot the calculated P_k values against the assumed J_1 values on an arithmetic graphical sheet, and obtain two curves; the point of intersection gives the solutions of P_k and J_1.
(e) From Eq. (50), Eq. (55) is derived.

$$b = \frac{\sum_1^n D - a\left[\sum_1^n \exp(J_1 X) - \sum_1^n \exp(J_2 X)\right] - D_0 \sum_1^n \exp(J_2 X)}{n - \sum_1^n \exp(J_2 X)} \qquad (55)$$

Equations (56) and (57) are then derived by substituting Eq. (55) into Eqs. (51) and (52):

$$a = \left\{\left[n - \sum_1^n \exp(J_2 X)\right]\left(\sum_1^n DX\right) + \left[\sum_1^n X \exp(J_2 X) - \sum_1^n X\right]\sum_1^n D \right.$$
$$+ D_0\left[\sum_1^n \exp(J_2 X)\sum_1^n X - n\sum_1^n X \exp(J_2 X)\right]\right\}$$
$$\div \left\{n\left[\sum_1^n X \exp(J_1 X) - \sum_1^n X \exp(J_2 X)\right]\right.$$
$$+ \sum_1^n \exp(J_2 X)\left[\sum_1^n X - \sum_1^n X \exp(J_1 X)\right]$$
$$\left. + \sum_1^n \exp(J_1 X)\left[\sum_1^n X \exp(J_2 X) - \sum_1^n X\right]\right\} \qquad (56)$$

$$a = \left\{\left[n - \sum_1^n \exp(J_2 X)\right]\left(\sum_1^n DX^2\right) + \left[\sum_1^n X^2 \exp(J_2 X) - \sum_1^n X^2\right]\sum_1^n D \right.$$
$$+ D_0\left[\sum_1^n \exp(J_2 X)\sum_1^n X^2 - n\sum_1^n X^2 \exp(J_2 X)\right]\right\}$$
$$\div \left\{n\left[\sum_1^n X^2 \exp(J_1 X) - \sum_1^n X^2 \exp(J_2 X)\right]\right.$$

$$+ \sum_{1}^{n} \exp(J_2 X) \left[\sum_{1}^{n} X^2 - \sum_{1}^{n} X^2 \exp(J_1 X) \right]$$

$$+ \left[\sum_{1}^{n} \exp(J_1 X) \right] \left[\sum_{1}^{n} X^2 \exp(J_2 X) - \sum_{1}^{n} X^2 \right] \Big\} \quad (57)$$

(f) The procedures for solving the parameters of J_2 and a are similar to that for solving J_1 and P_k [described in procedures (c) and (d)].

(g) Obtain the b value by substituting J_2 and P_k values into Eq. (55).

It is important to know that the values of J_1, J_2, a, b, and P_k that can be calculated from Eqs. (48)–(52) are not real parameters of water quality. The real parameters are K_2, K_3, K_s, α, and E; these should be further evaluated by Eqs. (37)–(41).

From Eqs. (37) and (38), Eqs. (58) and (59) are derived.

$$K_3 - J_1^2 E = -(UJ_1 + K_d) \quad (58)$$

$$K_2 - J_2^2 E = -J_2 U \quad (59)$$

By simplifying Eqs (39)–(41), one can obtain Eqs. (60)–(62), respectively.

$$P_k K_3 - K_s = -P_k K_d \quad (60)$$

$$a K_2 - a K_3 = K_d(a + L_o - P_k) \quad (61)$$

$$b K_2 + \alpha = K_d P_k \quad (62)$$

Equations (58)–(62) are linear equations with five unknowns. It is easy to evaluate K_2, K_3, K_s, α, and E except when the following determinant equals zero:

$$\begin{vmatrix} 0 & 1 & 0 & 0 & -J_1^2 \\ 1 & 0 & 0 & 0 & -J_2^2 \\ 0 & -P_k & -1 & 0 & 0 \\ a & -a & 0 & 0 & 0 \\ b & 0 & 0 & 1 & 0 \end{vmatrix} = 0 \quad (63)$$

The solutions of Eq. (63) are

$$L_o = P_k = K_s/(K_d + K_3) \quad \text{and} \quad J_1 = J_2$$

i.e., the parameters cannot be determined by any method under these conditions.

V. PRACTICAL EXAMPLES

Example 1

Tam-Sui River, a tidal river located in Taipei, Taiwan, Republic of China, has been selected for field investigation. Figure 6 shows its geo-

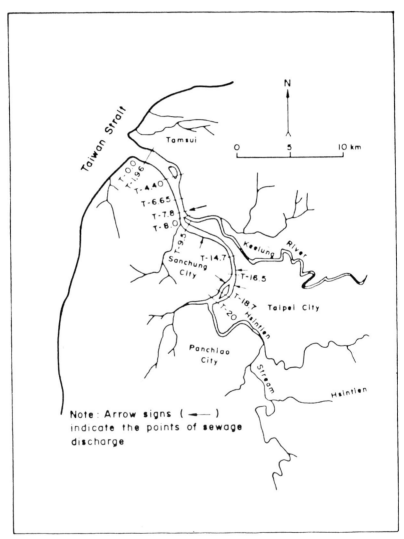

Fig. 6. Tam-Sui river and its sampling stations [15].

graphical location, sampling stations, tributaries, and point-sources of pollution. The length of the river is about 20 km. There are two main tributaries, Keelung River and Hsintien Stream, that discharge their flows into Tam-Sui River at 8 and 20 km, respectively, from Tam-Sui's mouth. The estuary is used mainly for recreation. Tam-Sui River is divided into two sections according to its hydraulic conditions. Section I is from Stations T-20.0 to T-8.0; while Section II is from Stations T-8.0 to T-0.0 (the numbers represent the distance in kilometers from the river mouth). The mean velocities (U) are 7.56 and 4.58 km/d for Sections I and II, respectively. Both sections are affected by tidal action.

Partial field data are listed in Table 8. Estimate the five water quality parameters, K_2, K_3, K_s, α, and E by a moment method [8, 14, 15].

Solution

It is seen from Table 8 that the initial concentrations of first-stage ultimate BOD (L_o) are 5.76 mg/L for Section I and 7.97 mg/L for Section II. The initial concentrations of dissolved oxygen deficit (D_o) are 5.6 mg/L for Section I, and 3.7 mg/L for Section II.

TABLE 8
Water Quality Data of Tam-Sui River During Summer, 1969 [15]

(1)	(2)	(3)	(4)	(5)	(6)	(7)	(8)	(9)
Section	Section	X	BOD[a] (5-day 28°C), mg/L	DO[a] (28°C), mg/L	k_d, 28°C (base 10) d^{-1}	Saturated DO[b] at 28°C, mg/L	L, mg/L	D, mg/L
I	T-18.70	0	5.30	2.30	0.22	7.90	5.76	5.60
	T-16.50	2.20	5.30	1.80	0.14	7.90	6.62	6.10
	T-14.70	4.00	6.00	2.35	0.12	7.90	8.02	5.55
	T-9.50	9.20	5.40	3.25	0.14	7.75	6.75	4.50
II	T-7.80	0	6.55	3.85	0.15	7.55	7.97	3.70
	T-6.65	1.15	5.05	4.05	0.15	7.52	6.14	3.47
	T-4.45	3.35	4.35	4.80	0.15	7.24	5.29	2.44
	T-1.35	6.45	2.70	5.00	0.15	7.14	3.29	2.14
	T-0.00	7.80	2.40	5.48	0.15	6.75	2.92	1.27

[a]Tidal average.
[b]Corrected by chloride concentration and atmospheric pressure.

The average k_d (28°C, base 10) values that are weighted by distance are 0.142 d^{-1} for Section I and 0.150 d^{-1} for Section II. Both are calculated with the data in Column 6, Table 8.

1. Tam-Sui River, Section I

By assuming J_1 values ranging from -0.01 to -1.00, two sets of P_k values are calculated by Eqs. (53) and (54), and listed in Table 9. The assumed J_1 values are then plotted against the two sets of the calculated P_k values, therefore, two curves (J_1 vs P_k) are obtained, as shown in Fig. 7. It is seen from the figure that two curves do not intersect each other; thus, there is no solution for J_1 and P_k for Section I.

There are no sanitary sewerage systems in the Cities of Taipei and Sanchung. The untreated raw sewage from the two cities is discharged into the Section I of Tam-Sui River (Stations T-20.0 to T-8.0) at several points, as shown in Fig. 6. It is concluded that the water quality of this section is not suitable for modeling unless the section is divided into several smaller parts.

2. Tam-Sui River, Section II

Section II is suitable for water quality modeling, and therefore is described in detail. Figure 6 shows that the major point source of organic pollution is from Station 7.8, where L_o and D_o have been estimated to be

TABLE 9
The Values of P_k with Respect to J_1 for the First Section of Tam-Sui River

Assumed J_1	Calculated P_k		Assumed J_1	Calculated P_k	
	Eq. (53)	Eq. (54)		Eq. (53)	Eq. (54)
−0.01	33.36878	25.53210	−0.55	7.35015	7.16679
−0.05	12.04294	10.36976	−0.60	7.31791	7.15017
−0.10	9.40154	8.50963	−0.65	7.29135	7.13674
−0.15	8.53574	7.91242	−0.70	7.26924	7.12574
−0.20	8.11150	7.62821	−0.75	7.25066	7.11662
−0.25	7.86239	7.46710	−0.80	7.23493	7.10899
−0.30	7.69990	7.36598	−0.85	7.22154	7.10255
−0.35	7.58631	7.29799	−0.90	7.21006	7.09710
−0.40	7.50292	7.24990	−0.95	7.20019	7.09245
−0.45	7.43946	7.21453	−1.00	7.19166	7.08846
−0.50	7.38982	7.18768			

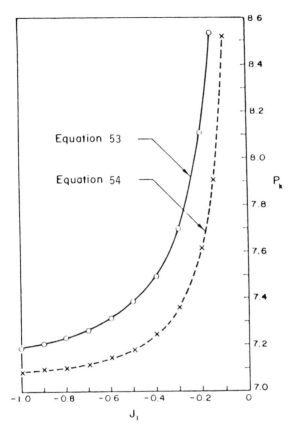

Fig. 7. The relationships of J_1 and P_k in Tam-Sui River, Section I (Stations T-9.5 to T-18.7).

7.97 and 3.7 mg/L, respectively. There are five sampling stations providing water quality data, as indicated in Table 8; therefore, n equal to 5. The average values of U and k_d (28°C, base 10) for this section are 4.58 km/d and 0.150 d^{-1}, respectively.

By substituting the observed data of X, L, L_o and n, and the assumed J_1 values into Eqs. (53) and (54), two sets of P_k values are calculated and listed in Table 10. For example, P_k values are calculated to be 2.21741 and 2.22578 by Eqs. (53) and (54) respectively, when J_1 is assumed to be −0.250.

With the data in Table 10, the relationship of J_1 and P_k in Tam-Sui River, Section II, is plotted as Fig. 8. The solutions for P_k and J_1 are

TABLE 10
The Values of P_k with Respect to J_1 for the Second Section of the Tam-Sui River

Assumed J_1	Calculated P_k		Assumed J_1	Calculated P_k	
	Eq. (53)	Eq. (54)		Eq. (53)	Eq. (54)
−0.050	−9.63830	−8.56550	−0.280	2.50776	2.47081
−0.100	−2.12766	−1.66788	−0.290	2.59008	2.53925
−0.150	0.62831	0.82950	−0.300	2.66642	2.60221
−0.200	1.52174	1.62268	−0.310	2.73737	2.66026
−0.210	1.68928	1.76946	−0.320	2.80346	2.71388
−0.220	1.84064	1.90134	−0.330	2.86515	2.76351
−0.230	1.97797	2.02029	−0.340	2.92284	2.80951
−0.240	2.10306	2.12798	−0.350	2.97689	2.85222
−0.250	2.21741	2.22578	−0.360	3.02761	2.89194
−0.260	2.32228	2.31488	−0.370	3.07531	2.92893
−0.270	2.41876	2.39627	−0.380	3.12022	2.96343

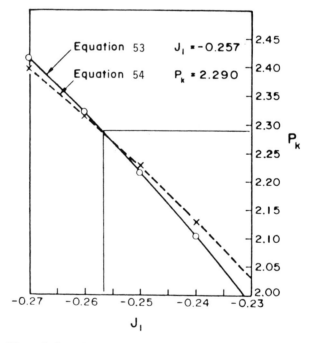

Fig. 8. The relationships of J_1 and P_k in Tam-Sui River, Section II (Stations T-0.0 to T-7.8).

2.290 and -0.257, respectively, obtained according to the procedure outlined in Section IV.C.3.

By substituting the observed data of X, D, D_o, and n, and the assumed J_2 values into Eqs. (56) and (57), two sets of "a" values are calculated and listed in Table 11. For example, when J_2 is assumed to be -0.130, the calculated "a" values are -2.37126 and -2.37069 by Eqs. (56) and (57), respectively.

The relationship of J_2 and a in the Section II of Tam-Sui River is then graphically illustrated in Fig. 9. The solutions of J_2 and a are -0.132 and -2.450, respectively, obtained according to the procedure outlined in Section IV.C.3.

By choosing the D, D_o, X, and n values from Table 8, and substituting the D, D_o, X, n, a, J_1, and J_2 into Eq. (55), the value of "b" can then be calculated to be -0.631 according to Eq. (55).

Knowing $L_o = 7.97$ mg/L, $U = 4.58$ km/d, k_d (28°C, base 10) = 0.15 d^{-1}, $J_1 = -0.257$, $P_k = 2.290$, $J_2 = -0.132$, $a = -2.450$, and $b = 0.631$, one could then determine the real water quality parameters using Eqs. (58)–(62):

$$K_3 - (-0.257)^2 E = -[4.58(-0.257) + 0.15 \times 2.3] \quad (58)$$

$$K_2 - (-0.132)^2 E = 0.132 \times 4.58 \quad (59)$$

$$2.290 K_3 - K_s = -2.290 \times 0.15 \times 2.3 \quad (60)$$

TABLE 11
The Values of a with Respect to J_2 for the Second Section of the Tam-Sui River

Assumed J_2	Calculated a		Assumed J_2	Calculated a	
	Eq. (56)	Eq. (57)		Eq. (56)	Eq. (57)
-0.090	-1.19604	-1.18092	-0.145	-3.03142	-3.03926
-0.095	-1.13073	-1.29701	-0.150	-3.29304	-3.30425
-0.100	-1.43281	-1.42060	-0.155	-3.58055	-3.59543
-0.105	-1.56315	-1.55248	-0.160	-3.89792	-3.91692
-0.110	-1.70239	-1.69352	-0.165	-4.25008	-4.27360
-0.115	-1.85167	-1.84464	-0.170	-4.64284	-4.67149
-0.120	-2.01197	-2.00698	-0.175	-5.08392	-5.11826
-0.125	-2.18463	-2.18178	-0.180	-5.58240	-5.62327
-0.130	-2.37108	-2.37055	-0.185	-6.15055	-6.19880
-0.135	-2.57293	-2.57499	-0.190	-6.80359	-6.86045
-0.140	-2.79230	-2.79710	-0.195	-7.56237	-7.62919

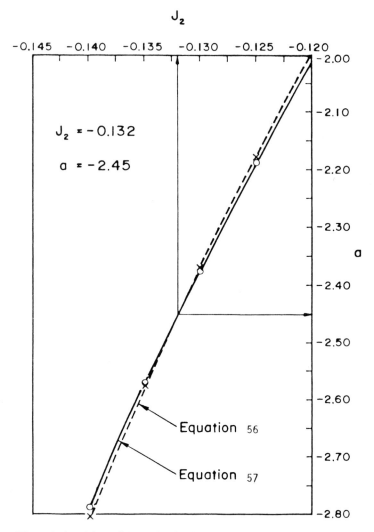

Fig. 9. The relationships of J_2 and a in the Tam-Sui river, Section II (stations T-0.0 to T-7.8).

$$-2.450K_2 + 2.450K_3 = 0.15 \times 2.3 \qquad (61)$$
$$(-2.450 + 7.97 - 2.290)$$

$$-0.631K_2 + \alpha = 0.15 \times 2.3 \times 2.290 \qquad (62)$$

The solutions of the above five simultaneous linear equations are:

$K_2(28°C$, base e$) = 0.69$ d^{-1}; $K_3(28°C$, base e$) = 1.14$ d^{-1}; $K_s = 3.40$ mg/L/d; $\alpha = 1.22$ mg/L/d; and $E = 4.7$ km^2/d.

The preceding computations yielded the following two specific mathematical models for Tam-Sui River, Section II:

Biochemical Oxygen Demand Model

$$L = 5.680 \exp(-0.257X) + 2.290 \qquad (35a)$$

Dissolved Oxygen Deficit Model

$$D = 6.781 \exp(-0.132X) - 2.450 \exp(-0.257X) - 0.631 \qquad (36a)$$

which are derived from Eqs. (35) and (36) because the values of L_o, K_s, E, K_d, K_2, K_3, J_1, J_2, α, D_o, and U are all known or identified.

After the Biochemical Oxygen Demand Model [Eq (35a)] and the Dissolved Oxygen Deficit Model [Eq. (36a)] for a target stream are developed, environmental engineers and planners can use them to determine the remaining concentration of first-stage ultimate BOD (L), and the concentration of dissolved oxygen deficit (D) at any downstream distance X. Columns 1, 2, and 4 of Table 12 indicate the stations, the observed BOD, and the observed DO deficit, respectively, that can also be found from Table 8. Columns 3 and 5 indicate the BOD and DO deficit values estimated by Eqs. (35a) and (36a), respectively. It should be noted that Station T-7.80 is considered to be a starting point, thus $X = 0$ mile at Station T-7.80 and $X = 7.8$ miles at Station T-0.00. It can be seen that the estimated L and D values are in close agreement with the observed L and D values. The BOD Model and the DO Deficit Model are accepted under 5 and 1%, respectively, of significance levels according to the chi-square test.

TABLE 12
Evaluation of Biochemical Oxygen Demand and Dissolved Oxygen Models[a]

Station	Biochemical oxygen demand, mg/L		Dissolved oxygen deficit, mg/L	
	Observed L	Estimated L	Observed D	Estimated D
T-7.80	7.97	7.97	3.70	3.70
T-6.65	6.14	6.52	3.47	3.37
T-4.45	5.29	4.69	2.44	2.69
T-1.35	3.29	3.37	2.14	1.80
T-0.00	2.92	3.06	1.27	1.46

[a]Source: Refs. [14] and [15].

It is then concluded that the moment method is an effective method in estimating the characteristic parameters of steady-state stream models. The errors are very small if the number of parameters in the models are five or less. The procedure for system identification is mathematically simple. Water quality parameters can be estimated easily by using an electronic computer [14]. When necessary, the sensitivities of various water quality parameters can also be determined [16,17].

Example 2

A low flow analysis for a continuous point source discharge has been conducted. The following are the given information:

Freshwater low flow (Q) = 50 ft^3/s
Mean water depth (h) = 10 ft
Average cross-sectional area (A) = 2000 ft^2
Average tidal velocity (U_T) = 0.6 knots
Dispersion coefficient (E) = 2 mi^2/d
$K_1 = K_r = 0.25$ d^{-1}
$K_{12} = K_d = K_r$
$K_2 = K_a$
$K_b = 2$ d^{-1}
$K_L = 4$ ft/d
PS (i.e. point source) effluent flow = 10 ft^3/s
PS effluent total nitrogen = 20 mg/L
PS effluent UOD = 120 mg/L
PS effluent total coliform = 1000 MPN/100 mL

Figure 10 shows both actual and idealized estuarine systems. Determine the following important water quality parameters and models under low freshwater flow conditions:

1. Total nitrogen concentrations at

$X = 0$, $X \leq 0$, and $X \geq 0$

2. Total coliform concentrations at

$X = 0$, $X \leq 0$, and $X \geq 0$

3. Dissolved oxygen deficit at

$X = X_c$, $X \leq 0$, and $X \geq 0$

where X = downstream distance from the source of pollution, and X_c = critical distance where the DO deficit is maximum.

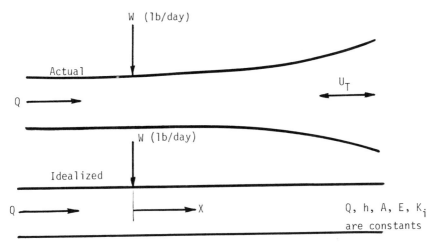

Fig. 10. Example of an estuary with a single-point source of pollution.

Solution

1. Determination of Total Nitrogen Concentrations:
   Average freshwater velocity

$(U) = Q/A = 50/2000 = 0.025$ ft/s $= 0.41$ mi/d
1 ft³/s-mg/L $= 1$ cfs-mg/L $= 5.4$ lb/d
$W_{tn} = W = (10)(20)(5.4) = 1080$ lb/d
$C_{max} = C_o = W/Q$ (7)
    $= 1080/(50 \times 5.4)$
    $= 4.0$ mg/L at $X = 0$
$C = C_o \exp(UX/E)$ (12)
    $= 4 \exp(0.41X/2)$
$C = 4 \exp(0.205X)$ at $X \leq 0$
$C = C_o = 4$ mg/L at $X \geq 0$ (17)

2. Determination of Total Coliform Concentrations

Estuary number $(N) = K_b E/U^2$ (2)
$N = (2$ d$^{-1})(2$ mi²/d$)/(0.41$ mi/d$)^2 = 23.8$
$m_1 = (1 + 4N)^{0.5} = (1 + 4 \times 23.8)^{0.5} = 9.81$ (9)
$j_1 = (U/2E)(1 - m_1)$ (19)
$j_1 = (0.41/2 \times 2)(1 - 9.81)$
   $= -0.903$ mile^{-1}

$$g_1 = (U/2E)(1 - m_1) \tag{14}$$
$$= (0.41/2 \times 2)(1 + 9.81)$$
$$= +1.108 \text{ mile}^{-1}$$

Total coliform discharge $(W_c) = W$

$W = (10 \text{ ft}^3/\text{s})(1000 \text{ MPN}/100 \text{ mL})$

$$L_{max} = L_o = W/(Qm_1) \tag{8}$$
$$= (10 \text{ ft}^3/\text{s})(1000 \text{ MPN}/100 \text{ mL})/[(50 \text{ ft}^3/\text{s})(9.81)]$$
$$= 20 \text{ MPN}/100 \text{ mL at } X = 0$$

$$L = L_o \exp(j_1 X) \tag{18}$$
$$= 20 \exp(-0.930 X) \text{ at } X \geqq 0$$

$$L = L_o \exp(g_1 X) \tag{13}$$
$$= 20 \exp(1.108 X) \text{ at } X \leqq 0$$

3. Determination of Dissolved Oxygen Deficit Concentrations

Average tidal velocity $(U_T) = 0.6$ knots

$U_T = (0.6 \text{ naut.mi/h})(1.15 \text{ stat.mi/naut.mi}) [88 \text{ ft/s}/(60 \text{ mi/h})]$
$= 1 \text{ ft/s}$

Mean water depth $(h) = 10$ ft

$K_L = 4 \text{ ft/d}$

$$K_2 = K_a = K_L/h = 4/10 = 0.4 \text{ d}^{-1} \tag{1}$$
$$N_1 = K_1 E/U^2 = (0.25 \times 2)/(0.41)^2 = 2.974 \tag{2}$$
$$m_1 = (1 + 4 N_1)^{0.5} = 3.591 \tag{9}$$
$$N_2 = K_2 E/U_2 = (0.4 \times 2)/(0.41)^2 = 4.759 \tag{2}$$
$$m_2 = (1 + 4 N_2)^{0.5} = 4.476 \tag{11}$$
$$j_1 = (U/2E)(1 - m_1) \tag{19}$$
$$= [0.41/(2 \times 2)] (1 - 3.591)$$
$$= -0.266 \text{ mile}^{-1}$$
$$j_2 = (U/2E)(1 - m_2) \tag{21}$$
$$= [0.41/(2 \times 2)] (1 - 4.476)$$
$$= -0.356 \text{ mile}^{-1}$$
$$g_1 = (U/2E)(1 + m_1) \tag{14}$$
$$= [0.41/(2 \times 2)] (1 + 3.591)$$
$$= 0.471 \text{ mile}^{-1}$$
$$g_2 = (U/2E)(1 + m_2) \tag{16}$$
$$= [0.41/(2 \times 2)] (1 + 4.476)$$
$$= 0.561 \text{ mile}^{-1}$$

$W_{UOD} = W = \text{UOD discharged} = (10)(120)(5.4) = 6480 \text{ lb/d}$

$$X_c = \frac{\ln[m_1 (1 - m_2)m_2^{-1} (1 - m_1)^{-1}]}{(U/2E)(m_2 - m_1)} \quad (22)$$

$$= \frac{\ln[3.591(1 - 4.476)4.476^{-1}(1 - 3.591)^{-1}]}{(0.41/2 \times 2)(4.476 - 3.591)}$$

$= 0.811$ miles

$L_o = W/Qm_1$ (8)
$= 6480/(50 \times 5.4 \times 3.591)$
$= 6.68$ mg/L

$K_{12} = K_d = 0.25$ d^{-1}

$D_c = [L_o K_{12}/(K_2 - K_1)] [\exp(j_1 X_c) - m_1 m_2^{-1} \exp(j_2 X_c)]$ (23)
$= [6.68 \times 0.25/(0.4 - 0.25)] \times [\exp(-0.266 \times 0.811)$
   $- 3.591(4.476)^{-1} \times \exp(-0.356 \times 0.811)]$
$= 11.14 (0.2049)$
$= 2.28$ mg/L at $X = X_c$ or critical point

$D = [L_o K_{12}/(K_2 - K_1)] [\exp(g_1 X) - m_1 m_2^{-1} \exp(g_2 X)]$ (15)
$= [6.68 \times 0.25/(0.4 - 0.25)] \times [\exp(0.471X)$
   $- 3.591(4.476)^{-1} \exp(0.561X)]$
$= 11.14[\exp(0.471X) - 0.802 \exp(0.561X)]$
at $X \leq 0$ or upstream

$D = [L_o K_{12}/(K_2 - K_1)] [\exp(j_1 X) - m_1 m_2^{-1} \exp(j_2 X)]$ (20)
$= [6.68 \times 0.25/(0.4 - 0.25)] \times [\exp(-0.266X)$
   $- 3.591(4.476)^{-1} \exp(-0.356X)]$
$= 11.14 [\exp(-0.266X) - 0.802 \exp(-0.356X)]$
at $X \geq 0$ or downstream.

Example 3

An estuary with multiple waste sources is being investigated. The following are the given information:

Freshwater low flow $(Q) = 50$ ft^3/s
Freshwater summer average flow $(Q) = 300$ ft^3/s
Freshwater summer storm flow $(Q) = 900$ ft^3/s
Mean water depth $(h) = 10$ ft
Average cross-sectional area $(A) = 5000$ ft^2
Average tidal velocity $(U_T) = 0.2$ knots
Dispersion coefficient $(E) = 2$ mi^2/d
$K_1 = K_r = 0.25$ d^{-1}

$K_{12} = K_d = K_r$
$K_2 = K_a$
$K_b = K_1 = 2 \; d^{-1}$
PS (i.e., point source) effluent flow = 20 ft^3/s
PS effluent total coliform = 10,000 MPN/100 mL
PS effluent UOD = 160 mg/L
NPS (i.e., nonpoint source) UOD = 80 mg/L
NPS total coliform concentration = 3 × 10^5 MPN/100 mL
NPS summer average flow = 20 ft^3/s
NPS summer storm flow = 100 ft^3/s
NPS distribution = uniformly distributed over 4 mi

Figure 11 shows the actual and idealized estuarine systems. Determine the following important water quality parameters and models under the stated freshwater flow conditions:

1. Total coliform and UOD of PS and NPS under three freshwater flow conditions: low, summer average, and summer storm flow.
2. Total coliform concentrations at Milepoints 38–60 under summer storm conditions (see Fig. 11 for milepoints and definitions of X and X_p).

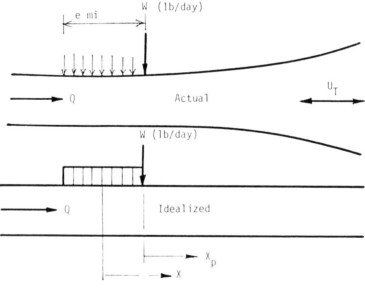

Fig. 11. Estuary with both point source and nonpoint sources of pollution.

WATER QUALITY IN RIVERS AND ESTUARIES   97

3. Maximum total coliform concentration and its location in terms of milepoint (MP) under summer storm conditions.
4. Dissolved oxygen deficit (D) at Milepoints 27–60 under summer average flow conditions.
5. Maximum dissolved oxygen deficit and critical distance.

Solution

1. Determination of Total Coliform and UOD under Various Flow Conditions. Total coliform of PS (ft^3/s-MPN/100 mL) = (flow, ft^3/s) × (concentration, MPN/100 mL)
Total coliform of NPS (ft^3/s-MPN/100 mL/mi) = (flow, ft^3/s) (concentration, MPN/100 mL)/(distributed length, mi)
UOD of PS (lb/d) = 5.4 (flow, ft^3/s) (concentration, mg/L)
UOD of NPS (lb/d/mi) = 5.4 (flow, ft^3/s) (concentration, mg/L)/(length, mi)
The answers are summarized below:

Flow conditions	Point source, W	
	Total coliform, ft^3/s-MPN/100 mL	UOD, lb/d
Low	20 × 10000	20 × 160 × 5.4 = 17,280
Summer average	20 × 10000	20 × 160 × 5.4 = 17,280
Summer storm	20 × 10000	20 × 160 × 5.4 = 17,280

Flow conditions	Nonpoint source, w	
	Total coliform, ft^3/s-MPN/100 mL	UOD, lb/d/mi
Low	0	0
Summer average	—	20 × 80 × 5.4/4 = 2160
Summer storm	100 × 3 × 10^5/4 = 7.5 × 10^6	—

2. Determination of Total Coliform Concentrations at Various Milepoints under Summer Storm Conditions
Q = 900 ft^3/s
U = Q/A = (900/5000)16.4 = 2.95 mi/d
E = 2 mi^2/d

$K_b = 2 \text{ d}^{-1}$
$N = 2 \times 2/(2.95)^2 = 0.46$ (2)
$m_1 = (1 + 4 \times 0.46)^{0.5} = 1.685$ (9)
$g_1 = [2.95/(2 \times 2)](1 + 1.685) = 1.98 \text{ mi}^{-1}$ (14)
$j_1 = [2.95/(2 \times 2)](1 - 1.685) = -0.505 \text{ mi}^{-1}$ (19)

Coliform concentrations from PS.

$X_p = 0$ at MP 52 (see Fig. 11)
$W = (20 \text{ ft}^3/\text{s})(10{,}000 \text{ MPN}/100 \text{ mL})$
$L_o = W/(Qm_1)$ (8a)
$\quad = (20 \times 10000)/(900 \times 1.685)$
$\quad = 132 \text{ MPN}/100 \text{ mL at } X_p = 0$
$L = L_o \exp(g_1 X_p)$ (13a)
$\quad = 132 \exp(-0.505 X_p)$ at $X_p \leq 0$
$L = L_o \exp(j_1 X_p)$ (18a)
$\quad = 132 \exp(-0.505 X_p)$ at $X_p \geq 0$

Using Eqs. (13a) and (18a), the PS data in the first three columns of Table 13 can be completed. The maximum coliform concentration caused by the PS is 132 MPN/100 mL at $X_p = 0$ according to Eq. (8a).

Coliform concentrations from NPS

$X = 0$ at MP 54 (see Fig. 11)
$a = 2 \text{ mi}$
$w = (100 \text{ ft}^3/\text{s})(3 \times 10^5 \text{ MPN}/100 \text{ mL})/(4 \text{ mi})$
$\quad = 7.5 \times 10^6 \text{ ft}^3/\text{s-MPN}/100 \text{ mL/mi}$

L at $X \leq -a$ (or ≤ -2)
$\quad = [w/(Qm_1 g_1)] \{\exp[g_1(X + a)] - \exp[g_1(X - a)]\}$ (28)
$\quad = [7.5 \times 10^6/(900 \times 1.685 \times 1.98)] \times \{\exp[1.98(X + 2)] - \exp[1.98(X - 2)]\}$
$\quad = 2498\{\exp[1.98(X + 2)] - \exp[1.98(X - 2)]\}$ (28a)

L at $X \geq a$ (or $X \geq 2$)
$\quad = [w/(j_1 Q m_1)] \{\exp[j_1(X + a)] - \exp[j_1(X - a)]\}$ (31)
$\quad = [7.5 \times 10^6/(-0.505 \times 900 \times 1.685)] \times \{\exp[-0.505(X + 2)] - \exp[-0.505(X - 2)]\}$
$L = -9793\{\exp[-0.505(X + 2)] - \exp[-0.505(X - 2)]\}$ (31a)

L at $-a \leq X \leq a$ (or $-2 \leq X \leq 2$)

$$L = \frac{w}{Qm_1} \left[\frac{\exp[j_1(X + a)] - 1}{j_1} - \frac{\exp[g_1(X - a)] - 1}{g_1} \right] \quad (25)$$

TABLE 13
Total Coliform Analysis of an Estuary with Multiple Waste Sources under Summer Storm Flow Conditions[a]

MP	PS X_p	PS L	NPS X	NPS L	Total L, MPN/100 mL
60	−8	0	−6	1	1
58	−6	0	−4	48	48
57	−5	0	−3	345	345
56	−4	0	−2	2498	2498
55	−3	0	−1	6375	6375
54	−2	3	0	8677	8680
53	−1	18	1	9794	9812
52.81	−0.81	26	1.19	9834[b]	9860[b]
52	0	132[b]	2	8495	8627
51	1	80	3	5126	5206
50	2	48	4	3094	3142
49	3	29	5	1867	1896
48	4	18	6	1127	1145
47	5	11	7	680	691
46	6	6	8	410	416
45	7	4	9	248	252
44	8	2	10	149	151
42	10	1	12	54	55
40	12	0	14	20	20
38	14	0	16	7	7
Column: (1)	(2)	(3)	(4)	(5)	(6)

[a]Example 3.
[b]Maximum Concentration

$$L = \frac{7.5 \times 10^6}{900 \times 1.685} \left[\frac{\exp[-0.505(X + 2)] - 1}{-0.505} - \frac{\exp[1.98(X - 2)] - 1}{1.98} \right]$$

$$L = 4946 \left[\frac{\exp[-0.505(X + 2)] - 1}{-0.505} - \frac{\exp[1.98(X - 2)] - 1}{1.98} \right] \quad (25a)$$

Using Eqs. (28a), (31a), and (25a), the NPS data in columns 1, 4, and 5 of Table 13 can be completed. Finally, the total coliform concentration caused by both PS and NPS are calculated and listed in column 6 of the same table.

3. Determination of Maximum Total Coliform Concentration and Its Location

Maximum total coliform from PS

$$L_{max} = L_o = 132 \text{ MPN}/100 \text{ mL at } X_p = 0$$

according to Eq. (8a)

Maximum total coliform from NPS. L_{max} is located at (see Fig. 5):

$$X_1 = a/m_1 \tag{64}$$

where X_1 is the location of L_{max}, and

$$0 \leq X_1 \leq a$$

Therefore,

$$X_1 = 2/1.685 = 1.19 \text{ mi} = X$$
$$L_{max} = 9834 \text{ MPN}/100 \text{ mL} \tag{25a}$$

Maximum total coliform from PS and NPS   From Table 13,

$$L_{max} = 9860 \text{ MPN}/100 \text{ mL}$$
$$\text{at } X = 1.19 \text{ mi or MP} = 52.81$$

4. Determination of Dissolved Oxygen Deficit at Milepoints 27-60 under Summer Average Conditions

h = 10 ft
Q = 300 ft^3/s
U = 0.984 mi/d
E = 2 mi^2/d$_1$
$K_1 = K_r = 0.25$ d^{-1}
K_L = 4 ft/d
$K_2 = K_a = K_L/h = 4/10 = 0.4$ d^{-1}
$K_{12} = K_d = K_r = 0.25$ d^{-1}

N_1 = 0.516
m_1 = 1.750
g_1 = 0.677 mi^{-1}
j_1 = -0.1845 mi^{-1}
N_2 = 0.826
m_2 = 2.075
g_2 = 0.756 mi^{-1}
j_2 = -0.264 mi^{-1}

D from PS:
W = 17,280 lb/d from Step 1
L_o = 17,280/(300 × 1.75 × 5.4) = 6.095 mg/L (8)
D at $X_p \leq 0 = [6.095 \times 0.25/(0.4 - 0.25)] \times$

$$[\exp(0.677X_p) - 1.75 \times 2.075^{-1} \exp(0.756X_p)] \tag{15a}$$
$$D \text{ at } X_p \geq 0 = [6.095 \times 0.25/(0.4 - 0.25)] \times$$
$$[\exp(-0.1845X_p) - 1.75 \times 2.075^{-1} \exp(-0.264Xp)] \tag{20a}$$
$$X_c = \ln[1.75(1 - 2.075)2.075^{-1}(1 - 1.75)]/$$
$$[(0.984/2 \times 2)(2.075 - 1.75)] = 2.37 \text{ mi} \tag{22}$$

D from NPS
$w = 2160$ lb/d/mi from Step 1
D at $X \leq -2$

$$= \frac{2160 \times 0.25}{300 \times 5.4(0.4 - 0.25)} \left\{ \frac{\exp[0.677(X + 2)] - \exp[0.677(X - 2)]}{1.75 \times 0.677} \right.$$
$$\left. - \frac{\exp[0.756(X + 2)] - \exp[0.756(X - 2)]}{2.075 \times 0.756} \right\} \tag{29a}$$

D at $-2 \leq X \leq 2$

$$= \frac{2160 \times 0.25}{300 \times 5.4(0.4 - 0.25)} \left\{ \left[\frac{\exp[-0.1845(X + 2)] - 1}{1.75(-0.1845)} \right. \right.$$
$$\left. - \frac{\exp[0.677(X - 2)] - 1}{1.75(0.677)} \right]$$
$$\left. - \left[\frac{\exp[-0.264(X + 2)] - 1}{2.075(-0.264)} - \frac{\exp[0.756(X - 2)] - 1}{2.075(0.756)} \right] \right\} \tag{26a}$$

D at $X \geq 2$

$$= \frac{2160 \times 0.25}{300 \times 5.4(0.4 - 0.25)} \left\{ \frac{\exp[-0.1845(X + 2)] - \exp[-0.1845(X - 2)]}{1.75(-0.1845)} \right.$$
$$\left. - \frac{\exp[-0.264(X + 2)] - \exp[-0.264(X - 2)]}{2.075(-0.264)} \right\} \tag{32a}$$

Table 14 is established using Eqs. (29a), (26a), and (32a).

5. Determination of Maximum Dissolved Oxygen Deficit and Critical Distance

For PS
$X_c = 2.37$ mi from Step 4
$D_c = 1.98$ mg/L at $X_p = X_c$ using Eq. (20a)

TABLE 14
Dissolved Oxygen Deficit Analysis of An Estuary with Multiple Waste Sources under Summer Average Flow Condition[a]

MP	PS		NPS		Total D, mg/L
	X_p	D	X	D	
60	−8	0.02	−6	0.05	0.07
58	−6	0.08	−4	0.15	0.23
56	−4	0.26	−2	0.40	0.66
55	−3	0.45	−1	0.58	1.03
54	−2	0.73	0	0.75	1.48
53	−1	1.59	2	0.95	2.54
51.34	0.66	1.80	2.66	0.96[b]	2.76
51	1	1.87	3	0.95	2.82
50	2	1.97	4	0.92	2.89
49.63	2.37	1.98[b]	4.37	0.90	2.88
49	3	1.96	5	0.87	2.83
48	4	1.88	6	0.80	2.68
46	6	1.60	8	0.64	2.24
44	8	1.29	10	0.50	1.79
37	15	0.47	17	0.17	0.64
32	20	0.21	22	0.01	0.22
27	25	0.09	27	0.00	0.09
Column: (1)	(2)	(3)	(4)	(5)	(6)

[a]Example 3.
[b]Maximum concentration.

For NPS. Table 7 lists four important equations for determining the location of maximum concentration for distributed sequentially reacting waste, such as DO deficit. Initially the slope of the DO deficit curve(s) must be determined by Eq. (65):

$$s = [\exp(2aj_1) - 1]/m_1 - [\exp(2aj_2) - 1]/m_2 \quad (65)$$
$$= [\exp(-2 \times 2 \times 0.1845) - 1]/1.75 - [\exp(-2 \times 2 \times 0.264) - 1]/2.075$$
$$= +0.016$$

Since the slope is positive, the location of D_c is

$$X_c > 2$$

and the numerical value of X_c can be determined by Eq. (66):

$$\frac{\exp[j_1(X_c + a)] - \exp[j_1(X_c - a)]}{m_1}$$

$$= \frac{\exp[j_2(X_c + a)] - \exp[j_2(X_c - a)]}{m_2}$$

(66)

or

$$\frac{\exp[-0.1845(X_c + 2)] - \exp[-0.1845(X_c - 2)]}{1.750}$$

$$= \frac{\exp[-0.264(X_c + 2)] - \exp[-0.264(X_c - 2)]}{2.075}$$

from which $X_c = 2.66$ mi. Then, $D_c = 0.96$ mg/L by substituting $X = X_c = 2.66$ into Eq. (32a).

The maximum DO deficit (D_c) caused by both PS and NPS is estimated to be 2.89 mg/L at MP-50 according to Table 14.

NOMENCLATURE

a	$\dfrac{K_d}{K_2 - K_d - K_3}\left(L_o - \dfrac{K_s}{K_d + K_3}\right)$ in the moment method; a constant otherwise
A	Average cross-sectional area of a surface water system, L^2
b	$\dfrac{K_d}{K_2}\left(\dfrac{K_s}{K_d + K_3} - \dfrac{\alpha}{K_d}\right)$ in the moment method
C	Conservative substance concentration, M/L^3
C_o	Maximum concentration of a conservtive substance at the point source discharge loctaion, M/L^3
D	Concentration of dissolved oxygen deficit, M/L^3
D_c	Critical (e.g., maximum) dissolved oxygen deficit, M/L^3
D_o	Dissolved oxygen deficit at the point source discharge, M/L^3
E	Longitudinal dispersion coefficient, L^2/T
$\overline{E}$	Turbulent dispersion vector, with components $\overline{E}_x$, $\overline{E}_y$, and $\overline{E}_z$
g_1	$(U/2E)(1 + m_1)$
g_2	$(U/2E)(1 + m_2)$
h	Average depth of receiving water at mean tide, L

j_1 $(U/2E)(1 - m_1)$
j_2 $(U/2E)(1 - m_2)$
J_1 $\dfrac{U}{2E} - \left(\dfrac{U^2}{4E^2} + \dfrac{K_d + K_3}{E}\right)^{1/2}$
J_2 $\dfrac{U}{2E} - \left(\dfrac{U^2}{4E^2} + \dfrac{K_2}{E}\right)^{1/2}$
k_d CBOD deoxygenation rate, T^{-1}, (base 10)
K First-order reaction rate, T^{-1} (base e)
K_a Reaeration coefficient, T^{-1}
K_b Coliform die-off rate, T^{-1} base e)
K_d BOD decay rate, T^{-1} (base e)
K_L Surface transfer coefficient, L/T
K_n NBOD deoxygenation rate, or nitrification rate, T^{-1} (base e)
K_p Reaction rate of phosphorus in receiving water, T^{-1} (base e)
K_r CBOD removal rate in receiving water, T^{-1} (base e)
K_s Rate of BOD addition to the overlying water from the bottom deposits and/or from the local run-off, $M/L^3/T$
K_1 Decay rate, system 1, T^{-1} (base e).
K_{12} Reaction rate between systems 1 and 2 (e.g., deoxygenation rate K_d), T^{-1} (base e)
K_2 Reaction rate of system 2 (e.g., reaeration coefficient K_a), T^{-1} (base e)
K_3 BOD settling rate, T^{-1} (base e)
L Concentration of reactive substances, (e.g., UOD), M/L^3
L_{max} Maximum concentration of reactive substances, M/L^3
L_o Concentration of reactive substances (e.g., UOD) at the point source discharge, M/L^3
m_1 $[1 + 4 K_1 E/U^2]^{0.5} = [1 + 4 N_1]^{0.5}$
m_2 $[1 + 4 K_2 E/U^2]^{0.5} = [1 + 4 N_2]^{0.5}$
M_l longitudinal mass flux, $M/L^2/T$
M_t transverse mass flux, $M/L^2/T$
M_T total mass flux, $M/L^2/T$
M_v vertical mass flux, $M/L^2/T$
n Numbers of observations
N Estuarine number; N_1 for system 1 and N_2 for system 2
P_k $K_s/(K_d + K_3)$
q Lateral inflow per unit length, $L^3/T/L$
Q Freshwater flow, L^3/T
R Assimilation ratio

s	Slope of the dissolved oxygen deficit curve
S	Sources and sinks of substance C in the surface water system, $M/L^3/T$
t	Time, T
U	Mean velocity, or advective velocity, L/T
U_T	Average tidal velocity, L/T
w	Distributed source (e.g., nonpoint source) of pollution, M/T/L
W	Mass of point source pollution discharge rate, M/T, W_{tn} for total nitrogen, W_c for coliform and W_{UOD} for UOD
x	Cartesian coordinate in the direction x
X	Downstream distance from the point source of pollution, L
X_c	Critical distance downstream where the critical dissolved oxygen deficit D_c occurs, L
X_1	Location of L_{max}, which is a distance measured from the center of a distributed source of pollution, L
X_p	Downstream distance of point source of pollution, L
y	Cartesian coordinate in the direction y
z	Cartesian coordinate in the direction z
α	Production rate of oxygen by photosynthesis, $M/L^3/T$
β	Order of moment = 0, 1, 2, . . .

REFERENCES

1. US Federal Water Pollution Control Act Amendment (PL 92-500), 1972, US Government Printing Office, Washington, DC.
2. US Environmental Protection Agency, *Areawide Assessment Procedures Manual*. EPA-500-9-76-014, US Department of Commerce, National Technical Information Service, Springfield, VA, 1976.
3. US Environmental Protection Agency, *Urban Runoff Pollution Control Technology Overview*, US Department of Commerce, National Technical Information Service, Springfield, VA, 1976.
4. A. N. Diachishin, *ASCE* **89**, No. SA1, 29, (Jan. 1963).
5. A. N. Diachishin, *Sanitary Eng. Div. ASCE* **89**, No. SA4, 39 (1964).
6. H. B. Fischer, *Int. J. Air Water Poll.* **10**, 443.
7. E. L. Thackston, J. R. Hays, and P. A. Krenkel, *Sanitary Eng. Div. ASCE* **93**, No. DA3, 47 (1967).
8. C. G. Wen, J. F. Kao, and L. K. Wang, *J. Env. Managment* **10**, 1 (1980).
9. A. T. Ippen, (ed.) *Estuary and Coastline Hydrodynamics*, McGraw-Hill Book Co., NY, 1966.
10. A. James (ed.) Mathematical Models in Water Pollution Control, Chapter 8 in *Estuarine Dispersion* (by L. Gallagher and G. D. Hobbs), Wiley, NY, 1978.

11. C. G. Wen, J. F. Kao, L. K. Wang, and M. H. Wang, *J. Env. Management* **14**, 17(1982).
12. C. G. Wen, J. F. Kao, L. K. Wang, and M. H. Wang, *Practical Application of Sensitivities of Stream Water Quality Parameters*, PB-80-141377, US Department of Commerce, National Technical Information Service, Springfield, VA, 1980, p. 56.
13. D. J. O'Connor, Estuarine Analysis, Manhattan College Summer Institute in Water Pollution Control, New York, 1977.
14. L. K. Wang, G. Gainforte, C. G. Wen, and M. H. Wang, Mathematical Modeling of Stream Water Quality by a New Moment Method: Computer Analysis, in *Proceedings of the Institute of Environmental Sciences*, vol. **26**, p. 259–267 (1980).
15. C. G. Wen, J. F. Kao, and L. K. Wang, *J. Env. Management* **12**, 127 (1980).
16. C. G. Wen, J. F. Kao, L. K. Wang, and M. H. Wang, *Civil Eng. for Practicing and Design Engrs.* **2**, No. 4, 425, Pergamon Press, NY (1983).
17. C. G. Wen, J. F. Kao, L. K. Wang, and M. H. Wang, *Civil Eng. for Practicing and Design Engrs.* **2**, No. 6, 537, Pergamon Press, NY (1983).
18. C. G. Wen, J. F. Kao, L. K. Wang, and C. C. Liaw, "Effect of Salinity on Reaeration Coefficients of Receiving Waters." Intern. Assoc. on Water Pollution Conf., York, UK (July 1983).

APPENDIX

A practical method of determining reaeration coefficients would greatly aid design engineers in determining the degree of wastewater treatment required for a proposed effluent discharge. The reaeration coefficient in saline water, K_{2s} (day^{-1}, base e) at any chloride concentration CL(g/L) and at 20°C, can be expressed by

$$K_{2s} = K_{2f} \exp(0.0127 CL)$$

in which K_{2f} is the reaeration coefficient in fresh water at 20°C [18].

3
Cooling of Thermal Discharges

William W. Shuster
Department of Chemical and Environmental Engineering, Rensselaer Polytechnic Institute, Troy, New York

I. INTRODUCTION

The discharge of heated condenser water from thermal power plants together with other heated effluents from various industrial processes represents an increasingly important contribution to stream pollution. More stringent water quality standards as well as increased pressure from local, state, and federal authorities have necessitated the consideration of means for heat removal from such emissions.

The amount of heat discharged to the environment by way of heated effluents can be reduced by cooling such streams either through the use of cooling ponds or of cooling towers. Cooling ponds are usually designated as either recirculating or once-through types. Cooling towers may be classified as natural or mechanical draft, and as wet or dry types, depending upon whether the air and water streams come into direct contact or are kept completely separate.

Cooling ponds are usually less costly than cooling towers, but because the cooling process occurs by natural means with limited contact between the water and its surroundings, the process is relatively slow, and consequently quite large heat exchange areas are usually required. On the other hand, the ponds may have the potential for a wide variety of recreational uses and can serve as settling basins for effluents containing settleable solids. The quantity of water lost through evaporative processes is usually less for cooling ponds than for cooling towers.

II. COOLING PONDS

A. Mechanism of Heat Dissipation

Basically, a cooling pond represents a holdup in the flow stream of the heated discharge of sufficient magnitude to allow the exchange of heat with the surroundings. The heated water both loses heat and gains heat in its passage through the pond by combined mechanisms of conduction, convection, radiation, and evaporation. The rates at which these exchanges take place will depend upon such factors as the temperature difference between atmosphere and water surface, the surface area, and various meteorological conditions, such as geographical location, air humidity, wind speed, and so on. When the heat inputs equal the heat outputs, the system will be in equilibrium and the water surface will be at an equilibrium temperature.

Edinger and Geyer [1] have identified the heat inputs and outputs and presented typical ranges of values. These values, expressed as BTU/d ft^2, together with recalculated values expressed in metric units of kilocalories per day meter^2 are given as follows:

Inputs
1. q_s = short-wave solar radiation, 1085–7588 kcal/d m^2 (400–2800 BTU/d ft^2)
2. q_a = long-wave atmospheric radiation, 6504–8672 kcal/d m^2 (2400–3200 BTU/d ft^2)

Outputs
3. q_{sr} = reflected portion of q_s, 109–542 kcal/d m^2 (40–200 BTU/d ft^2)
4. q_{ar} = reflected portion of q_a, 190–325 kcal/d m^2 (70–120 BTU/d ft^2)
5. q_{br} = long-wave back radiation from water surface, 6504–9756 kcal/d m^2 (2400–3600 BTU/d ft^2)
6. q_e = evaporative heat loss, 5420–21,680 kcal/d m^2 (2000–8000 BTU/d ft^2)

Input or output (depending on circumstances)
7. q_c = conduction–convection heat loss, -867 to $+1084$ kcal/d m^2 (-320 to $+400$ BTU/d ft^2)

The net heat flux across the air–water interface can be expressed as the difference between inputs and outputs, or

$$q_t = (q_s + q_a) - (q_{sr} + q_{ar} + q_{br} + q_e) \pm q_c \qquad (1)$$

At equilibrium, q_t will be zero. Under this condition the water surface temperature is called the equilibrium temperature.

It may be noted that q_{bn}, q_e, and q_c are dependent on the water surface temperature, while q_s, q_a, q_{sr}, and q_{ar} are independent of this temperature. The discharge of heated effluents adds another heat quantity to the input side of the above equation. This quantity is indicated as q_d.

Because of changing meteorological conditions, the temperature of most natural bodies approaches the equilibrium temperature asymptotically. Equilibrium temperatures are usually calculated on the basis of average daily conditions.

B. Design of Cooling Lakes

The design of cooling lakes is sometimes based on rules of thumb, such as 1–2 ac/MW of installed capacity or 200–400 kcals of heat dissipated per hour per square meter. These rules must almost always be tempered with engineering judgment. Local meteorological and topographical conditions may greatly influence the rate of heat transfer and, thus, the size of the pond required to produce a given amount of cooling. Therefore, it has become necessary to resort to a more rigorous approach to the design and construction of cooling lakes, both recirculated and once-through types [2].

The basic design equations are presented in this section for both completely mixed recirculated cooling lakes and completely mixed once-through cooling lakes. Pertinent calculations and the effects of longitudinal mixing and short-circuiting are given for both types of cooling lakes. Seasonal and geographical effects on cooling lake performance are also outlined. Lastly, other aspects of cooling lake design that are of interest are listed and described briefly.

I. Evaluation of Heat Dissipation at Completely Mixed Recirculated Cooling Lakes

In the design of a recirculated cooling lake for a thermal power plant, a parameter of interest is the amount of lake surface area needed to dissipate the surplus heat from a cooling water outfall. It is assumed that there will be a specific surface temperature difference above the equilibrium temperature for the lake. The lake surface required to dissipate this surplus heat from a discharge is intimately related to the mean surface heat dissipative capacity of the lake [1].

a. *Capacity of a Completely Mixed Lake to Dissipate Added Heat Loads*   As demonstrated by Edinger and Geyer, the capacity of a completely mixed lake to dissipate heat added from a heated effluent may be expressed in the following form [1]:

$$q = KA(T_s - E) \qquad (2)$$

where

q = heat dissipated, kcal/d
K = heat transfer coefficient at the surface, kcal/d meter^2 °C
A = surface area available for heat transfer with the air, m^2
T_s = surface temperature of the lake, °C
E = equilibrium temperature of the lake for zero added heat load, °C

Equation (2) may be rearranged as follows:

$$T_s = \frac{q}{KA} + E \qquad (3)$$

In terms of the heat flux, expressed as q_{rj} kcal/d m^2, the equation becomes:

$$T_s = \frac{q_{rj}}{K} + E \qquad (4)$$

It may be noted from Eq. (4) that in the design of a cooling lake, a balance between the lake surface area, and its temperature rise above equilibrium become involved. Equation (2), or a modification thereof, forms the fundamental basis of water surface temperature prediction techniques.

b. *Surface Heat Transfer Coefficient*   Brady, Graves, and Geyer have derived an expression for the surface heat transfer coefficient, K, by utilizing a balance of rates of heat exchange at the water surface. After mathematical manipulation of the heat balance and several approximations, the following expression for K results [3]:

$$K = 92.5 + (C_1 + \beta) f(U) \qquad (5)$$
$$\underset{\text{back radiation}}{} \quad \underset{\text{conduction}}{} \quad \underset{\text{evaporation}}{}$$

where

K = surface heat transfer coefficient, kcal/d m^2°C
C_1 = Bowen's conduction–evaporation coefficient
   = 0.468 mm Hg°C

β = slope of the saturated vapor pressure curve between the dew point temperature and the water surface temperature, mm Hg°C

$f(U)$ = evaporative wind speed function, kcal/d m² mm Hg

The surface heat transfer coefficient is expressed in Eq. (5) in terms of three components approximately representing the relative contributions to the total dissipation capacity arising from back-radiation, conduction, and evaporation, respectively [3].

c. *Relationship for the Slope of the Saturated Vapor Pressure Curve, β*   Appropriate values of β may be determined by reference either to vapor pressure tables, or to a vapor pressure chart (see Fig. 1). Values of β are defined from the relation

$$\beta = \frac{(e_s - e_a)}{(T_s - T_d)} \qquad (6)$$

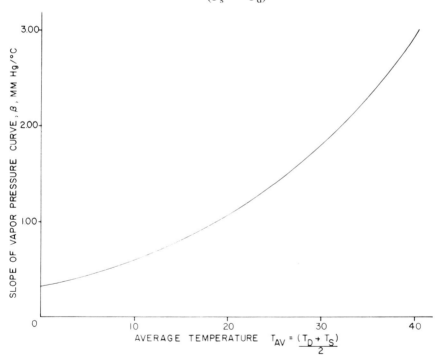

Fig. 1. Plot of the slope of the vapor pressure curve for water versus the average of the dew point and water surface temperature.

where

e_s = saturated vapor pressure of water at the surface temperature, mm Hg
e_a = air vapor pressure, mm Hg
T_s = surface temperature, °C
T_d = dew point temperature, °C

   d. Relationship for Wind Speed Function, f(U)   Brady, Graves, and Geyer [3], through the utilization of curve-fitting and plotting techniques, as well as multiple regression analysis on data gathered from three cooling-lake field sites, have proposed a model for predicting the wind speed function. This expression given in metric units is as follows:

$$f(U) = 189.8 + 0.784U^2 \qquad (7)$$

where

$f(U)$ = wind speed function, kcal/d m^2
U = wind speed, km/h

An important feature of Eq. (6) is that the wind speed function, $f(U)$ is relatively insensitive to wind speed when the wind speed is low. This, in turn, not only reduces the precision of wind speed data for low wind speed conditions, but also reduces the sensitivity of the heat transfer coefficient in the event of very calm conditions. The presumed explanation for this apparent insensitivity of K to calm conditions is related to the cooling effect of vertical atmospheric convection currents [3].

   e. Relationship for Equilibrium Temperature, E   An expression for lake equilibrium temperature has been derived by Brady, Graves, and Geyer by utilizing a balance of rates of heat exchange at the water surface. After mathematical manipulation of the heat balance and successive approximations, the following expression for E is derived [3]:

$$E = T_d + \frac{H_s}{K} \qquad (8)$$

where

E = equilibrium temperature, °C
H_s = gross solar radiation, kcal/d m^2
K = heat transfer coefficient, kcal/d m^2 °C
T_d = dew point temperature, °C

As a test of Eq. (8), values of E were evaluated directly using data for dew point temperature and gross solar radiation at three recirculated cooling lake field sites. The results of this evaluation were plotted, and it was found that the equation was usually accurate to within a few degrees when the heat transfer coefficient was greater than about 135 kcal/d m² °C [3].

2. Prediction of Surface Temperature T_s for a Completely Mixed Recirculated Cooling Lake

Both the surface heat transfer coefficient, K, and the equilibrium temperature, E, may be estimated from known meteorological conditions using Eqs. (5)–(8). However, since the average water surface temperature, T_s, is unknown initially, an iterative procedure must be employed. A trial value of T_s is assumed, and is then corrected through successive approximations.

The only meterological information required is (1) the dew point temperature, (2) the gross solar radiation, and (3) the wind speed, in order to calculate the surface temperature, T_s, for a specified heat loading [3].

a. Calculation of Lake Exit Temperature

Example Problem

A fossil fuel-fired power plant is using water for its condensers that is recirculated through a completely mixed cooling pond having an area of approximately 4×10^6 m². The heat rejected by the plant amounts to 8×10^9 kcal/d. With a flow of cooling water equal to 9.255 L/s, the temperature rise in the cooling water as it passes through the condenser is 10°C. It may be assumed that meteorological conditions found in southern US in the month of July will prevail for design purposes. Such conditions include a mean value for the dew point of 23°C, a mean daily value of solar radiation of 5750 kcal/m², and a mean wind speed of 13.5 km/h. The water surface temperature and the equilibrium temperature are to be calculated.

Solution

The water surface temperature, T_s, is assumed to be 36°C. Since T_d is given as 23°C

$$T_{ave} = \frac{T_d + T_s}{2} = \frac{36 + 23}{2} = 29.5°C$$

From Fig. 1, the value of β at this temperature is

$$\beta = 1.77 \text{ mm Hg °C}$$

For a given wind speed of 13.5 km/h

$$f(U) = 189.8 + 0.734U^2 = 189.8 + 0.734(13.5)^2$$
$$= 323.6 \text{ kcal/d m}^2$$

The surface heat transfer coefficient is calculated as

$$K = 92.5 + (C_1 + \beta)f(U)$$
$$= 92.5 + (0.468 + 1.77)323.6$$
$$= 816.5 \text{ kcal/d m}^2 \text{ °C}$$

The mean solar radiation, H_s, is given as 5750 kcal/d m². Therefore, since the equilibrium temperature, E, is given as

$$E = T_d + \frac{H_s}{K}$$

$$E = 23 + \frac{5750}{816.5} = 23 + 7.0 = 30.0 \text{°C}$$

The surface temperature may then be calculated from Eq. (4)

$$T_s = \frac{q_{rj}}{K} + E = \frac{8 \times 10^9}{4 \times 10^6 \times 816.5} + 30.0$$
$$= 2.5 + 30.0 = 32.5 \text{°C vs 36°C assumed}$$

Assume $T_s = 33 \text{°C}$

$$T_{ave} = \frac{T_d + T_s}{2} = \frac{33 + 23}{2} = 28 \text{°C}$$

From Fig. 1, $\beta = 1.65$ mm Hg °C

$$f(U) = 323.6 \text{ kcal/d m}^2$$

The value of K is calculated as

$$K = 92.5 + (0.468 + 1.65)323.6 = 778.5 \text{ kcal/d m}^2 \text{ °C}$$

Therefore

$$E = 23 + \frac{5750}{778.5} = 23 + 7.4 = 30.4 \text{°C}$$

The calculated surface temperature is

$$T_s = \frac{8 \times 10^9}{4 \times 10^6 \times 778.5} + 30.4 = 2.6 + 30.4 = 33.0°C$$

b. Design Chart To assist in design calculations, a chart has been prepared (see Fig. 2) after the manner of Brady et al. [3]. This chart gives values of the surface heat transfer coefficient, K kcal/d m² °C in terms of the wind speed, U km/h, and the average temperature $T_{ave} = (T_d + T_s)/2°C$. The chart is convenient in eliminating a part of the calculations.

Meteorological data may be obtained from the "Climatic Atlas of the United States" [4], which contains maps for dew point temperature,

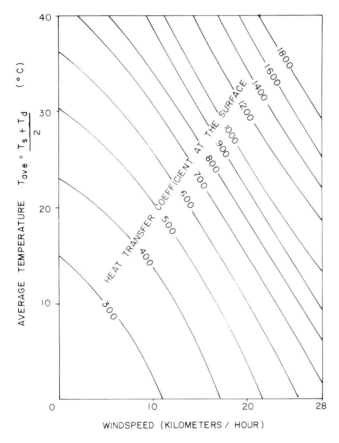

Fig. 2. Values of wind surface heat transfer coefficients in terms of wind speed and average temperature.

gross solar radiation, and wind speed for any region of the United States. The Atlas contains maps for each month of the year, plus annual averages for these and many other meteorological parameters.

3. Longitudinal Mixing Effects on Recirculated Cooling Lake Performance

The two main objectives of a recirculated cooling lake are to dissipate the heat rejected by a power plant, and also to yield the lowest intake temperature back to the power plant. These objectives are not necessarily the same. It is entirely possible to dissipate the heat rejected by the power plant, without, at the same time, achieving the lowest possible plant intake temperature.

The performance of a cooling lake in terms of the lake exit temperature (plant intake temperature) for a given set of specified meteorological conditions, is related to the amount of longitudinal mixing that takes place, and also the uniformity of flow distribution through the lake. Short-circuiting results in the direct transport of warm water entering the lake to the point of discharge, lowering the overall performance of the lake as far as providing the lowest possible temperature for return to the power plant is concerned.

When longitudinal mixing is a problem, improvements in operation may be obtained by a number of techniques. To increase the length of flow path, transverse baffles may be installed. Vertical skimmer walls may also be installed to reduce the possibility of wind-induced short-circuiting. When high ratios of length to width exist, the effects of wind on circuiting is usually minimized. Relatively shallow depths tend to minimize vertical velocity gradients.

4. Evaluation of Heat Dissipation in Once-Through Cooling Lakes

In this type of lake, heat effluents enter the lake at one region and exchange heat with the surroundings by the various mechanisms previously described. As a result of these heat transfer mechanisms, the water is cooled and leaves the lake at a lower temperature. Various models that will allow the prediction of discharge temperature for a given lake, or the specification of lake area required for a given temperature change, have been proposed to describe this exchange of heat.

A method originally described by Edinger and Geyer [1] and more recently summarized in a publication of the Federal Water Pollution Control Administration [5] allows the prediction of design parameters with a reasonable balance between accuracy and facility. The model used assumes complete mixing in the depth and width of the flowing stream, but no longitudinal mixing in the direction of flow. It further assumes constancy of crosssection to flow and constancy of flow velocity and the absence of flow channel curvature. It is presumed that the heated stream mixes completely with the lake water at the point of entrance and then progresses through the lake as a heated slug of water losing heat to the surroundings in an exponential manner. An expression for predicting the water surface temperature as a function of longitudinal distance through the lake may be derived within the restrictions described above.

The net rate of surface heat exchange can be expressed in differential form as

$$\frac{dq_t}{dt} = -K(T_s - E) \qquad (9)$$

where

dq_t/dt = net rate of surface heat exchange, kcal/d m^2
K = heat transfer coefficient, kcal/d m^2 °C
T_s = water surface temperature, °C
E = equilibrium temperature, °C

This exchange of heat results in a change in the water temperature with respect to distance traveled. This may be expressed as:

$$-K(T_s - E) = \rho c_p du \frac{\partial T}{\partial X} \qquad (10)$$

where

ρ = density of water, 1.0 g/cm^3
c_p = specific heat of water, 1 cal/g °C
d = average depth of flow path, m
u = average flow velocity, m/d
$\partial T/\partial X$ = longitudinal temperature gradient, °C/m

X = distance from the point of heat input, m

Integrating this equation and imposing the boundary conditions that $T = T_x$ at $X = X$ and that $T = T_o$ at $X = 0$ results in the following

$$\frac{(T_x - E)}{(T_o - E)} = e^{\frac{-KX}{\rho c_p du}} = e^{-\alpha} \tag{11}$$

where

$$\alpha = \frac{KX}{\rho c_p du} = \frac{KA}{\rho c_p Q}$$

where Q = volumetric flow rate, m³/d.

Using local meteorological data and the relations presented in previous sections, the cooling lake area required for a given heat load may be estimated, or for a given lake area the exit water temperature may be predicted.

a. Calculation of Lake Exit Temperature For the example previously cited values of $K = 778.5$ kcal/d m² °C, and $E = 30.4$°C, were found to be appropriate. Also, from the given data the volumetric flow rate, Q, may be calculated as follows:

$$Q = 9.255 \text{ L/s}$$
$$= 9.255 \times 3600 \times 24 = 8.0 \times 10^8 \text{ L/d}$$

Assuming that the inlet temperature to the condensers is 28°C, then a 10°C rise in condenser water will result in a discharge temperature of 38°C. The following additional data will also apply:

$$c_p = 1.0 \text{ kcal/kg °C}$$
$$\rho = 1.0 \text{ g/cm}^3 = 1000 \text{ kg/m}^3$$

then

$$\alpha = \frac{KA}{\rho c_p Q} = \frac{778.5 \times 4 \times 10^6 \times 10^3}{1.0 \times 10^3 \times 1.0 \times 8 \times 10^8} = 3.89$$

$$\frac{T_x - E}{T_o - E} = e^{-\alpha} = e^{-3.89}$$

$$\frac{T_x - 30.4}{38 - 30.4} = 0.0204$$

solving for T_x

$T_x = 30.56°C$, the temperature of the water leaving the cooling lake

To assess the effect of the surface area on the exit water temperature, the above calculation may be repeated for various values of A. Or conversely, for a specified exit water temperature, the required lake surface area may be determined.

5. Relationship of a Completely Mixed to a Completely Unmixed Lake for Once-Through Cooling Lakes

The objectives of a once-through cooling lake are essentially the same as those of a recirculated cooling lake; that is, to dissipate the heat rejected by a power plant, and also to yield the lowest effluent temperature of the cooling water. As mentioned earlier, these objectives are identical, and are largely a function of the degree of longitudinal mixing and short circuiting present in the lake.

a. Comparison Between Net Plant Temperature Rise of a Completely Mixed Lake to that of a Completely Unmixed Lake

The ratio of the net plant temperature rise at the exit from a completely mixed lake to that at the exit from a completely unmixed lake has been derived [1] and may be expressed as follows:

$$\frac{T_{SM} - E}{T_{SU} - E} = \frac{e^\alpha}{1 + \alpha}$$

where

T_{SM} = completely mixed lake's actual surface temperature, °C

T_{SU} = completely unmixed lake's actual surface temperature, °C

For the same flow rate Q, surface area A, and meteorological conditions, the completely unmixed lake will provide quicker cooling than the completely mixed lake. This happens because the hot water in the unmixed lake is not initially mixed or diluted with any cooler water. The driving force for cooling, $T_s - E$, is thus maintained at the highest possible value.

b. Area of Completely Unmixed Lake Necessary to Provide the Same Surface Temperature as the Completely Mixed Lake The area of a completely unmixed lake necessary to provide the same surface temperature as the completely mixed lake may be derived using Eq. (11). Since the surface temperatures T_{SM} and T_{SU} are now identical, $T_{SM} - E$ is equal to $T_{SU} - E$, or

$$\frac{T_{SM} - E}{T_{SU}} = \frac{e^\alpha}{1 + \alpha_M} = 1 \qquad (14)$$

where

$\alpha_M = \alpha$ evaluated for the completely mixed lake
$\alpha_U = \alpha$ evaluated for the completely unmixed lake

However, the only difference between α_M and α_U is the area of the respective lakes, since for both lakes the flow rate through the lake Q, the specific heat of the water c_p, the water density ρ, and the surface heat transfer coefficient K are identical. Hence, the ratio of the required areas for the two cases may be calculated.

6. Effect of Seasonal Variations on Cooling Lake Requirements and Performance

The equilibrium temperature E for most locations throughout the continental United States shows a low about December 31 and a high in mid or late July. The pattern of variation of E is customarily quite smooth and regular. The difference between high and low for the year and the relative sharpness of the summer peak and winter valley tend to increase with latitude [4].

The surface heat exchange coefficient K exhibits a pattern of variation almost exactly opposite that of the equilibrium temperature E. The highest heat exchange coefficients occur in mid or early July and the lowest occur about December 31. Deviations from the smooth pattern of annual variation are more common for the heat exchange coefficient than for the equilibrium temperature. Higher wind speeds during certain parts of the year at coastal locations may cause the highest heat exchange coefficient to fall at times other than mid or early July.

7. Effect of Geographical Location on Cooling Lake Performance

Latitude, which controls solar radiation, is the main influence on the equilibrium temperature. Topographic conditions, which strongly influence wind speed and the wet-bulb temperature, have a strong influence on the heat exchange coefficient. The result is that the best cooling conditions exist on the southern great plains between the Rocky Mountains and the Mississippi River. The second best cooling conditions exist along the Gulf and Atlantic coasts, all the way to Cape Cod. On the other hand, poor cooling conditions for lakes exist across the northern part of the United States and in the vicinity of the Appalachian and Rocky Mountains. The area best suited to cooling lakes seems to be Texas, while the least suitable area seems to be between the Sierra Nevada and Rocky Mountains. In any case, it should always be recognized that local topographic and meteorological conditions can exert great influence on cooling lake performance and requirements [4].

8. Other Factors Related to Cooling Lake Design

Topics of interest in the design of cooling lakes are enumerated and briefly described as follows:

a. Diking and Baffling Cooling lakes have customarily been constructed by placing dikes and embankments to form an artificial water body. However, to make the use of cooling lakes economically feasible, their construction is limited to locations where dikes or dams could take advantage of the natural topography. The cost incurred by excavating a complete lake is normally too high [4].

    b. Evaporation of Water In some areas, such as southwestern United States, evaporation from cooling lakes can be a serious problem. Since replenishment must come from outside sources, the design of lakes on such areas must take into account the availability of water and the overall plans for its management.

    c. Infiltration The entry and movement of water through soil is termed infiltration. Infiltration may affect the design and location of a cooling lake because (1) the recharging of the local groundwater structure may raise the water table elevation (therefore affecting the use of the surrounding land), and (2) there is a potential of contamination of the groundwater reservoir [4].

    d. Fogging One disadvantage to using cooling lakes is the potential increase in fog formation, which is undesirable and in some instances may be dangerous. The fog results from the condensation of atmospheric water vapor into water droplets. These droplets may remain suspended in the air, and if they are in sufficient concentration, can reduce visibility and create local hazards [4].

    e. Recreation Swimming has become one of the most popular outdoor recreational activities. Boating and fishing are also in the top ten most popular activities. Some states strongly recommend that cooling lakes be used as public recreational facilities. In these cases, quality of the water used must conform to state water quality standards for the specific activity intended [4].

    f. Economics Cooling lakes are generally a viable alternative when land is available at a reasonable price and the local topography is suitable. Incremental costs associated with the construction and operation of a cooling lake is less than the cost of constructing a wet tower for these cases. However, it should be noted that cooling ponds operate at temperatures that are usually 5–10°C in excess of the equilibrium temperature. This, in turn, means a higher condenser temperature, and a reduction in the capability of the turbine [4].

III. COOLING TOWERS

The use of cooling towers for removing heat from various liquid streams, particularly the high quantities of hot condenser water discharged from power plants, has received increased attention in recent years. Because of the very large quantities of heat and mass that must be handled, the size of the towers required is often quite large and can represent a major capital investment. Because of the massive size of these structures, the design and construction of cooling towers has become a rather specialized business. It is, however, important that engineers be familiar with the basic features of the mechanisms involved in tower operation, the types of towers available, their advantages and disadvantages, and some of the problems associated with their use, including the most important problems of environmental impact.

A. Mechanism of Heat Dissipation

In cooling towers heat is transferred from a warm water stream to a relatively cool air stream through a combined mechanism of latent heat exchange and sensible heat exchange. Provision is made to bring the water into intimate contact with unsaturated air in such a way that water is evaporated into the air resulting in cooling of the water and humidification of the air. The water, of course, cannot be cooled below the wet-bulb temperature.

Since the operation of cooling towers involves the simultaneous transfer of both mass and heat, the generation of heat and material balances, combined with the use of rate equations is useful in the development of suitable design equations. This has been shown in detail in numerous chemical engineering texts, as well as the excellent book of McKelvey and Brooke [6], and others.

To properly appreciate the mechanisms involved in cooling tower operation, it is helpful to examine the conditions existing in a tower with respect to changes in heat and mass flow. Figure 3 illustrates a typical operation at the top and at the bottom of a tower used for cooling. As can be seen in this diagram, the conditions in a cooling tower may change depending upon the relative magnitude of the water temperature, and the air dry and wet-bulb temperatures. In the upper part of a counterflow tower, for example, conditions may be represented as shown in Fig. 3a. In this case the water temperature is higher than the dry-bulb temperature of the air. Hence, the water is cooled by both evaporation and by the

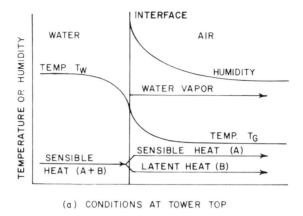

(a) CONDITIONS AT TOWER TOP

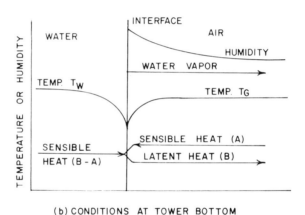

(b) CONDITIONS AT TOWER BOTTOM

Fig. 3. Temperature and humidity gradients in cooling towers.

transfer of sensible heat from the water to the air. Since evaporation takes place at the interface between water and air, the concentration of water in air (humidity) will vary as shown. Likewise the temperature of the air will decrease through the gas film away from the interface. The total temperature gradient in the water film must reflect the total heat effect of sensible heat change in the gas plus the heat associated with the evaporation producing the humidity gradient.

In Fig. 3b the conditions that might exist in the lower part of a counter flow tower are shown. Here the water temperature is above the wet-bulb temperature, but below the dry-bulb temperature. Since the water is being cooled, the water–gas interface is at a lower temperature than the water bulk temperature and a temperature gradient in the water exists in the direction of the interface. At the same time the air temperature is greater than the interface temperature and hence, a temperature gradient in the air exists in the direction of the interface. The sum of sensible heat flows through both water and air to the interface will exactly equal the latent heat flow from the interface into the air stream caused by evaporation at the interface.

B. Definitions

As an aid to understanding the language of cooling towers, the following definitions may be useful:
1. *Approach:* the temperature difference between the water out of the tower, and the dry-bulb temperature of the air entering the tower.
2. *Basin:* the bottom part of the tower in which the cooled water is collected after passing through the tower.
3. *Blow-down:* a portion of the water discharged to waste in a recirculating system for the purpose of preventing the buildup of dissolved solids in the water.
4. *Cooling range:* the temperature difference between the hot water entering and the cold water leaving the tower.
5. *Drift* (or windage): small droplets of water carried out of the cooling tower in the exit air streams; usually expressed as a percent of the water flow rate.
6. *Drift eliminators:* baffles placed in the exit air stream to change the direction of air flow and cause entrained water droplets to be separated from the air by centrifugal force.
7. *Dry-bulb temperature:* the air temperatures as read by a standard thermometer.
8. *Heat load:* the amount of heat exchanged in a tower per unit of time. It can be expressed as the product of the quantity of water circulated per unit of time, and the cooling range.
9. *Make-up water:* water added to the system to replace the amount lost through evaporation, drift, blow-down, and system leaks.
10. *Packing:* material placed within the tower over which the

water flows to provide uniform flow distribution and to provide surface area for water–air contact.
11. *Performance:* a measure of the ability of the tower to cool water, usually expressed in terms of the amount of cooling a quantity of water will undergo at a specific wet-bulb temperature.
12. *Relative humidity:* the ratio of the partial pressure of water vapor in the air, to the vapor pressure of water, expressed at the same temperature.
13. *Wet-bulb temperature:* the temperature that an ordinary thermometer would read when its bulb is covered with a film of water that is evaporating into a moving air stream. If the humidity of the air is low, evaporation can take place at a high rate, with associated cooling. If the air is saturated, no evaporation takes place and no cooling is accomplished, and the wet-bulb and the dry-bulb temperatures are the same.

C. Types of Towers

Towers may be classified in a variety of ways, but are usually designed as either natural-draft or mechanical-draft types. Natural-draft towers use the difference in density created by the warm wet air emerging from the tower and the cooler dry air outside the tower, as the driving force for air flow. Mechanical draft towers use some type of mechanically driven fan to move air through the tower. The majority of cooling towers in large-scale operations involve direct contact of air and water and are designated as "wet" towers. In some cases units have been designed in which air and water are kept separated and are designated as "dry" towers. Towers are designed for either counterflow or for crossflow. In counterflow, the water droplets fall through the tower in a direction opposite to the rising air flow. In crossflow, air flows into the sides of the tower in a more-or-less horizontal direction while contacting the falling water stream. Various combinations of the features described have been incorporated in recent designs. Some of the more common designs are shown diagrammatically in Figs. 4–7 and are described in the following sections.

1. Natural-Draft Atmospheric Towers

In this type of tower (see Fig. 4) air enters through open louvres on the sides and contact falling water droplets sprayed into the tower at the

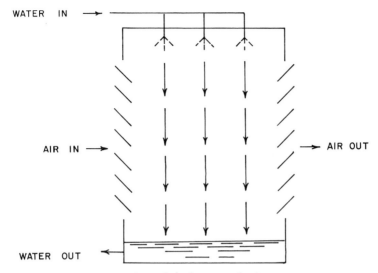

Fig. 4. Natural-draft atmospheric tower.

top. Air flow is essentially horizontal and is dependent on wind direction and other atmospheric conditions. The tower may be open and contain only the falling water spray, or it may contain some type of open packing such as a very open splash board arrangement to increase the surface area for air–water contact. Except for the hyperbolic type (a special case of natural draft towers), these towers are quite limited in cooling capacity and are used only for relatively small installations such as refrigeration units. They have virtually no application for cooling power plant discharges. Some advantages and disadvantages for this type of unit include the following [6]:

Advantages
    (a) No moving or mechanical parts.
    (b) Relatively small equipment maintenance.
    (c) Is not subject to recirculation of exit air.

Disadvantages
    (a) Has a comparatively massive size for a given heat load.
    (b) Substantial initial investment cost.
    (c) High pumping costs.
    (d) Requires unobstructed access to winds.
    (e) Amount of cooling is dependent on wind direction and velocity.
    (f) Windage losses are often high.

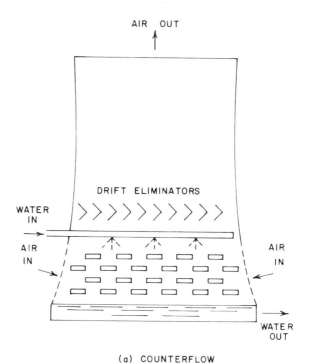

Fig. 5. Natural-draft, wet, hyperbolic towers.

2. Natural-Draft, Wet, Hyperbolic Towers

This type of tower takes its name from the hyperbolic shape of the stack through which air passes after contacting the warm water sprayed over packing in the lower part of the tower. The towers are often mamouth structures being as much as 150–180 m high and 100–135 m in diameter. The tower shell is usually made of reinforced concrete. The flow of air through the tower is generated by the density difference between the heated leaving air and the heavier, colder external air. Wind flow across the top of the tower may aid the air flow, but tower operation is not dependent on wind flow.

Intimate air–water contact is provided in the lower 3–6 m of the tower by packing arranged to promote either counterflow or cross-flow of air with respect to water. In a counterflow arrangement, the warm water is distributed over the top of the packing and flows downward counter to the upward flow of air. This is shown in Fig. 5a. The packing can also be located around the outside part of an enlarged section at the tower bottom

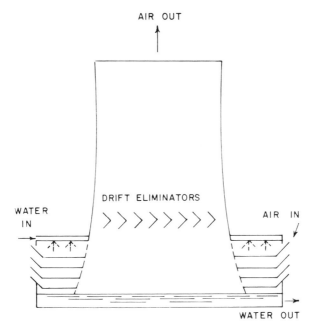

(b) CROSSFLOW

such that the air enters the tower horizontally and flows across the downward flowing water stream into the empty central section which it then turns upward to the tower top. This arrangement is shown in Fig. 5b.

The counterflow design provides more efficient heat transfer, since the coolest water comes in contact with the coolest air initially. For the same transfer of heat usually less packing is required for the counterflow arrangement. However, because of the geometrical arrangement, it is often possible to promote better air and water distribution with the crossflow tower, and for the same capacity less pressure drop is encountered in this unit.

The natural-draft hyperbolic tower is attractive from the standpoint of being free, for the most part, of mechanical and electrical components. This is reflected in both the initial investment and also the operating and maintenance costs. It can be built to handle large quantities of water and is relatively efficient as a heat transfer device. Its height, however, which is necessary to produce the desired draft, presents problems in terms of esthetics and public acceptance. Although its performance is not depend-

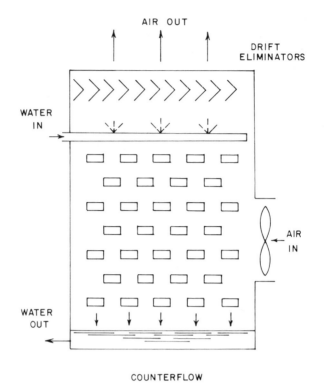

Fig. 6. Mechanical, forced-draft, wet towers.

ent on wind velocity, other atmospheric conditions do influence its operation. The air dry-bulb temperature, for instance, must be below the inlet water temperature to induce the necessary draft. Also, the exit water temperature is dependent on atmospheric conditions and may be difficult to control precisely.

The natural-draft hyperbolic tower appears to be most applicable to situations where the heat load is large, particularly during winter months. Also desirable are locations in which low wet-bulb temperatures and high relative humidity prevail, particularly in combination with relative hjigh inlet and exit water temperatures.

3. Mechanical, Forced-Draft, Wet Tower

This type of unit utilizes motor-driven fans to force the air through the tower. Since practical considerations may limit fan size, multiple units may be built to satisfy the total air requirement. The air paths may be either crossflow or counterflow, but are usually counterflow. The essential features of this type of unit are shown in Fig. 6.

An important feature of this type of unit is that the air flow may be controlled exactly, which in turn allows control of the exit water temperature. It also allows a closer type of packing with greater surface area per unit volume of tower since air flow resistance is not a major consideration. This means a smaller tower both with respect to height and cross-sectional area. The smaller height, in turn, reduces the pumping head and reduces pumping costs. In addition, the overall investment cost is usually lower, particularly for the smaller units. Usually a greater cooling range and a smaller approach is possible with this type of unit.

Power costs for the fans required for the mechanical draft tower may be quite high. In addition, this type of tower also has relatively high associated maintenance costs. Total operating and maintenance costs for the mechanical draft tower are usually higher than natural draft tower for the same cooling capacity. Because of the relatively low height of the mechanical draft tower, this type of unit is often plagued with problems of recirculation of discharged air into the inlet air stream, and also of local fogging. These problems usually necessitate that units be separated appreciably from one another which, in turn, means that land costs and equipment piping costs are high for this type of tower. The noise level associated with this tower can also present problems, particularly for the larger units.

Recently a design has been proposed that appears to combine some of the best features of the mechanical, forced draft, wet tower, and the natural draft, wet, hyperbolic tower. This unit, to be marketed by Research Cottrell, is called the fan-assisted hyperbolic cooling tower. It is constructed with a hyperbolic shaped-concrete shell with a number of motor-driven fans located around the wall at the base of the structure. Such an arrangement allows for much better control of the air flow than does the natural draft unit, but does not have the high power requirements of the mechanical draft towers. The structure is not as high as the natural draft and doesn't occupy as great a land area. Recirculation does not appear to be a problem because of height and exit air velocity. It also can be operated without fans when climatic conditions allow.

4. Mechanical, Induced-Draft, Wet Tower

Many of the characteristics of this type of tower are similar to the forced-draft mechanical tower. It differs primarily in having the fans located on the exit side of the packing section, and hence, tends to pull the

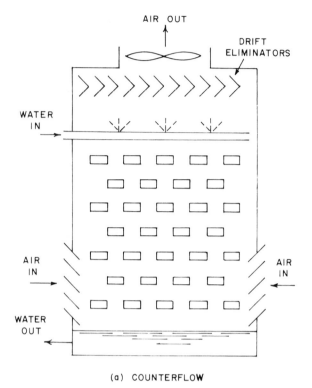

(a) COUNTERFLOW

Fig. 7. Mechanical, induced-draft, wet towers.

air through rather than force the air through. Again, the air path may be crossflow or counterflow with respect to the water path. This tower is shown in Figs 7a and 7b.

Compared to the forced-draft tower, this unit has a somewhat lessened problem with recirculation because of higher exit velocities. Air distribution is usually better and often larger fans can be used with this type of tower.

The induced-draft tower has a disadvantage that the moist warm air comes in contact with the fan, which increases corrosion. It also has a higher noise level associated with it.

The choices between forced- or induced-draft types, and between crossflow or counterflow, depend on the particular circumstances associated with the problem at hand. At present, many designers seem to favor induced-draft towers having a crossflow air path.

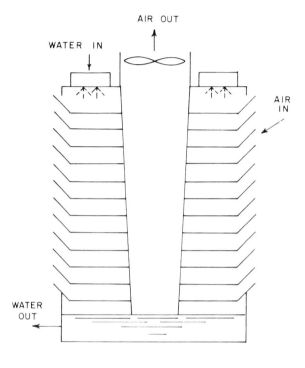

(b) CROSSFLOW

5. Dry Cooling Towers

Dry towers depend primarily upon convection for the transfer of heat and not upon evaporation. Heated water usually passes through the tower inside tubes, outside of which air passes. Heat is exchanged between the water and air, which are kept separated by the tube wall. The towers may be either of the natural-draft or mechanical types. There have also been some applications in the power industry, where exhaust steam from turbines has been condensed directly by air in mechanical draft type towers [7]. Another design utilizes a jet condenser using cooling water of feedwater quality to condense turbine exhaust. The condensate is then handled in an air-cooled tower and a part of the condensate returned to the power system as feed water [7].

This type of tower finds application where water temperatures are unusually high. It is not limited by the temperature to which the air is heated. It also allows flexibility in the siting of power plants, since the use of such towers is not dependent upon large sources of cooling water. Since there are no evaporative losses, such problems as fogging, icing, salt fallout, chemical treatment, and corrosion are either eliminated or markedly reduced.

The efficiency of the dry tower is usually decidedly lower than the corresponding wet tower. This leads to larger investment costs in most cases. It is also limited in the temperature to which the water may be cooled, by the dry-bulb air temperature.

Many design engineers are attracted by the possibility of combining the best features of wet towers and dry towers in the so-called wet/dry cooling tower. This unit provides for the passage of outside air through both wet and dry sections of the unit in parallel paths. These streams are mixed before being discharged to the atmosphere. This results, usually, in a much less visible plume. The hot water passes first through the dry section and then the wet section. Such a unit is conserving of water and is particularly applicable for situations where the amount of water is limited or where the amount available is variable with the season. Provision is often made to regulate the split between wet and dry operation.

Because of the increased attention being given to the impact of cooling towers on the environment, many engineers are looking with more favor on the use of dry towers. Despite higher initial costs, the elimination of the direct discharge of evaporated water is the atmosphere, and the attendent savings in make-up water requirements, stands as major attractions for this type of unit.

D. Problems Associated with Cooling Tower Operations

Although the use of cooling towers offers a number of distinct advantages for the handling of heated liquid streams, there are, nevertheless, a number of distinct problems that arise through their use. A number of these have already been alluded to, but will be briefly summarized.

In recent years, concern has been expressed with regards to the effect of thermal discharges on the global climate. Although opinions differ about future effects, most investigators agree that well-managed discharges to the atmosphere are preferable to discharges to most water bodies. The present very considerable quantity of heat being discharged is increasing each year and many people feel that it can markedly effect atmospheric and ocean dynamics. On a local scale, cooling towers have a slight effect on local climatic and weather conditions in the opinion of some investigators. Some modifications to local weather patterns have been reported, and changes in vegetation growth in the immediate vicinity of towers have been noted. Such effects are strictly limited to local areas and opinions differ with regards to their impact.

Under some conditions local ground fog may be created in the immediate vicinity of cooling towers when the water-carrying capacity of the air is exceeded. Depending upon local circumstances, it may be possible to locate and design towers so that the impact of such fog (and possibly icing in cold weather) can be minimized. Numerous investigators have studied the problem of fog occurrence and dispersion.

In many mechanically induced draft cooling towers, the generation of noise is a serious problem. Large quantities of air must be moved through restricted spaces by mechanical devices resulting in generally high sound levels. This problem is being studied and modifications to design and being developed to minimize this problem. Such features as changes in wall thickness, use of centrifugal fans instead of propeller fans, use of deflecting barrier walls, installation of discharge baffles, modification to the direction of tower discharge, and the use of acoustically lined plenums have been effective in dealing with this problem.

One of the most evident impacts of wet cooling towers is the tremendous plume that emerges from the top of the tower. Composed of water droplets of varying size as well as minute particles of salts plus any added chemicals introduced for control of algae and/or corrosion, the plume can produce problems of visibility, as well as problems of salt and water fallout. Plumes tend to have peculiarities related to the locality, and their formation, dissipation and general behavior are most difficult to predict. Though the problem is being studied, good solutions for control have not yet been generated.

The formation of scale on heat transfer surfaces and piping is a problem commonly encountered in cooling tower operation. Natural waters contain various levels of salts which become concentrated through evaporation until the limit of solubility is reached. The use of chelating agents together with the practice of bleeding off a portion of the water as blowdown can usually keep salt formation to a minimum.

Control of corrosion is obtained through the addition of additives to the water, which form protective coatings on metal surfaces. Various formulations of chromates, phosphates, and organic compounds are effective in many cases. Some of these materials, however, are toxic to biological life in waters that receive the blow-down stream. Stricter controls on the discharge of these materials can be expected in the future.

Biological growths tend to form on the surfaces in cooling towers where they promote corrosion and inhibit heat transfer. Such growths, which include algae, slimes, and so on, can be controlled by periodic slugs of chlorine that kill them. Again, the toxic properties of chlorine are cause for concern to many public health officials, and other less toxic biocides are being considered for this service.

E. Costs

It becomes almost impossible to generalize with respect to the costs associated with thermal problems requiring cooling equipment. Each problem involves many parameters that vary with such factors as the type of system, site characteristics, climatic conditions, desired level of cooling, materials of construction, local environmental restrictions and laws, local zoning ordinances, chemical and physical characteristics of the water,

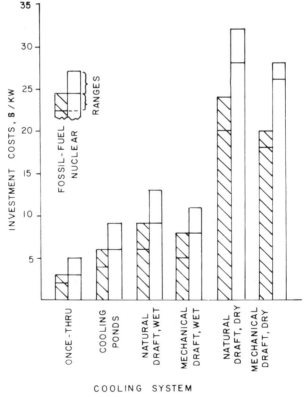

Fig. 8. Unit costs for cooling water systems.

and so on. It is probably more important for engineers to appreciate the general level of costs for specific cases, and to note the relative trends in costs as a function of process type, than to attempt to rely on average costs. One must determine exactly what components of the power system are considered to be included in the cooling section of the system, and what portion of total costs are assigned to the cooling operation before meaningful comparisons can be made. One type of cost comparison that illustrates general levels of unit costs for cooling systems associated with both fossil fuel and nuclear fuel power plants of capacities greater than 600 MW is illustrated in Fig. 8. These costs are based on data of Jimeson and Adkins [8]. It should be noted that although the cost of cooling systems is a large sum overall, it represents a relatively small part of the total investment of a power plant (perhaps 4–7%). This added cost, in turn, is reflected in an increase in the cost of power of probably less than 1 mill/kWh. Based on data of Hauser, Fig. 9 shows the total cost additions for

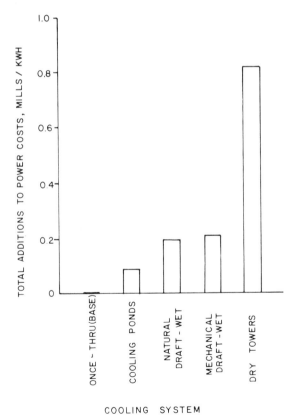

Fig. 9. Total cost additions for cooling water systems.

various cooling systems considered for a 1000 MW nuclear power plant at one particular location [9].

REFERENCES

1. J. E. Edinger and J. C. Geyer, *Edison Electric Institute Publication No. 65-902*, 1965.
2. L. D. Berman, *Evaporative Cooling of Circulating Water*, Pergamon Press New York, New York, 1961.
3. D. K. Brady, W. L. Graves, Jr., and J. C. Geyer, *Edison Electric Institute Publication No. 69-901*, 1969.
4. U.S. Department of Commerce, "Climatic Atlas of the United States," Environmental Data Service, Environmental Science Services Administration, U.S. Government Printing Office, Washington, D.C., 1968.
5. U.S. Department of the Interior, FWPCA, "Industrial Waste Guide on Thermal Pollution," U.S. Government Printing Office, Washington, D.C., 1968.
6. K. K. McKelvey and M. E. Brooke, *The Industrial Cooling Tower*, Elsevier, Amsterdam, 1959.
7. Special Report, "Cooling Towers," *Power*, March 1973.
8. R. M. Jimeson and G. G. Adkins, "Factors in Waste Heat Disposal Associated with Power Generation," paper presented at 68th National Meeting of AIChE, Houston, Texas, 1971.
9. L. G. Hauser, "Cooling Water Requirements for the Growing Thermal Generation Additions of the Electric Utility Industry," paper presented at American Power Conference, Chicago, Ill., 1969.

4
Control of Reservoirs and Lakes

Donald B. Aulenbach
Department of Chemical and Environmental Engineering, Rensselaer Polytechnic Institute, Troy, New York

I. INTRODUCTION

Pollution in general is a consequence of both the increased numbers of and the gregarious nature of humans. When there were only a few humans on earth, their bodily wastes were widely dispersed and created little problem. There were no concentrated industries and no industrial wastes. When an animal was killed for its food, its hide was also used for clothing, the fine bones were used as needles, the intestines and blood vessels were used as thread, the larger bones were used for weapons and tools, and any part that was not used by man was utilized by other parts of the immediate environment. This latter could include rapid consumption by other carnivorous animals or slow biodegradation by microorganisms, worms, termites, and so on.

In earliest times, human pollution abatement was most likely influenced by the presence of foul odors. This could be considered an esthetic problem or an air pollution concern. Body wastes and other decomposing matter generally had an undesirable smell and because of this there was an aversion to it. Thus, decomposing matter was usually removed as far as possible from the cave or other housing. Although the early humans did not know it, this was one of the best possible protections against the spread of diseases. However, as humans increased in numbers and moved into communities that grew into cities, the problems of waste disposal increased. Many of the great plagues of the middle ages have been traced to improper disposal of human wastes. The commonly accepted first incidence of relating human wastes to disease is attributed to studies of a cholera epidemic in London in 1849 by Dr. John Snow. It was through an epidemiological study that he traced the incidence of cholera to a particular well that was contaminated by the waste materials from a block of housing. Whether or not the story that he solved the problem by removing the handle from the pump is true, Dr. Snow did show the relationship between waste materials and disease. These findings were independent of the development of the microscope by Pasteur in France, only a few years later. His work confirmed the existence of many microscopic organisms, including bacteria of both desirable and disease-producing types. With this information available, societies became able to eliminate many diseases from the environment and increase the chances of human survival through reduction of diseases. Thus, early pollution control efforts were related to public health.

As cities grew, so did trade and manufacturing. Most frequently, cities were built along rivers to utilize the water power for running the factories, and to use the water itself for manufacturing processes, for transportation of raw and manufactured goods, and as a means of conveyance of the waste materials of that society from the location where they were generated. Soon we found that the increased organic load to the stream from waste discharges imposed on the stream an oxygen demand that exceeded the reaeration capacity. Subsequently, fish and other aquatic life in the stream suffered and some of the more desirable species died or disappeared completely. At the same time, city inhabitants initiated systems for water carriage of wastes. Sewers were almost essential to prevent or lessen the spread of diseases in high density population areas. The wastes were frequently conveyed directly to nearby streams adding to their pollution load. Some streams became open sewers with the majority of their flow sewage and/or industrial wastes from large communities. Soon it was realized that the load of organic materials on the streams was rendering them unihabitable for desirable species, particularly certain fish that provided food for many people. Gross pollution of streams produced other undesirable results such as production of sulfides and the destruction of the stream as a water supply for the city. Thus began the second phase of pollution control, in which methods were developed for removal of the organic material before the liquid was returned to the water course, and the era of the sanitary engineer. In general, pollution abatement methods proliferated around the turn of the past century.

Although much effort was expended on the removal of oxygen demanding material (primarily organic material) from waste waters, in general the soluble inorganic nutrients were overlooked. This was not necessarily the fault of the sanitay engineer at the time, probably unaware of the potential impact of these soluble nutrients—nor were methods readily available for their removal. However, the presence of soluble inorganic nutrients in a stream or a lake along with adequate sunlight can result in the production of excessive algal growths. Algae are organic material in the form of living cells. They have a finite life and when they die they rupture or lyse and contribute their organic matter back to the surrounding waters. This organic matter is the same as that which the sanitary engineer spent so much effort in removing. As more and more such interrelationships were observed, we had the development of the so-called environmental engineer who understood these interrelationships and the fact that removal of inorganic nutrients may be just as essential as the removal of the original organic matter. If nothing else, the cell growth had to be removed in order to prevent the cells from exerting the same amount of oxygen demand as the initial wastes. Although this in general did not present a public health hazard, it could result in depleting dissolved oxygen levels in a stream and rendering it unfit for fishing, which is still a potential source of work and food for many people.

Most recently, despite our concerted efforts to remove the inorganic nutrients from the waste before discharging to the receiving stream, a new threat has appeared in the form of toxic and hazardous materials. We are constantly striving to increase our standard of living and, therefore, constantly demanding more material items. As we develop more such better products, we find that they frequently require the use of many unnatural chemicals, chemicals that are often difficult for natural biological systems to degrade. Some of these have been discharged directly into the aquatic environment, whereas others have been landfilled and covered only to have the leachate from rainfall and/or runoff pass through the landfill area and carry them through the groundwater either into adjacent wells or ultimately to some surface water supply. In any event the threat of pollution is serious and in some instances has been shown to have extremely deleterious effects upon life. These effects arise not only from direct discharges of the hazardous or toxic materials into the water supply, but also from the vaporization of some of the organic chemicals, creating a hazardous atmosphere for local residents. Thus, we continuously find that we are subjecting ourselves to ever increasing problems in the environment.

It may be seen that pollution of the environment, particularly in the aquatic environment, is a threat to humankind and indeed to the entire earth. We do not yet know of the ultimate consequences of increasing the

CO_2 levels in the atmosphere from combustion of fossil fuels. It is speculated that an increase in CO_2 levels may increase the temperature of the earth sufficiently to change the climate and melt the polar ice caps. On the other hand, increased particulate discharges to the atmosphere could cool the earth by interfering with the sun's rays reaching the earth. Not only would this cool the earth, but also it would result in lower productivity because of the reduction of photosynthesis. This would drastically upset the total energy balance on earth. The problem includes the possible destruction or reduction of the ozone layer in the upper atmosphere by the Freons (chlorofluorocarbons) that are used in many refrigerants, and were formerly used in aerosol propellants. Another potential problem is the increased production of acid rain, which conjecture has traced to the combustion of fossil fuels, particularly coal. The nitrogen and sulfur oxides may be carried great distances and impact areas or nations many miles away. Acid rain results in the production of acid lakes, and many lakes have become sterile because of such acid rain. When these acid rains are traced to sources across national boundaries, additional political problems are created. Moreover, acidic bodies of water such as lakes or streams may dissolve more minerals out of the rock and soil materials they contact, thus potentially increasing the heavy metals content of the water and, therefore, further reducing the usability of this water even for drinking.

Many of the effects of pollution affect only humans. However, the results of pollution also affect numerous other terrestrial biological systems. Sometimes these have a direct impact upon humans, such as the shrinking of the fishing industry as their fish fish stocks died away, but in other ways the impact may be more indirect. The impact of DDT provides an excellent example of the roundabout effect of pollutants on the environment that ultimately touches humankind. DDT was found to create thin egg shells for many birds, including birds of prey. Thus, because birds of prey were reduced in numbers, the populations of rodents and other ground animals subsequently consumed more of the crops that DDT was intended to preserve. Because the total balance of nature can be upset, it is up to humankind to try to see to its preservation and not allow it to be destroyed by pollution. It is sufficient that we have already destroyed much of the natural balance by our high rate of reproduction and our aggregation into large cities and industrial sites. If we are to survive on earth, we must control the things which we discharge into the environment.

This chapter will concern itself primarily with waer pollution problems. This is not to say that air and soil pollution problems are not also of concern. There is a definite interrelationship between the pollution of air, water, and soil. However, most pollutants either ultimately end up in the aquatic environment or pass through the aquatic environment on their way from one phase to another. Thus it was considered within the limits of this chapter to be concerned primarily with the impact of pollutants on the aquatic environment.

As examples of this, pollutants discharged into the atmosphere are very frequently returned to earth in the form of precipitation. This may be in the form of rain or one of its frozen forms such as snow or ice. The transfer of air pollutants to the precipitation is usually accomplished by two means: one is by the washing out of particulates by the precipitation and the other is by the solution of the gaseous phases into the liquid water droplets. Both carry pollutants out of the atmosphere and into the aqueous environment. In terms of pollution of the soil, it is possible to place materials in the earth or the soil in such a manner that they do not ever reach the aquatic environment. However, land disposal systems very frequently allow a certain amount of infiltration of precipitation, or in some cases of groundwater, and both infiltrations result in the leaching of disposed materials from the land. In some instances this gets directly into the groundwater, but the groundwater in turn may flow into surface streams or lakes. Thus, hazardous materials stored in the earth, even when subjected to reasonable precautions, may still end up contaminating the aquatic environment. The other source of pollutants in the aquatic environment is the direct discharge of waste material, both treated and untreated, into flowing streams and into lakes. In general today, most direct discharges have been treated before discharge, but regardless of the degree of treatment, there is always some discharge of materials into the aquatic environment. It is for these reasons that this chapter will deal primarily with the impact of pollutants in water.

II. SPECIAL FEATURES OF WATER

Water is quite an unusual substance. It has many physical and chemical properties that do not follow the normally expected trends. Scientists have generally found the reasons for these seeming anomalies and have been able to explain them. However, this does not change these peculiar factors.

The observation of the peculiar properties of water date back at least as far as L. J. Henderson in 1912 [1], in which he describes the properties of water that are most important. First of all was that ¾ of the earth's surface is covered by water and that, if it were not for the mountains of land, the entire earth would be covered by 4 kilometers (2.5 miles) of water. Chemically, water is very stable; very few substances react directly with it, although many reactions take place within it.

Perhaps its thermal properties are the most anomalous properties of water. The first of these is its specific heat or heat capacity. Water is the standard, having been assigned a value of 1.000 in the range of 0–1°C. This is a relatively high value and enables water to maintain a relatively constant temperature. It is important in the stabilization of biological and chemical reactions in the aquatic environment, and serves to moderate the earth's climate by providing a reservoir of heat in the tropical oceans and a reservoir of cold in the polar ice caps. These effects in turn help aquatic organisms maintain a constant temperature and reduce the amount of heat regulation necessary for mammals.

A second important thermal property is water's latent heat of fusion and vaporization. The latent heat of fusion, 88 cal/g, is quite high. The latent heat of vaporization, 536 cal/g, is the highest known for any substance. The latent heats represent the amount of heat that must be provided to a given weight of water to convert it from either the solid to the liquid or the liquid to the vapor state at the same temperature. This is very important in the evaporation of water and represents a large transfer of energy from the tropical oceans where energy is required to evaporate the water and to release this energy elsehere on the earth in the form of condensation. The latter effect is also very important since it regulates or moderates temperatures throughout the world, preventing the tropics from becoming excessively hot and the poles extremely cold. It is similarly important in physiological heat regulation, in terms of both perspiration in animals and transpiration in plants.

The thermal conductivity of a substance represents the time rate of transfer of heat by conduction through a unit thickness across a unit area for a unit of temperature in a unit of time. The value for water of 0.0125 cal/s-cm-cm^2-°C, is quite high compared to other liquids. Ice, however, has a higher thermal conductivity, which results in a lake's more rapid cooling once the first layer of ice has formed.

Probably the most significant thermal anomaly is thermal expansion. Because of orientation of the water molecules, the most compact configuration occurs at 4°C. Thus, the most dense water will be 4°C. This results in deep lakes having a farily constant temperature of 4°C on the bottom. Furthermore, ice, the solid state of water, is less dense than water and therefore floats on the surface. Water is the only substance whose solid state floats on its liquid surface. This is very important to our environment since if this were not the case ice would form at the bottom of all deep lakes and this ice would never thaw. Such a circumstance would severely restrict biological activity in the bottom of the lake and result in a buildup of layers of sediments and ice in the bottom of all lakes.

There are several nonthermal properties of water that are also peculiar to it. The first one to be considered is its solvent power. Scientists are always looking for the universal solvent, and water comes closest to being just that since more elements and compounds have been recovered from it than from any other liquid. This is very important in making nutrients available to biological life in the water, whose food must be waterborne. In addition, the materials dissolved in the water help to maintain the osmotic pressure of the aquatic organisms.

Water also has a very high ionizing power as a result of its high dielectric constant, which helps to reduce the force of attraction between atoms or molecules in the liquid. This is important in providing soluble nutrients to aquatic organisms and also affects cell permeability.

Only mercury has a higher surface tension than water. The value for distilled water is 7.5×10^{-6} N/m (75 dynes/cm), whereas for mercury it is 43.6×10^{-6}. This creates a tough, thin elastic film on the surface of the liquid and results in the formation of drops. It also effects capillarity, which controls the rise of liquid in the soil. The surface film of a lake actually forms a habitat for certain organisms known as pleuston. Microscopic pleuston are called neuston. Those living on the upper surface of the interface film are called epineuston, and those on the underside are called hyponeuston. The pleuston include the water striders, the mosquito larvae, certain snails, and other organisms. Some organisms that do not live in the surface film nonetheless deposit their eggs there. Thus the surface film of water is a very special habitat to certain organisms.

The viscosity or resistance to flow of water varies with its temperature. This affects lake mixing, particularly during the summer, when the surface of the lake is warm and more readily mixed by the wind. It also effects the flotation of organisms in the euphotic zone (the area through which sunlight penetrates). This is important in maintaining balanced algal growths and in differentiating between the species that may be present at certain times of the year as a function of temperature.

Water is considered a noncompressible fluid, although precise measurements show that it is very slightly compressible. For all intents and purposes it may be considered a noncompressible fluid for either measurement purposes or for the aquatic organisms living therein.

From all these physical and chemical parameters it may be seen that water is very important in sustaining life. It is the author's belief that water is the most important factor in maintaining life on earth and that if scientists want to determine whether or not life could exist on any other planet, they must measure its presence there. In short, all life is dependent upon water and its unique properties.

III. HYDROLOGY

Hydrology is the science of the phenomena of water in all its states; its distribution and occurrence in the earth's atmosphere, on the earth's surface, and in the soil and rock strata; and the relationship of these phenomena to the life and activities of humankind. It is concerned with both the quality and the quantity of water available. Since water is essential to the preservation of life on earth, we need a good understanding of hydrology.

The amount of water present on the earth today is essentially the same as that which was present when the earth was formed. It may be in a different physical state and some minor additions may have occurred through chemical reactions, but for the most part, the total quantity of water on the earth has not changed. It does, however, undergo many physical transformations from the vapor phase to the liquid and the solid phases. This is a constant and ongoing cycle, commonly called the hydrologic cycle. Figure 1, from Osborn and Harrison [2], depicts the various phases and locations of water on the earth. It may be seen that water vapor occurs in the atmosphere, condenses, and forms precipitation that falls over the face of the earth, both the land and the water. A portion of the water evaporates in falling, another amount evaporates after striking objects on earth such as trees or buildings. Another portion flows along the ground, forming streams and/or lakes, whereas still another part infiltrates the soil to become groundwater. Both surface water and groundwater flow toward the lowest level, ultimately reaching the ocean. Here a large amount of the evaporation occurs. Evaporation occurs from all moist ground surfaces and all open waters. In addition, transpiration occurs from all plant life. Usually in evaluating the amount of water vapor formed, evaporation and transpiration are combined as evapotranspiration since it is difficult to differentiate one from another on the earth's surface.

Tremendous amounts of water pass through the hydrologic cycle. Some idea of this can be gained from Fig. 2 [3]. Large amounts of energy

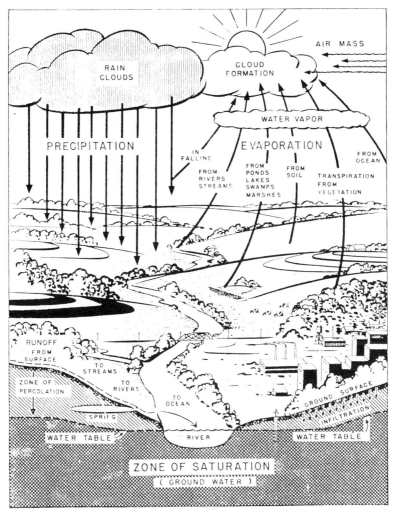

Fig. 1. The hydrological cycle [2].

are utilized in the evapotranspiration process in converting water to water vapor. Also, the amount of precipitation onto the land and the ocean is similar when one considers that approximately 75% of the earth is covered by the ocean. This does not imply, however, that the distribution of rainfall over the earth's surface is uniform. There are many localized factors that control the distribution of precipitation across the earth's surface.

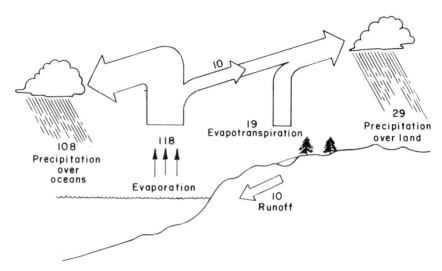

Fig. 2. World hydrological cycle. Units are 10^{15} gal/yr [3].

The main source of energy for evaportranspiration is the sun. Small increases in evaporation may occur as a result of human activities, particularly in the use of heat for power generation. However, this total amount is small compared to the energy of the sun and is significant only in localized areas. When water evaporates, the vapor is lighter than air; therefore, it rises and expands as the atmosphere rarifies. Expansion causes cooling, which ultimately results in condensation, releasing again the energy that was required to evaporate the water initially. Fortunately, the water vapors do not rise in the atmosphere indefinitely. About 50% of the water condenses before it reaches 1.6 km (1 mi) altitude. About 4% remains at 6 km (4 mi) and essentially none has been observed above 20 km (12 mi). Therefore, the water vapor does not escape the earth's atmosphere.

Precipitation occurs in many forms. Rain, in which droplets of liquid water fall to the earth, is the most common. Drizzle is the descent of very fine droplets of liquid water, usually with a diameter of less than 0.5 mm, which are small enough so that wind currents easily carry them both horizontally and occasionally upward. Snow is the precipitation of solid water, primarily in the form of hexagonal crystals. In North America, sleet is rain that falls through subfreezing air masses that cause the rain to freeze into ice pellets. In Britain, sleet is considered melting snow or a mixture of snow and rain. Freezing rain is caused when liquid rain falls on a subfreezing object coating it with ice. Whereas sleet occurs primarily in winter, hail is a form of frozen rain droplets that is found primarily in summer thunderstorms. In the formation of hail, the water droplets fall through a subfreezing air mass, forming small ice pellets. Depending upon the rising force of the air masses, these ice pellets may be carried upward into the warmer air mass where more condensation occurs, therefore, increasing their size. They fall again through the cold air mass, where the newly condensed water on the surface freezes forming a larger hail stone. Repeated rising, growing, falling, and freezing has been known to create hail stones as large as 5 cm (2 in). Dew forms directly from condensation on the ground, mainly during the night when the ground surface has been cooled by outgoing radiation. In some circumstances, dew can be a significant portion of the total precipitation.

All of precipitation is the result of cooling that causes the condensation of the water vapor. There are three main forms of cooling: dynamic, upward movement of air, and external cooling.

Dynamic cooling is caused by the expansion of the air as it rises to a higher altitude. In general the temperature of the air above the earth's surface decreases with altitude. This change in temperature with altitude is called the lapse rate. In the first 3–5 km above the earth's surface it is extremely variable. In some cases it is even negative (an increase in temperature with altitude), which is called an inversion. Generally in the range of 5–11 km of altitude, the decrease in temperature averages 0.7°C/100 m altitude (3°F/1000 ft). In the free atmosphere, where there are no other sources of heat or means of removal, the rising air cools at the dry adiabatic lapse rate. In the middle latitudes this decrease in temperature is 1°C/100 m (5.5°F/1000 ft). This decrease in temperature occurs until the dewpoint is reached. It must be pointed out that the dewpoint also decreases at a rate of 0.2°C/100 m (1.1°F/1000 ft). After the dewpoint has been reached, the wet or, as it is sometimes called, retarded adiabatic lapse rate occurs. This results in a decrease in 0.6°C/100 m (3.2°F/1000 ft). The difference between the wet and dry adiabatic lapse rate is caused by the latent heat of condensation, which releases energy as condensation occurs.

Under adiabatic conditions, three scenarios can occur, as shown in Fig. 3. The first scenario represents stable air in which the existing lapse rate is less than both the wet and dry adiabatic lapse rate. Therefore any rising air would be cooled more rapidly than the surrounding air and, since it would be cooler, it would not rise. Under this condition there would be no convective rising. This is depicted in the first diagram in Fig.

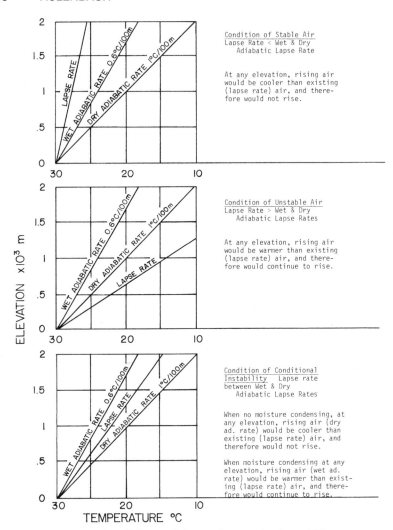

Fig. 3. Possible conditions of atmospheric stability.

3. The second scenario is one of unstable air. If the existing lapse rate is greater than both the wet and dry adiabatic lapse rates, any rising air will be cooled less rapidly than the surrounding air and thus would be relatively warmer than the surrounding air and would continue to rise. This creates the unstable conditions and is shown in the middle diagram in Fig. 3. If the existing lapse rate is less than the dry adiabatic lapse rate, but greater than the wet adiabatic lapse rate, the scenario is one of conditional

stability. The condition would be stable only when no moisture is condensing. However, when moisture is condensing, unstable conditions would develop and the water vapor would continue to rise. This condition is useful in artificial rainmaking, which relies upon initiating the precipitation using dry ice, silver iodide, or some other nucleating agent.

Upward movements of air that cause precipitation generally fall under three categories. The first is convective movement, which is typified by tropical rains and thunderstorms. The surface air is heated rapidly by a warm sun during the day. The water vapor rises rapidly, expanding to the point at which it then cools, condenses and precipitates. In tropical areas this is responsible for the routine daily precipitation that falls sometime in early afternoon. Thunderstorms may occur anytime of the day or night, but are most common late in the afternoon as a result of the rapid daytime warming by the sun.

The second upward movement of air is called orographic precipitation. This occurs when horizontal currents of warm air pass over large bodies of water such as the ocean where they pick up moisture. When they reach land, the air is forced upward by the coastal mountains where it is cooled and forms precipitation. Since the moisture is removed in the upward motion that produces this cooling, the air mass becomes warmer and hence drier as it proceeds down the other side of the mountain. Thus on the windward side of the mountain, areas of heavy rainfall occur, whereas on the leeward side deserts occur. As a typical example, in the Pacific Northwest, 190–300 cm/yr of precipitation occurs (75–125 in./yr) producing rain forests on the windward sides of the mountain ranges. On the leeward sides, only about 5% as much precipitation occurs. In some areas this is called the Chinook effect.

The third upward movement of air is caused by cyclonic motion. The air around low pressure areas flows in a counterclockwise direction. On the surface of the earth the air moves from areas of high pressure to areas of low pressure. This creates areas of differential temperature, and where such cooling occurs precipitation also occurs.

The third scenario for precipitation is caused by external cooling, usually the presence of cold air masses. When cold air comes down out of the Arctic (from the Antarctic below the Equator), it comes in along the surface of the earth, forcing the warm moist air upward. This creates cooling that results in precipitation.

In addition to the variation of precipitation with location and means of formation, precipitation within a storm also varies with time. As a general rule, the longer the duration of precipitation, the less is the rate of precipitation. Figure 4 depicts the intensity duration curves for New York

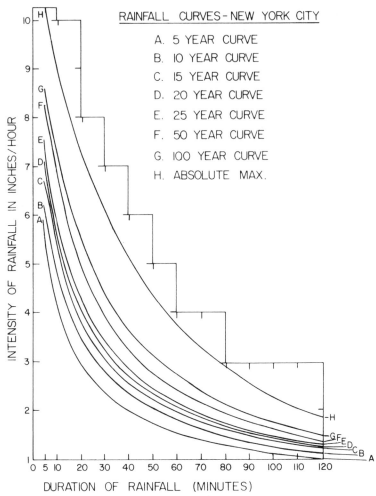

Fig. 4. Rainfall frequency–duration curves for New York City.

City. The figure also shows the recurrence interval or the number of years that a given precipitation would not be exceeded. For design parameters, the cumulative worst possible storm is also plotted. It is clearly evident from this figure that the longer the duration of precipitation, the less intense will be the storm.

In addition to the variation of precipitation with time, there is a very definite geographic variation in precipitation, particularly in considering a large area such as a drainage basin. Numerous methods have been presented to determine a representative value for precipitation over a large drainage basin area. The first method for determining a representative value would be to take the arithmetic mean of several precipitation stations within the basin. Values can be based on monthly or annual precipitation, but usually are not based on shorter periods of time. In the arithmetic method, only values from stations that are located within the stated drainage basin area are used. No results from adjacent stations are employed, even where a station may be very close to, but outside of, the drainage basin under study.

This has been corrected significantly by utilization of the Thiessen [4] method for determining precipitation over an area (Fig. 5). In this method stations located close to the basin boundary may be considered, but only a weighted average as a function of the area within the given

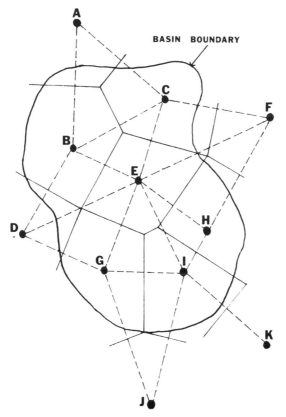

Fig. 5. Thiessen Method for average precipitation over an area.

basin that is represented by that particular station. To put it another way, if a station adjacent to the basin is so close to the basin that the perpendicular bisector of a line between a station immediately outside of the basin and its closest station inside the basin falls within the basin, then the area closer to the outside station, but within the basin is considered representative of the results of the station located outside the basin. In order to depict this, a stylized drainage basin is chosen as shown in Fig. 5. The points indicate the location of precipitation measuring stations within and adjacent to the basin. In order to determine the portion of the total area represented by each station, construction lines (shown as dotted lines in the figure) are drawn between each adjacent station. Then perpendicular bisectors of each of these dotted lines are drawn. The perpendicular bisectors are extended until they meet a similar perpendicular bisector of an adjacent line. With careful construction all lines should meet in points and it can be determined whether or not an area within the basin is closer to a measuring station outside of the basin or not. The perpendicular bisectors are drawn as solid lines on the figure. Then by means of a planimeter the area represented by each station may be measured. The total precipitation in the basin is then derived from the sum of all of the individual values of precipitation at the representative measuring stations multiplied by the fractions of the area of influence within the basin from that measuring station. A typical example of the calculations involved is shown in Table 1.

A third method for measuring precipitation in a large basin is the isohyetal method. Here the values of precipitation for each station are indicated on a map showing the location of each station. Smooth lines are drawn through areas of equal precipitation, as shown in Fig. 6. Now the area between each contour line is measured with a planimeter and the average precipitation between the two lines is multiplied by the ratio of that area to the total basin area, the sum of the individual values giving the total precipitation over the draniage basin. An example of the calculation for Fig. 6 is shown in Table 2.

All of these methods have some sources of errors. However, the use of the isohyetal method can also take into account such known conditions as mountains, lakes, and so on. Thus, the method may be improved by some subjective interpretation of the data.

Of importance to the engineer is the frequency of recurrence of a storm of a given intensity or the maximum or minimum rainfall over a given period of time. This normally requires extended periods of record that frequently are not available. In lieu of extended records, information

TABLE 1
Calculations for Thiessen Method

(1)	(2)	(3)	(4)	(5) or	(5a) omit (4)
Location	Ppt'n, cm	Area, km²	Fraction (or %) of total area[a]	Weighted ppt'n, cm. (2) × (4)	Ppt'n × A, (2) × (3)
A	1.65	7	0.01	0.0165	11.55
B	3.71	120	0.9	0.7049	445.20
C	4.88	109	0.18	0.8784	531.92
D	3.91	20	0.03	0.1173	78.20
E	6.83	120	0.19	1.2977	819.60
F	7.16	0	—	—	—
G	7.57	92	0.15	1.1355	696.44
H	11.43	76	0.12	1.3716	868.88
I	12.70	82	0.13	1.6510	1041.40
J	4.44	0	—	—	—
K	4.95	0	—	—	—
Σ		626	1.00	7.17	4492.90

$$\frac{4492.90}{626} = 7.18$$

[a] $(4) = A/\Sigma A = (3)/\Sigma(3)$ (× 100 for %)

may be obtained from a similar location or a nearby location where long-term data are available. Where sufficient data are available the probability of any event occurring can be determined by plotting the individual values on probability paper. Either an arithmetic normality or a geometric normality may be plotted. This is accomplished by arranging the individual data points in ascending (or descending) order of magnitude. The probability of recurrence of any value then is the order of magnitude of that event divided by the total number of events plus 1. These values can then be plotted on a graph similar to Fig. 7. To determine the probability of any event exceeding (or being less than) a given value, one needs only to refer to the probability curve. The mean value is that which is exceeded every other year as a statistical average and this produces a mean value at the 50% occurrence. For a value that would be exceeded once in 10 yr, one would refer to the 90% level and for once every 100 yr refer to the 99 yr probability, and so on. It must be remembered that these are statistical values. The Fig. 7 was plotted from 26 actually measured values. It may

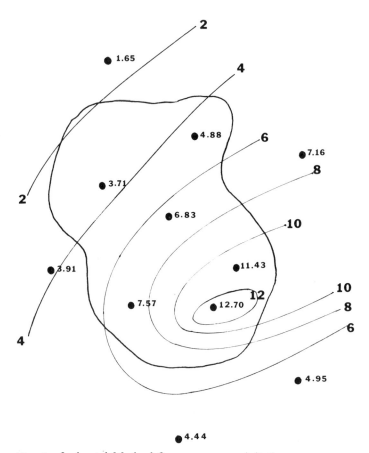

Fig. 6. Isohyetal Method for average precipitation over an area.

be seen that the upper end of the curve does not fit the data points. Furthermore, there is nothing to prevent a 100-yr occurrence from happening 2 yr in a row. The information merely provides data upon which the engineer can design a facility or a structure to handle the worst anticipated problem in a given period of time. The value of the structure and the value of the potential property damage must all be taken into account. For example, a dam should be constructed to handle the worst possible conditions, whereas a small storm drain could be designed to overflow on an average of once every 20 yr if there would be no significant damage caused by such flooding. Obviously, social concerns must also be considered, not just the financial aspect.

TABLE 2
Calculations for Isohyetal Method

(1) Isohyet, cm	(2) Area (A), km²	(3) Fraction (or %) of total area, $A/\Sigma A = (2)/\Sigma(2)$ (× 100 for %)	(4) Avg. ppt'n, cm	(5) Weighted ppt'n, cm (3) × (4)
12	13	0.021	12.2	0.2562
10–12	77	0.123	11.0	1.3530
8–10	116	0.185	9.0	1.6650
6–8	145	0.232	7.0	1.6240
4–6	150	0.239	5.0	1.1950
2–4	105	0.168	3.0	0.5040
2	20	0.032	1.8	0.0576
Σ	626	1.000		6.6500

usually (isohyet$_1$ + i_2/2)

or can find $\Sigma[(2) \times (4)]/\Sigma(2)$ = same value as column 5

IV. EVAPORATION

Evaporation is the formation of water vapor. Strictly speaking, it is the formation of water vapor from a liquid surface; however, for the sake of this discussion, evaporation here will also include evaporation from the solid state or ice, which technically is called sublimation.

The most important factor causing evaporation is the amount of solar radiation. Indirectly, other factors affecting evaporation are also related to solar energy. The vapor pressure of the air controls the relative humidity; the lower the vapor pressure, the greater the evaporation. The temperature of both the water and the overlying air affects the vapor pressure. The higher the temperature, the greater will be the evaporation. Wind is very important in creating evaporation in that it carries away the saturated monolayer of water vapor immediately above a body of water. This brings in fresh air with a lower vapor pressure, thereby allowing increased evaporation to occur. An increase in the barometric pressure decreases evaporation. Thus, the lower pressure at greater elevations would be expected to result in increased evaporation. However, this effect is generally small and is exceeded by those from the temperature differential and the wind velocity, each of which has a greater effect upon evapo-

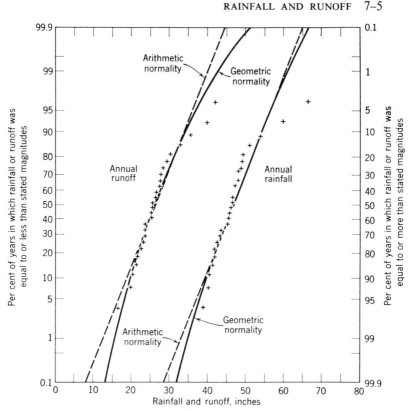

Fig. 7. Frequency distribution of annual rainfall and runoff.

ration at higher altitudes. The greater the quantity of dissolved salts in the water, the less evaporation occurs. Thus, slightly less evaporation occurs over oceans than over fresh water bodies. The difference becomes significant only in the evaporation of salt brine. The motion of water also increases the evaporation as a result of the increase in surface area created by waves, and of course by the formation of spray droplets from water falls. Also important is the shape of the bottom of the body of water, particularly in lakes. The greater the relative area of shallow water to deep water, the higher will be the rate of evaporation.

Several means are available for determining the amount of evaporation from open water surfaces. The simplest is the use of the water budget, following the equation:

$$E = P_t - R_o \pm S \pm L_a \tag{1}$$

in which

E = evaporation
P_t = precipitation
R_o = the runoff
S = the storage
L_a = the leakage and refers to groundwater flow into or out of the given basin

The precipitation may be fairly readily meausured by means of rain gages. The runoff can be determined by a flow measuring device on the outlet stream of the body of water. Storage merely refers to the change in elevation during the period of reference. It relates to the volume of water involved in any change in elevation. Normally on an annual cycle using the water year beginning October 1, the storage is approximately 0 from year to year. The groundwater inflow or outflow is rather difficult to determine. In actuality, this equation is often used to determine the amount of leakage based upon some measured amounts of evaporation.

Evaporation (E) may also be determined by the use of an energy budget. In this case, the equation may be written as:

$$E = \frac{Q_s - Q_r - Q_b + Q_v - Q_\theta}{\rho H_v (1 + B)} \qquad (2)$$

This equation gives E in cm when the energy values of Q_s, Q_r, Q_b, Q_v, and Q_θ are measured in cal/cm^2.

Q_s = the solar and sky shortwave radiation incident at the water surface
Q_r = the reflected solar and sky shortwave radiation incident at the water surface
Q_b = the net energy loss by longwave radiation to the atmosphere
Q_v = the net energy advected to the water which is the difference between the energy of the inflow and the outflow of the given basin
Q_θ = the increase in energy stored in the water
ρ = the density of the water
H_v = the latent heat of vaporation which is assumed to be 536 cal/g
B = the ratio of heat loss by conduction to heat loss by evaporation and is commonly known as the Bowen ratio.

The Bowen ratio (B) may be written as:

$$B = 0.61 \frac{T_s - T_a}{e_s - e_a} \cdot \frac{P_a}{1000} \qquad (3)$$

in which

T_a = the temperature of the air in degrees C
T_s = the temperature of the water surface in degrees C
e_s = the vapor pressure of the water surface in millibars
e_a = the vapor pressure of the air in millibars
P_a = the atmospheric pressure

There are numerous mass transfer equations that can determine evaporation mathematically. Most of these are empirical and have been designed to fit a specific location. Typical among these is Meyer's equation [5]:

$$E = C_e(V_s - V_a)\left(1 + \frac{w}{K}\right) \qquad (4)$$

in which

E = the evaporation, in/mo
V_s = the vapor pressure of the water surface in inches of mercury,
V_a = the vapor pressure of the air or dew point temperature, in inches of mercury
w = the wind velocity 30 ft above the surface in mi/h
K = 10
C_e = a coefficient which varies being 15 for small shallow lakes and ponds and for leaves and grass and 11 for large and deep bodies of water

Evaporation may also be measured by means of a pan. Here the basic principle is to place water in an appropriate pan and to measure the amount of water evaporated over a given period of time. There are numerous sizes, shapes, and depths of pans. However, they come under three basic types. First is the sunken pan, which is usually placed in an indentation on the shore of the body of water whose evaporation is to be determined. By placing these in the ground there is less heat loss through the sides and the pans may be somewhat larger; however, they have a higher cost and are not portable. The second type is the floating pan, which is placed directly in the body of water. These most nearly simulate evaporation from that water since there are no significant heat losses in this system. However, this type presents problems particularly from splashing and wave action, which can change the volume of water inside or outside of the pan. Also, it is difficult to read the water level when they may be sloshing back and forth and they are difficult to maintain in place. The third type of pan is a surface pan and is the most common type used. It is portable and relatively inexpensive. The standard weather bureau class A pan is 4 ft (1.2 m) in diameter and 10 in. (25 cm) deep. It is filled to 8 in. (20 cm) and evaporated to 7 in. (17.5 cm). It is placed above the ground adjacent to the body of water whose evaporation is to be measured. Nomographs and, of course, computer programs can be used to determine the pan coefficient, which relates the evaporation in the pan to the actual evaporation in the body of water.

In addition to evaporation from the water surface, there is evaporation of moisture from the soil. This is usually a function of the evaporation of recent precipitation, but where the groundwater level is close to the ground surface, capillary action may maintain moisture at the surface of the ground, where it may evaporate. As a general rule, ground cover reduces the amount of evaporation from the soil surface. Compared to the evaporation from bare ground, areas covered by vegetation may result in only 0.7–0.8 as much evaporation and a dense forest may result in only 0.2–0.4 as much evaporation as from the bare soil. In general, the same factors that affect evaporation from an open water surface also have similar effects upon evaporation of the soil moisture.

V. TRANSPIRATION

Transpiration is the vaporization of water from the breathing pores of leaves and other plant surfaces. This amount can be significant in forested areas and can be as much as 2 million pounds of water per acre per year (300,000 kg/ha-yr). The amount of transpiration increases significantly with an increase in temperature, and sunlight is necessary for the process to occur. It is relatively difficult to measure transpiration alone, but it can be determined experimentally in special plots, greenhouses, and so on. However, it is difficult to compare these experimental studies with actual field conditions. In general under field conditions, it is almost impossible to separate transpiration from evaporation from the soil surface.

VI. EVAPOTRANSPIRATION

Since it is difficult to determine transpiration alone, and since the total amount of vaporization of water is really the important parameter, evaporation and transpiration are usually combined in a term called evaportranspiration.

Evapotranspiration may be estimated by the inflow–outflow method, which is essentially the same as the method for measuring evaporation, where the two are combined and called evapotranspiration.

Considerable attention has been paid to the total of evapotranspiration by soil scientists. Meyer [5] has designed a series of charts that show the estimated amount of evaporation as a function of the rainfall during a month and the potential transpiration over the same period of time. These charts are available for different latitudes, but are basically designed for use in the middle latitudes. From the charts for evaporation and transpiration, the total evapotranspiration for a given area may be calculated.

Another method for determining evapotranspiration is the Blaney-Criddle method [20]. This method provides the potential consumptive use: that is, the amount of evapotranspiration that would occur if sufficient precipitation is available. It can be modified to provide the actual use based on the precipitation and a balance sheet. The basic equation for this method is:

$$U = \Sigma(f \times k) \tag{5}$$

in which

U = the consumptive use in inches per period (usually a month)
f = a coefficient equal to $T \times p_m/100$ and equal to the monthly consumptive use factor
T_m = the mean monthly temperature in degrees farenheit
P_m = the monthly percent of daytime hours of the year
k = the monthly consumptive use coefficient, which equals 0.55 for garden vegetables, melons, and winter grain, 0.65 for grasses, brambles, leaves, peas, potatoes, tomatoes, peppers and small grain, and 0.75 for corn, alfalfa, and orchard with cover.

Obviously with this method the operator must know the areas of each crop under production in the area being studied. The time of growth must be known and information must be available about the possibility of a second crop. This is a function of the length of the growing season for the particular area. The method gives more precise data, but obviously considerable calculation is involved in determining the areas of the various crops produced. Since U is merely the potential consumptive use, a balance sheet must be made up over a complete growing season or year, as appropriate. When precipitation exceeds potential evaportranspiration, the excess moisture may be stored. When evapotranspiration exceeds precipitation, the potential consumptive use may be fully utilized if there is sufficient storage from the previous month when there was excess precipitation. This may continue until all of the stored water has been utilized. Beyond this point, evapotranspiration cannot exceed the precipitation. The typical situation for the Northern latitudes is to have excess precipitation during the spring and insufficient precipitation in summer to allow for all of the potential consumptive evapotranspiration. One of the uses of this method is to determine when and how much additional irrigation would be needed to provide the optimum crop growth.

A more detailed method, based upon the potential evapotranspiration, for determining evapotranspiration has been developed by Thornthwaite [6]. The formula involves the relationship between temperature, latitude, and potential evaporation, and the actual use also depends upon the climate, soil moisture supply, plant cover, and land management. Since this method also determines potential evapotranspiration, a balance sheet similar to the Blaney-Criddle method must be utilized in order to determine the actual evapotranspiration over a growing season.

VII. INFILTRATION AND PERCOLATION

The term infiltration represents the entrance of surface water into the ground at the soil–water or soil–air interface. It is affected by topography, the surface conditions of the ground, the characteristics of the surface material, and the rate of precipitation. It is greatest with plowed fields, with vegetative cover, and on flat ground.

Percolation on the other hand represents the flow of water within the soil in both the horizontal and vertical direction. There is a fine line of distinction where infiltration ends and percolation begins. Percolation is controlled by gravity and the type of soil through which the water is flowing. Percolation will continue in a downward movement until an impermeable barrier is reached. This may be a very fine material such as clay or an impermeable substance such as rock. From this point the water will flow in more of a horizontal direction, but still downslope along the top of the impermeable barrier. The velocity of flow is a function of the soil material and the amount of water being applied to force the system in a downward slope as a function of gravity. This zone of water above the first confining layer is called the free or unconfined ground water. Since at various locations on the earth there are intermittent layers of porous and nonporous materials, water may enter a lower level and in flowing downward may pass under an impermeable layer and become a confined aquifer. Confined aquifers usually flow under pressure as a function of the connection to the free groundwater surface. Thus there may be numerous aquifers as one proceeds deeper into the earth. However, as a general rule, the deeper the aquifer, the more dissolved solids are present and the poorer the quality of the water. Since the free groundwater has the potential to be contaminated from the surface, the most desirable water from the standpoint of domestic use is usually that found in the first confining aquifer. A more detailed discussion of groundwater will be given in a separate section on groundwater.

VIII. RUNOFF

Precipitation which does not evapotranspire back into the atmosphere or infiltrate into the soil as groundwater runs across the surface of the earth as runoff. This is very important since it forms the lakes and streams on the earth. These may be used as a source of water for various uses including navigation, recreation, water power both directly and in the generation of electricity and many industrial uses including production of materials and cooling water. Surface waters serve as a water supply for many people particularly for large metropolitan areas. The animals in the water provide food for many people. When used for irrigation the water may be used for growing crops on land. Runoff represents the earth's freshwater resources. Once the water flows to the ocean, it becomes saline and takes on significantly different characteristics.

There are numerous factors that affect stream flow. These may be listed as follows:

1. Precipitation, particularly the amount of water equivalent, its distribution across a given watershed and the extent and duration of the storms.
2. The temperature is important particularly with regard to the accumulation of ice and snow during the winter months in temperate and polar climates.
3. Topography, specifically the slope and character of the ground.
4. Geology with regard to the relative imperviousness of the soil and the bottom of a stream. This represents the difference between surface water runoff and groundwater.
5. Ground surface covering, specifically the extent, character, and type of vegetation.

6. The storage of water, whether it be naturally in lakes, ponds, swamps, or marshes, or under artificial storage in dams, reservoirs or within dykes.
7. The watershed itself, particularly the size, the prevailing winds, and the orientation of the watershed in relation to storms and mountains.
8. Wind intensity which will carry precipitation into or out of a basin and which, under unusual conditions, may even control the outflow of a river.
9. Erosion and silt, which tend to fill the stream bottom causing the channels to change and to increase flooding potential with subsequent storms.

When studying a given watershed, items 3–9 are normally considered to be constant. There can be some allowable changes in the character and type of vegetation, and artificial storage can be provided. Within limits, erosion and silt pollution can be controlled. Also in studying runoff, conditions during the nonfreezing time of the year are usually considered. However, it must be recalled that the worst potential for flooding in a stream occurs when a heavy rainfall coincides with the spring snowmelt. This leaves the amount and distribution of precipitation as the most important factor to consider when evaluating storm runoff in a given drainage basin.

In evaluating storm runoff, the most important factor is normally the peak or maximum rate of discharge. Normally, the peak discharge occurs at the time of concentration, which is the longest time for flow from the farthest point in time in any given drainage basin to reach a specific location downstream. The development of the time of concentration, t_c, is shown in Fig. 8. An idealized drainage basin is depicted with a stream flowing through it. The point of determination of flow is designated as point A. The units of time are designated as the numbers 1 through 5. This indicates that in the first unit of time all of the water falling within line B will reach point A. These lines are not necessarily arcs of a circle, but may be contoured as a function of the slope and the character of the ground. However, for convenience they are shown as relatively smooth lines on the figure. By the end of the second time period, all areas within line C will contribute flow to point A. Continuing this analogy, finally, after somewhat more than 5 time periods, any rainfall that occurs within the given drainage basin area will flow to point A during that given amount of time. The time needed for the rainfall that comes down the farthest distance from point A in the basin to reach that point A is consid-

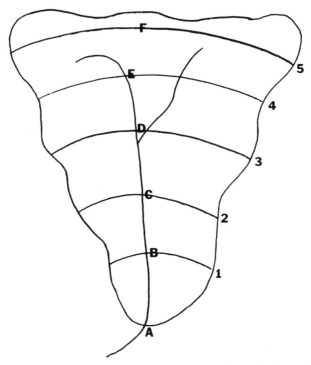

Fig. 8. Illustration of time of concentration, t_c, in a drainage basin.

ered t_c. As the area contributing to point A increases, the amount of runoff reaching point A increases significantly. However, it must be recalled from Fig. 5 that as the time of precipitation increases, the intensity of precipitation decreases. Thus, up until the time of concentration there are two factors that affect the rate of runoff: they are the increasing area with time, and the decreasing rate of precipitation with time. However, after t_c, the area contributing is constant and the only factor affecting runoff is the duration of precipitation, which results in a decreasing rate of precipitation within time. Thus, a curve of runoff vs time during a storm event is depicted in Fig. 9, which shows that the initial increase in runoff is affected by both the increasing area contributing and the decreasing intensity of precipitation up to t_c and, thereafter the decrease in runoff is caused by the decrease in intensity of the storm. As a general rule, for a given basin, the shorter the t_c, the higher will be the peak runoff rate; or conversely, if anything can be done to increase t_c, the peak flow can be reduced.

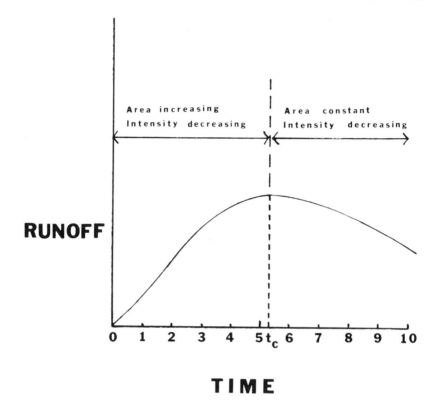

Fig. 9. Factors influencing a storm hydrograph.

The value of t_c can be determined for a small drainage area, such as that going into a storm sewer or culvert, or for a large area to determine the flow in a stream. The main difference is the time factor, whether this be in minutes or days. An empirical equation to calculate t_c has been given by Horner and Flint [7]:

$$t_c = 4.68 L^{0.332} O^{-0.675} s^{-0.281} \tag{5}$$

in which

L = the length of flow in feet
O = the excess rainfall in in./h
s = slope in ft/ft

For L in m and O in cm/h, the constant becomes 3.62. Excess rainfall is defined as that which produces runoff. It may be seen that this is also a function of prior wetting or the wetness of the ground surface at the beginning of the storm. The highest runoff occurs when a storm occurs shortly after a previous storm that has saturated the ground with water. Then a greater portion of the new precipitation occurs as runoff instead of infiltration.

There are numerous methods available for estimating storm runoff. The statistical method implies a Type III Pearsonian family curve. This is valid when large amounts of data are available from previous runoff data. Frequently insufficient data are available and other methods must be used to estimate the storm runoff.

The most common method for measuring runoff is the rational method using the equation:

$$Q = CiA \qquad (6)$$

Q = the rate of runoff at a given location and time

C = the coefficient of runoff, is equal to the ratio of runoff to precipitation for the individual area at a specific time, and usually averages about 0.3. However, if there was a recent antecedent storm that has saturated the ground, the value may approach 0.9, and it may be greater than 1 with melting ice and snow.

i = the average intensity of rainfall for the specified period of time and may be expressed as cm/h (in./h). It may also be calculated by the equation:

$$i = \frac{KF^x}{a + t^n}$$

in which $K = 5-50$, $a = 0-30$, $n = 0.5-1.0$, $x = 0.1-0.5$. All of the values K, a, n, and x vary with geographical location. F is the recurrence interval or frequency in years that a stated intensity is equalled or exceeded. Thus, F equals $(n + 1)/m$ where n is the number of events and m is the order of magnitude of a particular event; t is the duration of rainfall in minutes. The values expressed all present i in units of in./h.

A = the tributary drainage area in hectares or acres or mi^2. For ready conversion factors, there are 640 ac/mi^2. Also, if English units of area in acres are combined with precipitation in in./h, the value of Q becomes ac-in./h and this is approximately equal to 1 ft^3/s (cfs) (actual value, 1.002). If metric units of hectares are used for area, the value Q is given in ha-cm/h, which can be converted since ha-cm = 100 m^3. The intensity of the rainfall can also be determined by a precipitation gage and knowing the duration of precipitation. This will provide the average intensity of rainfall during the time of precipitation.

Numerous empirical relationships based upon the rational method have been proposed as shown in Fair et al. [8] and Gray [9]. As may be expected, the empirical equations apply most validly to the specific drainage basin being studied by the researchers who published their results.

The rational method is useful for mean storm runoff values over the period of time of a storm. For the maximum flow, the time of precipitation used for the equation should be the time of concentration, t_c. To determine runoff at any particular time during and after a storm occurrence, various values of time would have to be chosen and the total flow integrated over a longer period of time.

In order to avoid the multiple calculations in determining storm flow by the rational method, Sherman [10] has devised a system called the Unit Hydrograph. The hydrograph applies only to a specific drainage basin and is derived from data pertaining to runoff over a period of one complete storm event. Based on this storm hydrograph, the anticipated flow from other storms in that same drainage basin can be estimated using four basic assumptions. These are:

1. The effects of all physical effects of a basin are reflected in the shape of the storm runoff hydrograph. That is, all runoff in that specific basin will have the same characteristic shape of curve.
2. For a given duration of rainfall, the duration of runoff is essentially constant and independent of the magnitude of the rainfall. This indicates that the curve is the same length for the same time of precipitation.
3. The ratio of added runoff at any time is proportional to the total volume of storm runoff. This means that an increased magnitude of precipitation over the same period of time will increase proportionately all portions of the hydrograph curve.
4. Precipitation extending over several units of time, with or without interruption, creates a hydrograph that is a composite of all the hydrographs from unit time hydrographs. This indicates that the total runoff is a sum of the individual runoffs based upon the unit hydrograph value.

The unit of time in the unit hydrograph is the duration of rainfall and varies as a function of the size of the basin; for stormwater inlets it may be in the order to 10 min, whereas in large basins it may be in the order of 1 d. Frequently a 6-h duration of precipitation is a convenient value for calculating the unit hydrograph for a river.

In the use of the concept of the unit hydrograph, the first process is to determine stream flow as the result of an individual storm event. When possible, the beginning time for this sequence is long enough after a significant pre-occurring precipitation so that the stream will reach its base flow or relatively constant flow. The onset of precipitation and its total amount are recorded and correlated with the stream flow resulting from this storm. When possible, the flow data are continued until the flow returns to the initial base flow. This is depicted in Fig. 10. After the storm flow is plotted, the base flow is subtracted from the storm flow. Base flow is considered to be the value at the lowest point of the runoff curve immediately prior to the start of the increase in flow caused by the storm. The point at which the flow returns to base flow is determined by N as in Fig. 10. For large drainage basin area N is usually measured in terms of days. The value of N may be determined by one of several means, most of which are empirical. These are shown in Table 3. Using a determined value of N, which is time from the peak of the runoff until it is considered that the stream has returned to base flow, a line can be drawn representing base flow. Individual values of base flow are then subtracted from the storm hydrograph to provide the unit hydrograph as a function of the specific storm as shown in Fig. 11. There are numerous alternatives for determining base flow, but in general the use of a straight line as indicated in Fig. 10 provides results that are just as valid as any others. It may be seen that Fig. 11 represents a unit hydrograph for the duration of the

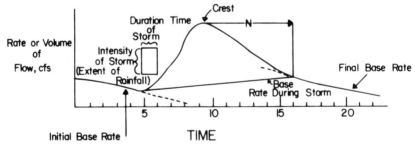

Fig. 10. Stream hydrograph showing a complete storm, hydrograph, and the means for separating base flow.

TABLE 3
Values of N as a Function of Drainage Basin Size

Area, mi²	N	$N = A^{0.2}$
100	2	2.5
500	3	3.5
2000	4	4.5
5000	5	5.5
10,000	6	6.3

storm studied. Very frequently it is desired to convert this to a storm of a uniform duration such as 6 h. Also the figure represents the amount of runoff from the measured amount of precipitation. For ease of calculation, it is frequently desirable to convert this unit hydrograph to a unit hydrograph of given precipitation such as 1 cm or 1 in. This may be calculated by multiplying or dividing the individual values in Fig. 11 by the ratio of 1 cm to the total cm of storm precipitation (1 in. to the total inches of precipitation). Whereas this step is not absolutely necessary, it is convenient in calculating the impact of future storms. For example, using a unit hydrograph for 1 cm of precipitation, if the actual precipitation were 2 cm, each individual value on the unit hydrograph would be multiplied by a factor of 2 to determine the actual storm runoff. If the precipitation continues for longer than the unit of time for which the hydrograph has been drawn, a subsequent curve starting 1 unit of time later can be calculated and a series of curves drawn until the end of the storm. The total runoff then caused by the storm would be the sum of the individual values of each of the unit time hydrographs. It must be kept in mind that, to

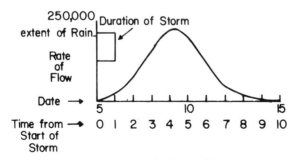

Fig. 11. Unit hydrograph.

determine the total stream flow, the value for the base flow must be added to the total flow in the stream.

The information obtained from this method is useful in predicting storm runoff from any intensity and duration of storm. One precaution, however, must be taken when using the unit hydrograph for determining peak flow. This is that the values obtained in the initial storm hydrograph may represent different conditions of groundwater moisture content at the start of the storm. This directly affects the ratio of the precipitation to the runoff. If the ratio of the total precipitation to the runoff is determined for the initial storm hydrograph, proportionate calculations can be made for any other ratio of precipitation to runoff.

The amount of flow and the rate of flow are important in the design of flood control facilities and for the design of dam spillways. The maximum height of a flood or the adequacy of a spillway design can be determined by a method called flood routing, per Creager et al. [11] and Pickels [12]. Basically this compares the amount of inflow to a reservoir during any short period of time with the rate of outflow through any control systems such as culverts, penstocks, or the spillway. When the inflow exceeds the outflow, the reservoir elevation will rise. As the reservoir elevation rises the rate of discharge through the control systems also increases. Thus, by plotting the incremental increases in water level vs the rate of discharge through the outlet devices, the maximum flood level and the duration and amount of runoff downstream can be calculated. This sequence can be determined utilizing a computer or by a graphical method, as shown here.

The first step in flood routing is to determine the initial reservoir stage at the start of the design flood. As a general rule, if the volume of the flood runoff is high compared to the storage capacity of the reservoir, it is assumed that the reservoir would be full at the beginning of the design flood. However, if the capacity of the reservoir is large compared to the volume of flood runoff, then the initial reservoir elevation would be that expected at the beginning of the design flood or the elevation obtained by a flood having an average frequency of once every 25 yr. The rate and volume of inflow into the system may be determined utilizing the unit hydrograph and projecting to the worst conceivable storm. The worst conceivable storm can be estimated if sufficient long-term data are available for the watershed in question. The other information needed concerns the nature of the discharge system. This may consist of any combination of regulating outlets, pentstocks, consumer demand, and so on. In a small dam, the discharge may be considered through a culvert. To determine the rate of flow through a culvert the equation used is:

$$Q = KA\sqrt{2gh} \tag{7}$$

in which

- Q = the discharge in m³/s (cfs)
- K = the coefficient of discharge
- g = gravity, 9.81 m/s² (32.2 ft/s²)
- h = the effective head on the center of the orifice in m (ft)

K may be calculated as

$$1/(1 + 2\ gL/C_c^2\ R_h)$$

in which

- C_c = the Chezy coefficient
- L = the length of the conduit in m (ft)
- R_h = the hydraulic radius which = A/w_p = the area (A) divided by the wetted perimeter (w_p). Once the pipe is full, the value of R_h = pipe diameter/4.

For discharge over a spillway, a sharpcrested weir with no approach of velocity may be assumed. Here

$$Q = cLH^{3/2} \tag{8}$$

in which

- Q = the rate of discharge in m³/s (cfs)
- c = the weir coefficient, which varies between 2.5 and 3.5
- H = the head above the spillway in m (ft)
- L = the length (width) of the spillway in m (ft)

The next information needed to evaluate the flood routing is the volume of storage in the reservoir at any elevation. This may be determined by measuring the area under water at a series of different elevations. The volume then becomes the sum of the calculations of the mean area times the difference in elevation between any two area measurements. The means of calculation is indicated in Fig. 12. The upper portion of the figure indicates the method for calculating the volume at a given elevation, and this may be applied to the total storage available at any given elevation, as shown in the lower diagram. In this figure, English units of area and volume are indicated because of the simplified relationship between ac-ft and cfs. Obviously, metric units may be used with the appropriate conversion factors. For convenience, the discharge through the control devices may be plotted on the same curve as the storage vs elevation, with the scale of the discharge in cfs double the scale for

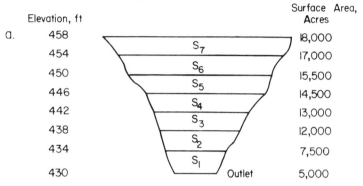
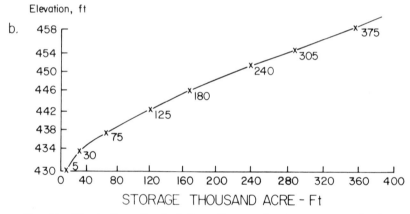

Fig. 12. Illustration of calculation of reservoir storage at any depth.

storage in ac-ft, as shown in Fig. 13. The reason for this correlation of scales is that 1 cfs is approximately equal to 2 ac-ft/d. However, in working with mean rate of discharge it is desirable to work with half of the mean rate of discharge at the beginning of the time period, and half of the mean rate of discharge at the end of the time period. That this is equivalent to the mean rate of discharge is shown by the comparison:

$$\frac{Q_1}{2} + \frac{Q_2}{2} = \frac{Q_1 + Q_2}{2} = Q_M \text{ or } \bar{Q} \qquad (9)$$

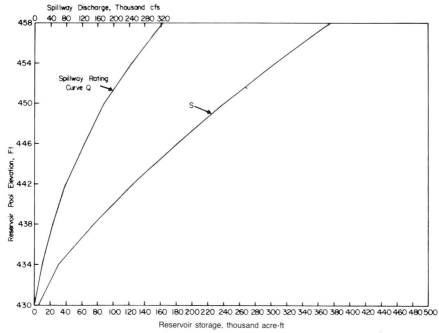

Fig. 13. Storage and discharge rate at any reservoir elevation.

which represents the mean rate of discharge over a given time period. Also, one normally works with only a 12-h period, which is one-half of a day. Thus to determine half of the outflow rate at a particular time with an inflow of 12 h (1/2 d) the calculation would be:

$$\frac{\text{cfs}}{2} \times 2 \, \frac{\text{ac-ft}}{\text{da-cfs}} \times \frac{12 \text{ h}}{24 \text{ h/d}} = \frac{\text{ac-ft}}{2}$$

Thus by setting the scale of discharge (in cfs) in Fig. 12 to ½ that of the ac-ft, we have effectively correlated the ac-ft half day and half of the mean discharge over that time period. For a 6-h interval of inflow, which is used over the peak period, we would have to divide the cfs by another factor of 2.

Next on a curve such as Fig. 13 are plotted the values of the storage plus and minus half of the outflow over a 12-h period and plus and minus half of the storage over a 6-h period. The storage minus half the outflow in a 12-h period is designated as

$$S - \frac{Q_{12}}{2}$$

or $B12$ indicating the beginning curve for a 12-h period and the plot

$$S + \frac{Q_{12}}{2}$$

is indicated by the $E12$ curve or the ending line for a 12-h period. The similar lines for a 6-h period are indicated as the $B6$ and the $E6$ curves. Since the discharge curve in its upper scale is equal to half of the discharge in a 12-h period at any elevation, a compass can be used to draw the $B12$ and $E12$ curves. The value of the spillway discharge Q is, respectively, added to and subtracted from the storage curve in ac-ft. The $B6$ and $E6$ curves then are the midpoints between the respective $B12$ curve and the storage curve and the $E12$ curve and the storage curve. These lines are indicated in Fig. 14.

The next step is to choose the inflow hydrograph and determine the values of inflow during a 12- or 6-h period. An example of computation for this is shown in Table 4. The first column indicates the corresponding lines that have been drawn in Fig. 14. The first few influent calculations are not indicated as lines because the lines would be too short and would not show in the figure. The values for the first five columns are the values to be used to enter the Fig. 14. It may be seen that 12-h intervals may be chosen for the noncritical portion of the curve; however, a 6-h interval should be chosen over the peak of the curve. This time interval may vary for different drainage basin sizes. The values for column 3 are the instantaneous rates of inflow in cfs determined from the storm hydrograph; and the values for column 4 are merely the mean rate over the time period. Column 5 simply converts these values of cfs to ac-ft during the time interval. As may be seen during the 12-h time interval, these are exactly the same values as for cfs, and for the 6-h period are half these values. Then, choosing the predetermined elevation of the reservoir at the start of the flood (for this example, the reservoir was considered to be empty), the amount of runoff from column 5 is added to the appropriate B ($B12$) line in Fig. 14. For example, line AB occurring 48 h after the beginning of the flood represents an inflow of 112,700 ac-ft and it is added to the $B12$ line at an elevation of 433.6 ft, which is the elevation at the end of the previous time period. The length of line AB represents the 112,700 ac-ft. Since this line extends beyond the $E12$ or ending curve for a 12-h period, line BC must be drawn in an upward direction in order to meet the $E12$ curve. This establishes a new reservoir elevation at the end of the time period and a new discharge rate that is equivalent to the Q value from the Q curve at the new elevation. From this point, a 6-h time interval was used and line *DE* therefore is drawn as the volume of flow of 98,600 ac-ft beginning with the *B*6 curve and extending to the right. Again, this line extends beyond the *E*6 curve, and therefore must be extended as line *EF* upward to the *E*6 curve. In all instances, when working with a 12-h influent one must begin and end with a 12-h curve and correspondingly when working with a 6-h influent one must begin and end with a 6-hr curve. This system is continued until the point where a line designating the influent does not extend beyond the ending curve, then a downward line would have to be drawn to the corresponding ending curve, indicating a drop in elevation of the reservoir and a decrease in the discharge rate. It may be seen from the values of Table 4, column 6, that the maximum reservoir elevation reached was 451.8 ft, at which time the spillway discharge equalled the volume of inflow.

The final results of the calculations in Table 4 are shown in Fig. 15, which plots the inflow vs the outflow and also shows the stage or elevation at the corresponding time periods. Note that the units for stage are in feet of elevation, whereas the inflow and outflow units are given in terms of flow. It may be seen that the peak of the stage occurs at the point where the inflow and outflow curves cross. Also the peak flow has been reduced by this reservoir. Increasing the time of concentration and reducing the peak flow both combine to reduce the possible flood damage downstream. The calculation will also indicate the maximum stage, and if this exceeds a predetermined elevation it will enable the design engineer to adjust the size of the spillway discharge to decrease the maximum stage. Conversely, the maximum stage can determine the limit of flooding and provide information on restrictions on building in the flood plain.

It should be noted that this type of calculation lends itself to a computer solution, since the rate of outflow at any elevation can be entered and the rate of inflow can be applied. The storage capacity would also have to be entered into the program. Once the storage vs elevation and the rate of discharge are determined for any particular system, the impact of any potential storm can be readily determined.

In some instances, not the maximum flood, but the maximum dependable supply is desired. This is essential where a dependable supply is required for any purpose such as drinking water supply, navigation, or today as an increasing source of hydropower. For this purpose a dam may frequently be constructed to create a reservoir for storage of water for periods when the flow is less than normal. The parameters in question include the stream flow, the demand flow, and the storage that can be provided. At any point in time, the demand flow cannot exceed stream flow

TABLE 4
Sample Flood-Routing Computations

Line on Fig. 14	Time from beginning of flood, h (1)	Length of interval t, h (2)	Instantaneous rate of inflow into reservoir, I, cfs (3)	Mean rate of inflow into reservoir during interval, $(I_1 + I_2)/2$, cfs (4)	Volume of inflow into reservoir during interval, $(4) \times 2 \times t/24$, ac-ft (5)	Reservoir elevation at end of interval, ft (6)	Spillway discharge rate at end of interval, cfs (7)
	0		2,000			430.0	0
	12	12	3,800	2,900	2,900	430.3	500
	24	12	10,400	7,100	7,100	431.0	2,000
	36	12	52,400	31,400	31,400	433.6	12,000

AB	48	12	173,100	112,700	112,700	440.5	64,000
DE	54	6	221,400	197,200	98,600	444.6	106,000
GI	60	6	245,800	233,600	116,800	448.2	147,000
	66	6	237,500	241,600	120,800	450.6	178,000
	72	6	211,800	224,600	112,300	451.8	193,000
	78	6	180,000	195,800	97,900	451.8	193,000
	84	6	148,600	164,200	82,100	451.1	184,000
	96	12	92,800	120,700	120,700	448.1	146,000
	108	12	54,100	73,400	73,400	444.5	105,000
	120	12	29,700	41,900	41,900	441.0	69,000
	132	12	16,500	23,100	23,100	438.2	44,000
	144	12	8,800	12,600	12,600	436.1	28,000
	156	12	4,000	6,400	6,400	434.5	17,000
	168	12	2,500	3,250	3,250	433.4	11,000
	180	12	2,000	2,250	2,250	432.6	8,000
	192	12	2,000	2,000	2,000	432.0	5,000

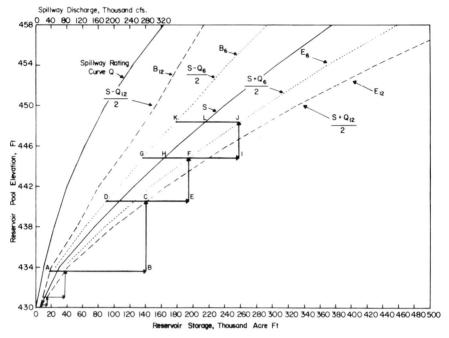

Fig. 14. Worksheet (partial) for determining reservoir elevation and discharge with a given inflow over a specified time period.

plus the storage capacity. Storage capacity must be provided prior to the demand's exceeding the stream flow. The limit of demand approaches the mean stream flow over an extended period of time.

The relationship between runoff, demand, and storage may be shown in a mass diagram using the Rippl method [13]. In this system, a long-term plot of runoff vs time is first made. It is desirable to keep the time integral for runoff to a maximum of a month, because a severe drought occurring during 11 months of a year could be missed if all of the annual precipitation occurred during 1 month and annual plots were made. From the plot of the long-term data, the period of the most severe drought can be extracted and plotted on an expanded scale. A typical mass diagram of flow vs time is shown in Fig. 16. Using this figure, several factors can be determined. If a desired draft is known, the rate of this draft can be plotted on the same scale as the mass diagram. In order to determine the practicality of this rate, a line parallel to the draft is drawn on the mass diagram tangent to the point at which the maximum drought begins. This line is shown as line D on Fig. 16. If this line intersects the

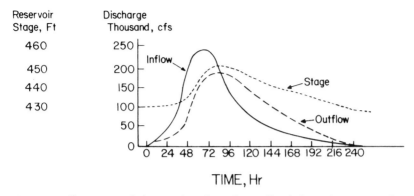

Fig. 15. Summary of the results of routing a flood through a reservoir.

mass diagram of the stream flow later in time, then this demand can be satisfied providing that a reservoir is constructed that represents the maximum difference between the stream flow and the desired draft. This is represented as the line S_D in Fig. 16. In order to determine the maximum draft that can be exerted on this stream during this drought, a line tangent to the start of the drought and tangent at the end of the drought could be drawn. This is designated as line M on Fig. 16, representing the maximum draft when the assumption is made that there are no physical limits on the size of the dam or the reservoir. In this case, the maximum storage to be provided is the maximum difference between this draft and the runoff and is depicted as S_M. Any demand that exceeds this rate could not be applied because it would be impossible to build a storage reservoir large enough to provide this water. Obviously, the maximum draft may also be limited by the maximum size of the dam and/or the storage reservoir. If this is the controlling factor, then the potential available storage should be drawn on the figure and a line tangent at the start of the drought and touching the top of the maximum storage capacity should be drawn. Thus, this simple diagram can be used to determine the maximum allowable draft and the maximum size of the reservoir needed to provide this draft. One other useful bit of information can be derived from this figure. If a line is drawn parallel to the chosen draft, but tangent at the lowest point of the drought and extended backward in time until it again meets the runoff curve, this will indicate the time at which the reservoir would have had to have been built in order for it to have filled prior to the onset of the design drought. Thus it may be seen that this figure represents a simplified balance between stream runoff as it varies, the amount of withdrawal from the stream, and the storage that must be provided prior to the

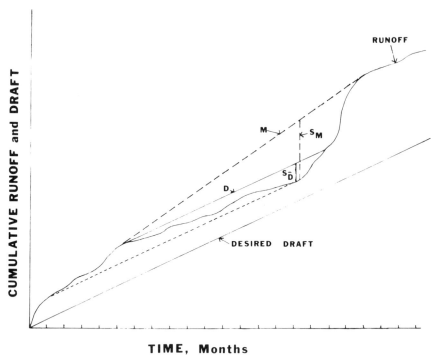

Fig. 16. Ripple Method for determining reservoir storage.

withdrawal exceeding the stream flow. It must be pointed out that the demand curve does not have to be a constant value; however, at no time can the total demand exceed the sum of the inflow and the storage available.

IX. GROUNDWATER

Groundwater represents a significant source of water supply. In the order of 30 billion gal/d (110 million m^3/d) of ground water is used in the United States. It is the prime source of water for drinking in rural areas. Many medium size cities and numerous industries use groundwater as their source of supply for drinking, processing, and cooling. One principle advantage of groundwater is its relatively cool and constant temperature. Furthermore, the soil acts as a filter, removing particulates including bacteria. Thus, under most circumstances groundwater is not subject to contamination as is surface water. There are a few potential sources of contamination such as landfill and chemical dump sites. These are discussed in more detail in a later section.

Groundwater results from the infiltration of precipitation into the soil. The water will proceed downward vertically until it reaches the first confining layer. Above this layer is a saturated layer of soil containing the groundwater. The upper edge of this saturated area is called the groundwater table. This will vary with the amount of precipitation and the time since the last precipitation. There is flow in this saturated area in a downhill gradient. At some point the water in the saturated zone may flow under a confining zone. This now produces a confined aquifer. An aquifer is defined as a geologic formation that contains water and transmits it from one point to another in quantities sufficient to permit economic development. This usually precludes a sand or gravel soil. An aquiclude is a formation that contains water, but cannot transmit it rapidly enough to furnish a significant supply. This is usually represented by clay, which holds much moisture but does not transmit it. An aquifuge is a formation such as rock that neither holds nor transmits water. There may be many intermittent levels of aquifers, aquicludes, and aquifuges through the soil. Water may occur at any depth and there are few locations where there is no water whatsoever beneath the soil surface. However, in general, as the depth increases, the total dissolved salts increase, with saline water being found at great depths. Thus, the useable quality of the water decreases with depth.

The volume of water in a saturated formation is a function of the pore space of the soil. The porosity, p, is equal to the ratio of the pore void or interstitial space to the total volume of the rock or soil as:

$$p = \frac{\text{void volume}}{\text{total volume}} \qquad (11)$$

concurrently, the voids ratio, e, is the ratio of the pore volume to the solid volume and equals:

$$e = \frac{p}{1-p} \qquad (12)$$

The safe yield of an aquifer is the rate at which water can be withdrawn from it without depleting the supply to such an extent that further withdrawal at this rate is no longer economically feasible. The safe yield thus is a function of the velocity of flow through the aquifer and the source of recharge of the aquifer. The velocity of flow through an aquifer is a function of the permeability of the soil. According to Darcy's Law $V = ks$, in which s is the slope of the hydraulic grade line in vertical units divided by horizontal units (m/m, or ft/ft), and k is a constant as a function of the type of soil. Some relationships between soils and various parameters are shown in Fig. 17 and Table 5. The quantity of flow through the soil, Q, is a function of $Q = Ps$, in which Q is the flow per unit area and P is the permeability. Standard English units for permeability have been gal/ft²-d at a slope of 1 ft/ft at 60°F; the equivalent metric units would be m³/m²-d at a slope of 1/1 at 15°C. To handle large areas, a standard field coefficient of permeability is sometimes used in which the slope is extended to 1 ft/mi. In metric units this translates to 1 m/km. It must be noted that the permeability P is through a unit area of cross-section. To determine the coefficient of transmissibility, $T = Pt_a$, in which t_a is the thickness of the aquifer and T then gives units of volume/day/unit of area of the aquifer for its entire depth. This is useful information in terms of determining the amount of water available in the soil.

It is useful to know the direction of flow of groundwater and the slope. This can be determined by the use of three or more observation

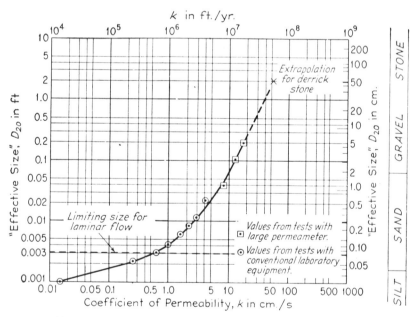

Fig. 17. Effect of transition from laminar to turbulent flow on permeability coefficient.

TABLE 5
Characteristics of Soils

Type	Size (in.)	Size (mm)	Porosity, %	Permeability, gal/ft²-d	Velocity
Quartzite, granite			1	0.1	
Limestone and shale			5	1	
Sandstone			15	700	
Gravel	> 0.08	1.5–8	25	100,000	1000 mi/yr
Very coarse sand	0.04–0.08	0.8–3	30	65,000	600 mi/yr
Coarse sand	0.02–0.04	0.5–2	35	8500	80 mi/yr
Medium sand	0.01–0.02	0.25–0.5	40	2000	20 mi/yr
Fine sand	0.005–0.010	0.05–0.25	45	80	1 mi/yr
Very fine sand	0.003–0.005	0.005–0.05	50	2	100 ft/yr
Silt	< 0.003	0.005–0.05	58	1	50 ft/yr
Clay		0.005–0.05	70	0.002	1 ft/10 yr
Colloidal clay		10Å–0.01	90	2×10^{-5}	1 ft/1000 yr

wells. The water level of each well is measured relative to some reference elevation. Frequently, mean sea level is used, but this is not important. The three (or more) observation wells are plotted in plan to scale (see Fig. 18). Choosing three adjacent wells, a line is drawn between the well with the highest water level elevation and the well with the lowest water level elevation. This line is then subdivided into equal increments assuming that for the relatively short distance between these two points the slope is constant. This line will then represent equal increments of slope such as meters or feet. A line is then drawn connecting the intermediate level well to the equivalent elevation on the line drawn between the highest and the lowest water levels. This is then considered a line of equal elevation. Any line perpendicular to this line will indicate the direction of flow from the highest to lowest elevation. Further if this perpendicular line is extended on the diagram to the location of the well with the highest water level, the distance from the well to this line can be determined. Knowing the change in elevation and the distance, the slope can be determined as the

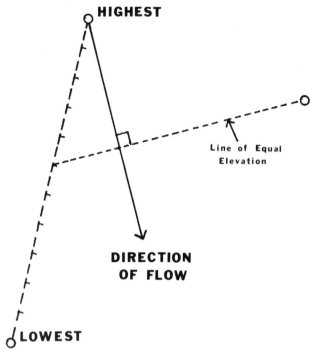

Fig. 18. Method for determining direction of flow and slope of ground water.

change in elevation with distance. In large areas where more than three wells are located, similar sets of wells are studied and the direction and slope may be determined for each area. In some instances both the slope and direction of the flow may change in relatively small distance.

It is also useful to know the time of flow so that the velocity between any two points may be established. This can be most readily accomplished by means of various tracers, but the dye Rhodamine WT has been found to be very useful for monitoring groundwater flow, Aulenbach and Clesceri [14]. The dye or other tracer is placed in the well with the highest water level elevation. Samples are then obtained continuously or periodically (usually daily) until the dye appears in the next downstream well. Depending upon the means and location of the injection of the dye, either the first appearance or the peak concentration of the dye in the observation well is used. Knowing the time until the tracer is observed in the observation well and the acutal distance within the soil, the velocity may be determined as the distance divided by time. It may be seen that the slope of the groundwater table is especially essential in this case since the flow in a nonhorizontal direction will increase the distance from one point to another.

Since wells are utilized to obtain water from the ground, it is necessary to determine how much water can be obtained from the soil without exceeding the safe yield. Consideration must be made for wells in an unconfined aquifer and those in a confined aquifer. Those in a nonconfined aquifer are frequently referred to as gravity or nonpressure wells, whereas those in a confined aquifer are referred to as pressure wells. If water is removed from a well at a constant rate until the conditions within the soil reach equilibrium, measurements can be made to determine the permeability of soil under a set of operating conditions. Knowing the permeability, the maximum safe yield can be determined. For both nonpressure and pressure wells, at least one observation well is needed within the zone of influence of the pumped well. If measurements of water level can be made in the pumped well, this can serve as the second observation well. If it is not possible to determine the water level in the pumped well, then a second observation well is needed for the determination.

For the nonpressure well in the surface aquifer, the relationship between the flow and the permeability may be determined by the equation:

$$Q = \frac{\pi K_p (h_1^2 - h_w^2)}{\ln r_1/r_w} \tag{13}$$

The terms of h_1, h_w, r_1, r_w for the equation are shown in Fig. 19, in which one observation well and the pumped well are used. K_p is a coefficient. If two observation wells are used (see Fig. 20), then h_1 and r_1 become the observation well farther from the pumped well and the h_2 and r_2 for the second observation well closer to the pumped well may be substituted for h_w and r_w, respectively.

For pressure wells in a confined aquifer, the equivalent equation is:

$$Q = \frac{2\pi K_p t (h_1 - h_w)}{\ln r_1/r_w} \tag{14}$$

The representative parameters for this equation are shown in Fig. 20. It may be seen that, in working with a pressure (artesian) well, the water level in the observation well will rise above the confining layer. This water level represents the piezometric surface or the elevation or pressure head to which the water will rise with no confinement. In this case, how-

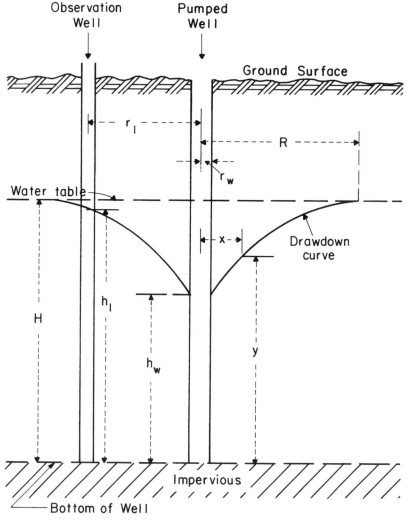

Fig. 19. Diagram illustrating the hydraulics of a nonpressure well.

ever, there is no actual cone of depression surrounding the pumped well in the confined aquifer. There is only a reduction in the piezometric head in the neighborhood of the pumped well.

The previous two equations assume that the zone or cone of influence is circular. Under actual conditions this may not be true, since there is some horizontal flow in the aquifer and, therefore, the cone of

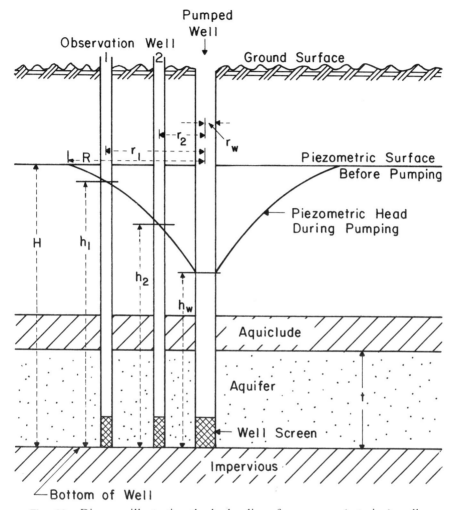

Fig. 20. Diagram illustrating the hydraulics of a pressure (artesian) well.

depression will be lower on the downstream side of the pumped well. Also, under certain conditions, it may take a long time (up to 10 yr) to establish equilibrium. Thus, other methods for determining the yield of an aquifer are necessary. One common method used is the Theis method. This considers the fact that an aquifer is somewhat elastic, so that removing water from it causes a low pressure in the area of the pumped well that causes more water to enter into the area. The Theis method is based upon four assumptions: (1) the soil of the aquifer is homogeneous, (2) the coefficient of transmissibility, T, is constant, (3) the coefficient of storage, S_c, is constant; the storage represents the yield of water per unit area of aquifer per unit drop in hydraulic gradient, and (4) the water is released from the aquifer instantaneously. Actually under normal conditions there is a lag in the release of the water from the aquifer. This Theis method makes use of the well function $W(u)$.

$$W(u) = \int_u^\infty \frac{e^{-u}}{u} du \qquad (15)$$

for metric units

$$d = \frac{QW(u)}{4\pi T} \qquad (15a)$$

$$u = \frac{r^2 S_c}{4Tt} \qquad (15b)$$

in which Q is then in m³/d. For English units

$$d = \frac{114.6 QW(u)}{T} \qquad (15c)$$

$$u = \frac{1.87\ r^2 S_c}{Tt} \qquad (15d)$$

In these equations

$W(u)$ = Theis' well function value
d = drawdown in m (ft) in the vicinity of the pumping well in which the observation well is r m (ft) from the pumped well
Q = the discharge in m³/d (gal/min)
T = the coefficient of transmissibility in m³/d unit gradient [gal/day (ft/ft)]
S_c = the coefficient of storage as a decimal
t = the time of pumping in days
u = a tabulated value in Theis method

In the English units of measurement the constants convert directly from time of pumping in days to a discharge in units of minutes. The Theis has prepared a table of values for u and $W(u)$; however, it is much easier to plot these values on a log–log scale, as shown in Fig. 21. This diagram is for English units of measurement and the time t must be in terms of days. To solve this system, a well is pumped at a constant rate and the level of water in an observation well is determined on a daily basis. On log–log

CONTROL OF RESERVOIRS AND LAKES 191

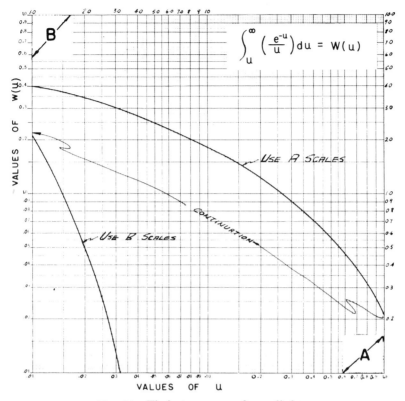

Fig. 21. Theis type curve for well data.

paper with the same scale as the type curve (Fig. 21) values of d in ft and r^2/t in ft^2/d are plotted. Then, overlaying this plot on the type curve and keeping the ordinates parallel, the one curve is moved over the other until they coincide or produce a curve of best match. Note that the two lines in the type curve (Fig. 21) are really one continuous line merely divided into two different ranges. With the two curves coinciding, any point on the two curves may be chosen and the values for u, $W(u)$, d, and r^2/t can be determined. Using these values the transmissibility T can be determined as:

$$T = \frac{QW(u)}{4d} \qquad (16)$$

and

$$S_c = \frac{4T + u}{r^2} \qquad (17)$$

For English units the value of T becomes

$$114.6 \frac{QW(u)}{d} \tag{16a}$$

and

$$S_c = \frac{uT}{1.87(r^2/t)} \tag{17a}$$

Jacob modified the Theis method to simplify it. His calculations for T are

$$T = \frac{264Q \log(t_2/t_1)}{d_2 - d_1} \tag{18}$$

and

$$S_c = \frac{0.3Tt_o}{r^2} \tag{19}$$

In Jacob's equations (Fig. 22),

d = the drawdown, ft, in the observtion well at time t, in days, from the start of pumping
Q = in gal/min

Values of the log t are plotted vs the drawdown, d. This may be done by plotting t on the log scale of semilog paper. To simplify the calculation, corresponding values of d over one log cycle of t may be taken, as shown in Fig. 22. This simplifies the calculations because t_2/t_1 will always be 10 and the log of 10 = 1. To determine the transmissibility, T, the values of

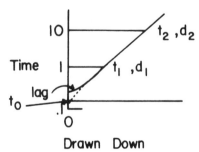

Fig. 22. Jacob's modification for determining storage and transmissibility of ground water.

the drawdown at the two times then may be substituted into the equation and t determined. For the calculation of the storage, t_o must be determined. This is determined by extending a curve of best fit of a straight line intersecting the origin of the drawdown line. This gives a projected value of t at $d = 0$. The equation then may be used for determining the storage of water in the aquifer.

The International Ground Water Modelling Center at Holcomb Research Institute, Butler University, Indianapolis, IN 46208 has an annotated database of approximately 400 groundwater models from all over the world.

X. IMPACT OF POLLUTION ON LAKES

It has frequently been stated that all lakes are born to die. This implies that over an extended period of time lakes will slowly fill in by erosion and increase in nutrient content to support ever greater biological systems, which will also die and settle to the bottom increasing the rate of sedimentation and filling in of the bottom. As the lake becomes shallower, rooted aquatic plants can grow in the nutrient-rich sediment, thereby increasing the production of organic matter and increasing the rate of sedimentation both by dying of plant life and increasing erosion sedimentation by slowing the velocity of the flow of water in the area where the rooted aquatic plants occur. Ultimately the lake will fill in and may become a bog in which very little water is found, but high productivity occurs. Ultimately the large rooted terrestrial plants take over in the moist soil and a high-moor or climax bog is produced as the climax of the lake. Whereas all lakes may eventually reach this termination, the rate at which this process proceeds is often influenced by humans, whose activities frequently result in increasing the erosion process that speeds the filling of the lake with sediment. Also, the discharge of sewage effluent and runoff from fertilized land may add to the nutrients in the lake, thereby increasing the rate of biological growth. This also adds to the increasing rate of sedimentation. Whereas little can be done to control the natural cycle of a lake, much can be done to prevent the increased rate of eutrophication that occurs from these human activities.

In terms of the natural rate of eutrophication in lakes, the morphology of the lake frequently has the greatest controlling influence. The morphology may even limit or restrict the impact of human activities until the morphology has been changed. As a general rule, a deep lake with steep sloping sides will have a much slower rate of eutrophication from all causes than a lake of similar volume that is shallow and has large shallow shoreline areas. Also the larger the total volume of the lake, the slower will be the rate of eutrophication.

Lakes in temperate climate normally go through a stratification pattern that results in dimictic conditions (equal mixing throughout the lake twice a year). This is primarily caused by the anomalous temperature–density relation of water. It may be recalled that the most dense phase of water is 4°C; thus, water at this temperature will tend to settle on the bottom of the lake, assuming that the external air temperature reaches 4°C or lower sometime during the year. During the winter, ice will form on the surface of the lake, with the temperature of the water immediately below the ice at or near 0°C. However, with a deep lake the temperature at the bottom approaches 4°C. As spring comes and the sun warms the water, the ice melts and then begins to warm the surface of the lake. When all of the lake is warmed to 4°C even a gentle wind will circulate the water from top to bottom of the lake creating the spring overturn. As spring continues and the sun warms the surface of the lake, the warmer surface waters will tend to float on the colder lower temperature water. If this heating occurs during a period of strong wind there may still be complete mixing of the lake and the entire lake is warmed to the temperature of the surface. However, if warming occurs on a calm day, the surface of the lake will become significantly warmer than the lower level. With several days of warming and little wind, a point is reached at which the wind does not have sufficient energy to mix the upper warmer layers of lower density and lower viscosity water with the colder lower levels of higher density and greater viscosity water. Thus a period of stratification begins. Frequently, in large temperate lakes the level of stratification is established near 10 m depth. In addition to the combination of warming and wind during the establishment of stratification, the shape and orientation of the lake with the wind have a great influence on the depth of the upper mixed zone. During the summer, there is an upper layer that is equally mixed by the wind, then a zone of rapidly decreasing temperature with depth, and then a third layer at the bottom with relatively constant and cold temperatures. The upper layer is called the epilimnion, the zone of great temperature vs depth change is called the thermocline or metalimnion, and the lower layer is called the hypolimnion.

Once this summer stratification is established, there is little to no mixing in the hypolimnion. Thus if large amounts of organic material reach the hypolimnion, the oxygen available will be utilized and the area will become anaerobic. If on the other hand there is little productivity in the lake, or if the lake is so clear that photosynthesis can occur in the hypolimnion, then there will not be oxygen depletion in the hypolimnion and this area will support a fine crop of cold water fish. Also this system helps to stabilize the nutrients that settle to the bottom and are normally precipitated under aerobic conditions. It can be seen that this is a critical condition, for once the hypolimnion becomes anaerobic the nutrients that are soluble under anaerobic conditions are released to the water column, allowing for the growth of more biological materials that will in turn die, settle to the bottom, and utilize more dissolved oxygen. Thus, in order to maintain the lake in an oligotrophic condition, management efforts should be made to maintain the hypolimnion in an aerobic state.

As fall approaches, the surface of the water is cooled and the cooler water circulates to a depth of equal temperature and/or density. This tends to lower the thermocline until the lake again becomes uniform in temperature and even a gentle wind will mix the lake from top to bottom. This is called the period of fall turnover. As the air temperature reaches $4°C$ and becomes colder, the surface of the lake will approach $0°C$, but the warmer $4°C$ water will remain on the bottom. Particularly after ice is formed, there is no wind effect on the lake to cause mixing of the water. Thus the bottom of a deep lake never freezes. However, again during ice cover there is no opportunity for reaeration of the lake and thus the biological activity in the water must be such that it does not utilize the available oxygen in the lake. Because of the cold temperature, this is usually fairly easy to maintain; however, in certain eutrophic lakes, oxygen depletion may occur under the ice. This period of stratification under the ice is called the winter stratification period. In evaluating a lake it may be seen that it is essential to evaluate it under all conditions of stratification and nonstratification.

A lake contains many biological communities. Within the water column are numerous organisms of microscopic size. These are generally termed plankton, which are microscopic floating organisms. The plankton may be subdivided into two general groups: the phytoplankton are the plant life and include the algae, the fungi, and the pollen that fall into a lake, and the zooplankton, which represent the animal forms. In another category, the plankton may be broken down into the nekton or free-swimming organisms, and the benthon, which are the microscopic organisms that exist on the bottom.

A prime concern are the algae, the microscopic green plants floating in the water column. These organisms represent the base of the food chain in that they can convert simple inorganic materials into organic materials with the aid of sunlight. In addition, during the daytime or sunlight hours these plants undergo photosynthesis, which is a process in which oxygen is liberated in the water. It has been estimated that ¾ of the world's supply of oxygen is generated by algae in the ocean. In terms of the food chain, the algae are normally consumed by the zooplankton, which are in turn consumed by larger animal forms, which may be consumed by small fish, which may be consumed by large fish, which may be consumed by larger vertebrates, including humans. But it is the small algae that are the start of this food chain.

All biological systems require the presence of the proper nutrients in order to grow and reproduce. For the larger organisms, the smaller organisms provide these nutrients. However, the algae, as the base of the food chain, must gain their nutrients in the inorganic form from the surrounding water. Organisms that rely on inorganic matter for growth are termed autotrophic, whereas those that require organic matter as a source of food are called heterotrophic. Growth is a function of the nutrients available (plus other limiting factors, such as temperature, light, and so on), but these nutrients may vary both with location and time within a lake. Specific organisms may vary in their specific nutrient requirements; however, in general these organisms have a certain demand for the essential elements. Carbon may be obtained from the solution of carbon dioxide. Hydrogen may be formed by hydrolysis of water, or from bicarbonates dissolved in the water. Oxygen may be produced in photosynthesis or may be secured from dissolved atmospheric oxygen. Nitrogen is secured from dissolved nitrogenous materials including both ammonia and nitrates. Phosphorus is usually derived from soil and animal excreta. Sulfur is usually derived from the soil, but is also present in animal excreta. The needed ratio of each individual material varies, but there is a general demand for approximately 60 parts of carbon to 15 parts of nitrogen to 1 part of phosphorus in most cellular material. In addition to these stated growth factors, there may be many other trace elements that are required in small concentrations in order to support adequate growth. Most frequently, the requirements for these elements are so low that there is an adequate amount available. Certain specific organisms may have special requirements. A typical example of this is the diatoms, which require the presence of silicon in order to manufacture their cell case, called a frustule.

Normally organisms such as algae will growth until one of the nutrients becomes limiting. Then growth may be retarded or eliminated entirely. The limits are in the ratio of the requirements; therefore, the limit is different for each element. This is referred to as Liebig's Law of the Minimum, which states that biological systems will grow until they are limited by the nutrient that is present in the limiting concentration. In lakes the most common limiting nutrients are nitrogen and phosphorus; however, in a few instances carbon has also been shown to be limiting. Most frequently, phosphorus is the limiting constituent; however, this is not exclusively so and there are many lakes in which nitrogen is the limiting factor.

Productivity is a measure of the utilization of inorganic material to produce plant growth. Very frequently, productivity in lakes is a measure of the amount of fish available in the lake. Since the number of fish is indirectly related to the growth of the other organisms lower in the food chain and, ultimately to the amount of algal growth, it may be seen that the productivity of fish is merely an indirect measure of the amount of algae growth in the first place. Whether or not productivity is desirable, is a function of individual taste. A lake that is low in productivity will be clear and conversely will have a small fish population. On the other hand, a productive lake will be turbid because of the high concentration of plankton, but this will support a large fish population. However, this fish population usually represents different species. A lake low in productivity tends toward the more preferred game fish such as trout and salmon, whereas a lake that is high in productivity is more commonly inhabited by bass, pickerel, and catfish. The term oligotrophic has been used to describe lakes of low productivity and eutrophic for lakes of high productivity. In between is an area called mesotrophic, indicating that there is a gradual transition between oligotrophic and eutrophic. However, these terms are not intended to imply that all eutrophic lakes are undesirable or that all oligotrophic lakes are desirable. The desirability of a specific level of productivity is a function of the specific use of the lake and what is needed is a balance of various types. The long-range problem is that as lakes age the nutrients accumulate within the lake. New nutrients are brought into the lake by allocthonous inputs, which are substances entering the lake from outside of the water body. This includes siltation, organic matter from decaying leaves and, of course, anthropogenic sources such as sewage or treated sewage effluent. Autochtonous inputs are those that are generated within the lake. This mostly represents a recycling of the nutrients within the lake since these are relatively constant. Thus, as there is an increase in allochtonous sources, more nutrients are made available to the organisms within the lake, and the productivity increases.

Based on Liebig's Law of the Minimum, it is possible to control excess productivity by eliminating one of the essential nutrients. It is not necessary to eliminate all of them. Since phosphorus is most commonly the limiting nutrient, most efforts have been expended towards controlling phosphorus inputs to lakes in order to control excess productivity. Whereas little can be done to control the natural process of eutrophication, there can be control of the anthropogenic sources of the nutrients that will encourage the unwanted higher productivity. In some instances, reduction of the anthropogenic sources of nutrients has been shown to reverse the eutrophication trend. In other instances the lowering of anthropogenic inputs merely slows down or delays the rate of eutrophication. What works in one lake may not necessarily work in another seemingly similar lake.

Since phosphorus is most frequently the limiting nutrient in a lake, more efforts have been directed toward finding means of reducing phosphorus inputs to a lake. This may include diversion of all storm runoff from gaining direct access into the lake, treatment of wastewaters for phosphorus removal prior to discharge, and use of land application wastewater techniques that have been shown to reduce phosphorus content significantly. Another reason phosphorus has been chosen as the nutrient to remove is that it may be fairly readily removed by chemical precipitation with iron, aluminum, or calcium. On the other hand the next most important nutrient for removal would be nitrogen. Removal of nitrogen is somewhat more difficult; however, it can be achieved by producing a nitrified treatment plant effluent and then creating denitrification (anaerobic) conditions. This will convert the nitrate nitrogen to atmospheric nitrogen, which will then escape to the atmosphere. However, there are certain organisms, particularly blue-green algae, that are able to utilize or "fix" nitrogen directly from the atmosphere into plant material. Since elementary nitrogen makes up approximately 80% of the air, it can be expected that there will be sufficient nitrogen available in the water body to provide for all of the nitrogen fixation that may be needed in most lakes. Thus phosphorus has received the greatest attention in terms of a method for controlling excessive productivity. It has been shown that controlling phosphorus to concentrations of less thatn 10 μg/L at the time of spring overturn will result in the control of excess productivity in most lakes.

There has been a direct correlation between the phosphorus concentration in a lake at the time of spring turnover and the amount of productivity. This has been shown by Sakamoto [15] (Fig.23). In this case, productivity is measured as the amount of chlorophyll-a present in the summer. It may be seen that there is a good correlation between the total phosphorus and the chlorophyll-a productivity. Correspondingly, the greater the chlorophyll-a content, which indicates the presence of algae, the greater the turbidity of the water, and therefore the lower the clarity of

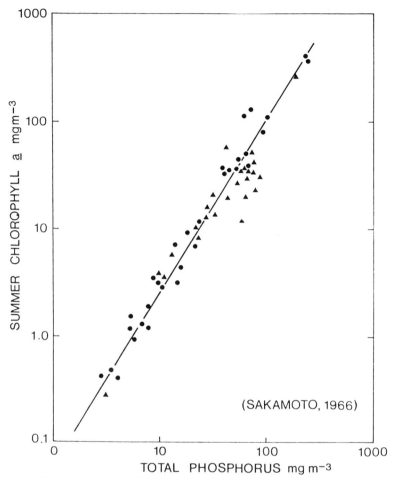

Fig. 23. Plot of total phosphorus concentration at spring overturn vs average chlorophyll α concentration in summer for a number of lakes. Circles represent data from Sakamoto (1966); triangles are for other lakes in the literature. Line is regression line for Sakamoto's points. Corrrelation coefficient (r) is 0.97.

the water as measured by the Secchi disc depth. Whereas there is good correlation between phosphorus content and chlorphyll-a, there is poor correlation between chlorophyll-a content and the clarity of the water. This is so because other substances in the water can reduce its clarity. This includes the zooplankton that feed on the algae, and can also include particulate matter, such as fine clay or silt, that is carried into the body of water from allocthonous sources. This is especially true at the mouth of a stream after a heavy rain that produces significant runoff. It may be seen that there is a poor correlation between the clarity of the water and the phosphorus or even the algae content, as measured by chlorophyll-a levels.

There have been many models derived to correlate certain specific parameters with the trophic state of a lake. However, two stand out as being quite reliable and simple. These both relate total phosphorus loading to the trophic state of the lake as a function of the body of water. In the original work of Vollenweider [16], there was a correlation between the total phosphorus loading and the mean depth of the lake. Vollenweider observed many lakes and compared their trophic state with the correlation between the phosphorus and the depth. In most cases there was a good agreement with the general observation of the trophic state of the lake and these parameters. Vollenweider worked with Dillon [17] to improve this model by comparing phosphorus loadings with the mean depth and the retention time of the lake. Using this comparison with numerous lakes around the world, an even better correlation was made with the actual trophic state of the lake and location where it fell on this graph (Fig. 24). It is conceded that in lakes where phosphorus is not the limiting nutrient this correlation will be poor. However, in lakes in which phosphorus is the limiting nutrient, which is the most common occurrence, this correlation has been shown to be extremely satisfactory. Information such as this is being used to indicate means of controlling the productivity in lakes. One of the largest efforts in this direction is the study of controlling nutrients in the Great Lakes of the United States. Knowing the phosphorus inputs to a lake and the morphology of the lake, it can be determined what amount of phosphorus control would be required to limit the productivity to a certain desired level. This is presently being done by varied means, including the removal of phosphorus from all wastewaters being discharged into the Great Lakes Basin, by controlling urban runoff through the use of holding tanks and subsequent treatment, and by placing buffer zones between farmland and streams into which the fertilizer would otherwise be leached. Based on these calculations, the most efficient and/or economical system for control of phosphorus discharges into lakes can be established.

XI. THERMAL IMPACTS ON THE AQUATIC ENVIRONMENT

With increasing human activities, and the accompanying increasing demand for energy, more and more waste heat is being produced. Where possible, means are being considered for the utilization of this waste heat.

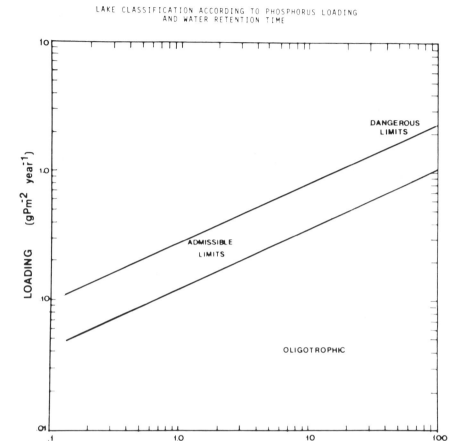

Fig. 24. Trophic state of a lake based upon its annual phosphorus loading vs its mean depth and hydraulic detention time.

In other instances, the heat is merely wasted to the environment. The two main categories of the environment receiving this waste heat are the atmosphere and the water. Discharge of heat into the atmosphere may even impact lakes or streams by changing precipitation patterns. If dry cooling towers are used, the atmosphere is heated and the relative humidity thereby decreases. On the other hand, if wet cooling towers are used, significant amounts of water vapor are passed into the atmosphere, thereby increasing the relative humidity and increasing the possibility of precipitation downwind. In terms of discharge into bodies of water, this can significantly increase the temperature of the body of water, resulting in increased evaporation and loss of water in the surface lakes and streams.

Excess heat in the aquatic environment may be considered to be no different from other forms of pollution: a resource out of place. Every human action produces heat in some form or other. It is the main source of energy and energy itself is usually expended in the form of heat: friction produces heat. The problem is compounded by the fact that industries are expanding and concentrating. When there were many small mills spread aorund the country, there was no serious problem. However, because of the economy of scale, there are now fewer but much larger mills located at a few choice locations. Thus, the same amount of heat may be produced, but it is discharged into a much more limited local area where it has a greater impact. One of the greatest sources of heat is from electrical power generation. Under price controls for oil, the amount of electric power generation doubled approximately every 10 yr. Also, it was not economical to utilize hydropower for generating electricity. Now with the increased price of oil, the expansion of electrical power has been reduced to approximately 4–5%/yr. Furthermore, it is becoming more economical to utilize hydropower even from small plants located on small streams where sufficient water fall is available. One of the major problems in a thermal power generation system is the efficiency of utilizing the heat. For fossil fuels, the efficiency is only about 40%, whereas for a nuclear fuel the efficiency is only 33%. Thus, over half of the heat that is generated to produce the electricity must be disposed of in some satisfactory manner. Some of this heat is dissipated into the atmosphere, whereas other portions are dissipated into the local aquatic environment. Even that which is dissipated into the atmopshere may have some influence on the hydrologic cycle since an increase in temperature can increase the evaporation in a local area. Much effort has been expended in finding uses for this waste heat. However, particularly with nuclear power reactors, the heat is produced in a remote location not close to a place where it can be readily used. The losses in transmitting such heat long distances preclude its use at distant locations. Thus large amounts of waste heat frequently reach the aquatic environment.

There are many problems that occur as the temperature of a lake or stream is increased. Some of these are obvious, but some represent problems that are difficult to assess. Probably the simplest problem that can be seen is the reduction in the saturation value of dissolved oxygen in a stream at an elevated temperature. At 20°C, approximately 9.2 mg/L of oxygen is the saturation value. At 30°C this is 7.6 mg/L, at 40°C it is 6.6

mg/L. Thus it may be readily seen that there is less oxygen available for biological respiration at the higher temperature. Another physical impact is the reduction of viscosity at a higher temperature. This is a benefit in waste treatment plants in that it allows a more rapid settling of the particulate matter. However, certain plankton require a specific viscosity of water to remain in suspension. At a lower viscosity they will sink and may not be able to remain in the euphotic zone. Obviously, greater evaporation occurs at a higher temperature, thereby removing water from the liquid to the vapor phase. This does result in an increased rate of cooling; however, the net effect is a greater loss of water. Rapid heating of water or pressure reduction, such as in passing through a power turbine, causes supersaturation of the nitrogen in the water, which has been shown to kill fish by the formation of nitrogen bubbles in their blood systems.

In terms of chemical effects, most rates of reaction are increased at a higher temperature. Unfortunately, this reduces the biological tolerance to change. This is similar to the reaction with enzymes or catalysts that normally are more effective at higher temperatures. Tastes and odors are also increased at higher temperatures. The BOD rate, which is both chemical and biological, also increases, thus satisfying the BOD in a shorter period of time. This can benefit an area farther downstream, but can result in serious oxygen depletion in the area immediately below a discharge containing at high BOD. Since there is less oxygen available at the higher temperature, it may be seen that the depletion is considerably more rapid and oxygen deficiency occurs in a shorter period of time.

From the standpoint of the biological system, there are many interrelated factors that result from the impact of higher temperatures. In general there is an increased rate of reaction at the higher temperature. Photosynthesis is somewhat increased at the higher temperature. However, the rate of BOD satisfaction, which is a function of respiration, may exceed the increase in oxygen production from photosynthesis. The combined impact of photosynthesis and respiration is shown in Fig. 25. The actual ordinates of the graph will vary for each individual organism or species, but the concept is valid. It may be seen that at a lower range of temperature, the increase in photosynthesis may exceed the increase in respiration; however, at higher temperatures, the respiration exceeds the increase of photosynthesis. The respiration increases to a maximum rate at the optimum temperature, and at only a slightly higher temperature reaches a critical or fatal temperature at which the organism is destroyed. Since individuals and different species have different optimum and critical temperatures, it may be seen that a change in temperature may seriously change the character of the biological systems present at different

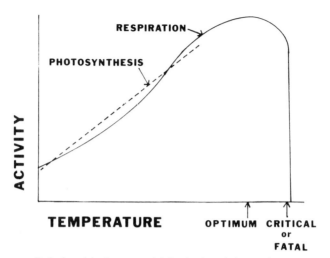

Fig. 25. Relationship between biological activity and temperature.

temperatures. As a rule of thumb, in the growth range, respiration doubles for each 10°C rise. The increase in the rate of respiration is correlated with the increased rate of metabolism. Also, the rate of reproduction is increased. This normally is recorded as a shortening of the time for the reproduction cycle to be completed. Thus, more organisms are produced that respire and metabolize at a higher rate, thereby utilizing the diminished oxygen reservoir in a stream at a higher rate. This, of course, will reduce the BOD rapidly, but with a commensurate reduction in DO. As all these factors increase, the total activity may be limited by the availability of food.

Some organisms have a temperature sensitivity in which they prefer warm or cold water. In general in cold streams, fish swim toward warm water, whereas in warm streams, fish swim toward cold water. However, there are also some fish that are quite temperature insensitive. This can result in an accumulation of fish and other swimming aquatic organisms at either a warm spot or cold spot in a stream. Here they may utilize all the food available, ultimately resulting in their demise.

The increase in temperature also adds to the synergistic effects of other harmful substances, such as toxicity, disease and so on. In general, organisms have a lower tolerance to these impacts at a higher temperature. In some instances, for example, disease may be controlled by a lower temperature. Thus, the higher temperature merely adds to other problems. Another important factor is the rate of change of temperature. Many aquatic organisms can tolerate a slow change in temperature, but not a rapid one. If there is a discharge of hot water into a stream, there can be a narrow zone of mixing that results in a rapid change in temperature within a short distance.

There are many combinations of impacts related to increasing aquatic temperature. For example, certain organisms will reproduce only at or above a certain temperature. If the body of water is maintained above this temperature they will reproduce for a longer period of time or constantly. However, their development is dependent upon the food available to them. If the food on which they rely is not available because of other temperature impacts, then they cannot survive. If they cannot survive, their predators may in turn disappear, because they are not available at the proper time for use by their predators. Thus, it may be seen that a change in temperature at one location can impact the entire food chain throughout a stream or ecosystem. This is especially true with algae in which it has been shown that the blue–green algae are more tolerant of higher temperatures. As a matter of fact, these have been shown to be present at up to 77°C in hot springs such as at Yellowstone. Most of the colors in the hot springs are produced by different species of blue–green algae. The blue–green algae, however, are less desirable as food for other organisms. Therefore, they will consume the nutrients available and produce organic matter that is not readily used by higher organisms in the food chain. This obviously distorts the entire system.

Probably one of the greatest problems is that of a rapid change in temperature. Almost all power plants must shut down at certain times for repairs and/or accidents. If the thermal discharge from a power plant has been keeping a stream warm through the winter, a sudden stoppage of the power generation would allow the stream to cool very rapidly in the winter. This would impact a significant thermal shock on the aquatic organisms. Therefore, most regulations call for prohibiting routine shutdowns during the winter months. Obviously, accidental shutdowns may still occur. Thus there are two main concerns: (1) the impact of the higher temperature and (2) the impact of sudden changes of temperatures because of operation of the thermal generating system.

The impact of a thermal discharge in a body of water will become dissipated with time and the water body will reach equilibrium after sufficient time. In a stream this implies sufficient flow time from even a constant discharge of heat. However, in a lake, since there is no flow time, the only time would be that from the batch discharge of a slug of heat. The temperature relationship in a stream during summer and winter is depicted in Fig. 26.

Although there is no space here for detailed description of control

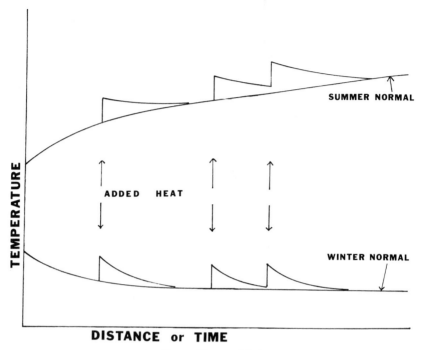

Fig. 26. Fate of heat added to a stream.

methods, some basic concepts should be mentioned. The first is the use of cooling ponds in which the water is recirculated through the cooling pond and reused for cooling. Crescent-shaped ponds have been shown to be quite economical. The heated water is discharged into one end and as evaporation causes cooling, the cooler water is taken from the opposite end. Various forms of cooling towers are also used for dissipating heat. There are various combinations of wet and dry towers using natural or forced draft. Detailed descriptions of these may be found elsewhere. (Suggested reading [18].)

Regulations on thermal discharges are the responsibility of the various states. It is impossible to cite regulations common to all states; however, some of the criteria for evaluation should be useful.

The first criterion is the definition of a thermal discharge, usually as one that is at a temperature above a set value such as 70°F (21°C) or if it causes a natural body of water to rise above a recommended limit, such as 3–5°F (1.5–3°C). It should be defined separately in terms of lakes, streams, and tidal waters. There is usually a definition for trout and nontrout streams in which the requirements are more stringent for trout streams. There may be greater temperature tolerances during the summer. Summer and winter may be defined as periods when the temperatures reach certain limits. Many criteria involve the degree of mixing in a stream with a description such as at least 50% of the cross-sectional area and/or volume of the stream, including a minimum of ⅓ of the surface, shall not be raised above a certain minimum temperature. This is to allow fish to pass a heated section of the stream. In a lake, frequently the regulations are to prevent a certain temperature rise in a circle or a plume from the point of discharge into the lake. Here it is recommended that a surface area be specified rather than a radius of a circle, which is frequently specified. This will compensate for any distortions in the shape of the plume other than circular. In a summer stratified lake, there should be no discharge of heated effluent into the hypolimnion. Withdrawal of water for cooling may be from the hypolimnion or the epilimnion, depending upon regulations. Discharging the heat to the epilimnion will help maintain the cool water in the bottom and maintain the thermocline relatively constant. These are just some of the generalities to be observed in regulations on thermal discharges.

It is most useful to find alternate uses for this wasted heat. When homes or factories may be located close to a source of heat, it may be used for heating or industrial processes. However, during the summer there is less need for this heat unless it can be converted to operating air cooling systems. The heat has been used for agricultural purposes. Here, again, this has benefit only during certain times of the year. Some aquaculture systems have been put into operation in which the waste heat is placed in ponds, lagoons, or bays to encourage the growth of fish or shellfish. This has been shown to be very beneficial in reducing the reproduction time and the time to reach maturity. However, again this has limited use and will not solve all the problems of the utilization of excess heat. As energy becomes more and more expensive, it is certain that more systems will be devised to utilize this excess heat.

XII. TOXICS IN WATER RESOURCES

The concern for toxic materials in water supplies has created one of the greatest controversies of recent time. The US Environmental Protection Agency issued a list of 129 priority pollutants that are of concern and should be eliminated from water supplies. However, there is uncertainty about what concentrations of these substances represent hazardous levels.

One problem is that individuals may have varying tolerances to these substances; the other problem is our present analytical methods. In some cases we do not have sufficient precision to measure concentration in ranges low enough to detect their presence, whereas in other cases we are improving our techniques by finding even lower concentrations that may or may not prove to be harmful. To add to this dilemma are the present requirements that in some instances tend to add to the problem of toxic materials. A typical example is the requirement for chlorination of wastewaters before discharge and of drinking waters prior to treatment. Chlorination of certain organic materials produces the trihalomethanes, which are on the list of priority pollutants. If chlorination of wastewater is not practiced there will be fewer trihalomethanes produced. However, the downstream area will have a higher disease potential because the pathogenic organisms have not been reduced sufficiently. Similarly, in the water intake, where poor quality water is used as the drinking water supply, prechlorination has been practiced to reduce the number of pathogenic organisms. This can result in an increase in the trihalomethane content. In order to reduce trihalomethane production, it is recommended that prechlorination not be applied, but that secondary chlorination, after the removal of the organics, be the only point for chlorination. This of course will reduce the production of the trihalomethanes, but may increase the potential for the carryover of pathogenic organisms in the water supply system. Similarly, several heavy metals are on the list of priority pollutants. Some of these metals have been shown to be essential growth factors in very low concentrations. However, at only slightly higher concentrations, they have been found to be detrimental. Thus, very close tolerance limits must be set, and it is difficult to establish safe levels and still prevent the occurrence of unsafe levels.

Another problem comes in the determination of the priority pollutants. The regulations call for the banning of any substance suspected of being carcinogenic or mutagenic. This usually is based upon feeding massive doses to a susceptible species of mice or rats. If any carcinogenic or mutagenic effects are found, that substance is placed on the priority pollutant list. There is, however, no guarantee that what is harmful to rats and mice will also be harmful to humans. Also it is difficult to extrapolate to humans the concentration limits that were harmful to the animals.

A regulatory agency must of necessity be extremely conservative. Therefore, any substance even suspected of being harmful must be included on this list. Environmental conditions are to be maintained so that even the most susceptible individual is protected from any harm. Thus, the most conservative values and the lowest levels are considered in listing the priority pollutants and the concentrations that must be avoided.

One of the major problems that comes in the evaluation of priority pollutants is the impact of long-term use of any substance. There may be no immediate symptoms, but symptoms may occur after long periods of time, in another generation, or in a very indirect manner. For example, the use of certain drugs (Thalidomide) for increased fertility has been shown to have an impact on the offspring of the second generation of persons who used these drugs. Findings such as these take 20–25 yr to evaluate. Another is the use of DDT, which has not been shown to be harmful to humans, but indirectly has particularly affected the eggs of certain fish and birds. In society's total concern for the environment, these impacts are also valid concerns. The problem then arises about how one evaluates a new substance that appears to have no deleterious impact at the moment, but may in the future. We certainly cannot ban every new substance; however, the answer probably lies in continued evaluation and monitoring of any new substance, particularly those that are suspected of being or related to other substances that have been shown to be deleterious. However, there must also be provision to remove a substance from the priority pollutant list if it has been proven over a long period of time not to be harmful to humans or the environment.

Another condition that has not been evaluated is the risk:benefit ratio. As an example, DDT has been shown to have saved on the order of 500 million lives, particularly in preventing malaria. It is now on the priority pollutant list primarily because of its impact on other organisms rather than humans. There must be some means of evaluating a substance to determine whether its adverse effects may be more than offset by its benefits.

Of some concern must also be cost; however, cost related to human health must be the final evaluation. A complete analysis of the 129 priority pollutant costs in the order of $1000 per sample. This represents a significant cost to a small water supply system. A question may arise about the need for such a complete analysis for every water supply system. Obviously some evaluation must be made of the potential harm or pollution of water supply systems. A water supply not subject to contamination may not need such a complete analysis; however, where potentially toxic chemicals are in the area or in the ground, it is advisable to monitor water supply systems to assure that such contaminants do not reach the public through the water supply system.

There is no simple solution to the concern for toxic matetrials in the aquatic environment. What is essential is continued monitoring for the presence of toxic materials and a rational evaluation of the impact of these potentially toxic materials upon humans.

XIII. GOALS OF WATER POLLUTION CONTROL

Goals should not be the conclusion of any discussion of water pollution control. They should be the reason for establishing some procedure for pollution control. Once proper goals are established, one can make some reasonable decisions in terms of what is necessary and how they may be accomplished. However, some goals have already been established, and it is assumed that the previous information will aid in reaching the prescribed goals.

The goals of water pollution control have been summarized briefly: to provide fishable, swimmable, and drinkable water. Public Law 92-500 passed by the US Congress and commonly known as the Amendments to the Federal Water Pollution Control Act, starts off with a "declaration of goals and policy." A summary of this section is in order. The main objective of this Act is stated to be "to restore and maintain the chemical, physical and biological integrity of the Nation's waters." In order to achieve this, stated goals include: (1) discharge of pollutants into the navigable waters should be eliminated by 1995, (2) an interim goal of water quality providing for the protection and propagation of fish, shellfish, and wildlife and providing for recreation should be achieved by July 1, 1983, (3) the discharge of toxic pollutants in toxic amounts should be prohibited, (4) there should be Federal financial assistance to construct publicly owned treatment works (POTW), (5) to assure adequate pollution control areawide waste treatment management processes should be developed and implemented, and (6) there should be major research and development to develop the technology necessary to eliminate the discharge of pollutants.

The successor to the above law was Public Law 95-217, commonly known as the Clean Water Act. Its main goals were similar to the previous legislation with some minor modifications in the timetable. Greater emphasis was placed on innovative and alternative technology, individual wastewater treatment systems, sludge disposal, and energy conservation. All of the laws and regulations that have been passed and promulgated by EPA and various other agencies have as their goal the protection of the quality of the water within the US. To this end, much money and time have been expended. Some improvement has been seen; however, there are some glaring cases where the efforts have not been successful. Some of these most likely will be taken care of in a matter of time; in other specific cases, there may be other problems that will have to be resolved.

One of the major concerns in any effort or program, whether it be that of the Federal or local government or of any other organization, is who is to be the beneficiary of the energy expended. There are two major directions to which pollution control efforts can be directed. One is to improve the quality of the environment for the sake of the environment and the other is to improve the quality of the environment for the sake of society. There is no clear-cut distinction between the two because in general what benefits the environment also benefits humankind. However, there are specific instances where benefits to the environment may not have any obvious benefit to humans; thus, certain individuals may question the expenditure of public monies that will not directly benefit the public.

It is the author's belief that there is room for both points of view. There are cases in which the environment is and should be the sole direct beneficiary of any control practice. For example, there are areas of the Adirondack Park in New York that are to be preserved as "forever wild." Even airplanes are not allowed into these areas to land on the lakes to provide for hiking, hunting, or fishing. Large areas of Alaska have been set aside as wilderness area in which humans are viewed as unwanted interlopers. Many areas of Africa are presently wildlife areas that will have to be preserved if we hope to salvage some of the rarer animals. In some instances it has been questioned whether we really need certain of these rare animals. They may have no direct benefit to humans and in some cases, aside from hunting expeditions that benefit only a very small number of individuals, they could probably be allowed to become extinct without any serious impact upon the environment. On the other hand, we must consider that humankind is also a part of the environment and not the only factor within that environment. Other species have just as much right to exist on this earth as do humans. Another concept that must be considered is the fact that the earth has maintained a blanace of biological systems that houses various species including humans. As the latter increase excessively in number, they crowd out the other species. This may result in a biological imbalance in which humans may ultimately destroy the earth by overpopulation. It may be necessary to control population so that the earth can continue to support a limited number of human beings. This concept of course rankles many people. Each person feels that he or she has a right to propagate as much as seems desirable and to fill the earth with his or her progeny. Certainly we do not want to have a dictator or government body dictating who may have how many children, when, and where. On the other hand, human beings must take responsible positions and realize that in the total environment, population control is essential. To accomplish this will take a considerable amount of education. It is made more difficult by the fact that traditions and religious beliefs have fostered the concept that humankind is the center of the universe. Thus, we can do whatever we want and everything was made for us. Even though everything may indeed have been made for us, if we utilize everything made for us there will be no more left and of course we will decline. It would be far better for us to control our own numbers than to have these numbers controlled by a large famine, pestilence, or other noxious factor.

In terms of the concept that the environment should be controlled for the benefit of society, there arises the concept that complete pollution elimination may not be a desirable situation. For example, an oligotrophic lake is very unproductive, particularly in terms of producing the large amounts of fish that could provide food for many. Just as adding fertilizer to the land increases the productivity of the terrestrial crop, adding nutrients to a lake will increase the total productivity of that lake. This will result in increased plant growth and ultimately increased animal (fish and shellfish) production. Obviously, limits of productivity are controlled in the aquatic environment by the amount of dissolved oxygen that is available. Here again, in terms of benefits to humans, we could install aerators in all lakes and use them to their maximum productivity. This may even be necessary one day if the population continues to increase as it has. Thus, a different set of goals would have to be established if all of the environment were to serve humankind directly.

There is probably some middle ground that offers the best solution. It is certainly desirable to maintain some wilderness areas undisturbed. It is desirable to have some bodies of water that are highly oligotrophic. The problem seems to be that, if these are preserved in their pristine purity, how will humans get to appreciate them. Possibly here limited access will have to be provided to some oligotrophic lakes. Other lakes however may be maintained at higher productivity levels. This will provide greater production of fish and similar related business and recreational activities. Thus, there could be a balanced, diverse environment that would satisfy both phases of the pollution control concept. A problem that arises here, of course, is how does one establish whether a particular area or lake or stream is to be preserved for the benefit of the environment and which other area is to be maintained for the benefit of society?

These are not simple questions to answer. Each individual has a private concept that has some validity. This will require a considerable amount of discussion and cooperation among individuals and obviously will result in various pressure groups trying to get a little more for themselves or a special cause. However, we must look at the total picture, not just at what affects a few individuals here and now. Considerations must be made for the future so that humankind can coexist with the environment on earth for many years to come. With an educated society, logical thinking, and good information, human beings can make good decisions about what efforts must be expended both now and in the future in order to improve and preserve the environment for the best use of all for now and for time to come.

REFERENCES

1. L. J. Henderson, "The Fitness of the Enviroment: An Inquiry into the Geological Significance of the Properties of Matter," in part delivered as lectures in the Lowell Institute, February, 1913. Macmillan, New York, 1913, XV, p. 317.
2. B. Osborn, and P. O'N. Harrison, "Water . . . and the Land," US Dept. of Agriculture, Soil Conservation Service, SCS-TP-147, 1965.
3. M. I. Lvovich, "Fresh Water—Resources of World," 1972, Vol. 70, pg. 72, Issue 11. Vestnik, Akademii Nauk, USSR.
4. A. H. Thiessen, *Monthly Weather Rev.* **39**, 1082 (1911).
5. A. F. Meyer, *Tans. Amer. Soc. Civil Engr.* **79**, 1 (1915).
6. C. W. Thornthwaite, *Geograph. Rev.* **39**, 55, (1948).
7. W. W. Horner and F. L. Flint, *Tans. Amer. Soc. Civil Engr.* **101**, 140 (1936).
8. G. M. Fair, J. C. Geyer, and D. A. Okun, *Water and Wastewaer Engineering*, Vol. 1, Wiley, New York, 1966, pp. 7–23.
9. D. M. Gray, *Handbook on the Principles of Hydrology*, Water Information Center, 1970, pp. 3.6–8.10.
10. L. K. Sherman, *Eng. News. Rec.* **108**, 501 (1932).
11. W. P. Creager, J. D. Justin, and J. Hinds, *Engineering for Dams*, Vol. I, *General Design*, Wiley, New York, 1945.
12. G. W. Pickels, *Drainage and Flood Control Engineering*, McGraw-Hill, New York, 1941.
13. W. Rippl, *Proc. Inst. Civil Engr.* **71**, 270 (1883).
14. D. B. Aulenbach and N. L. Clesceri, *Water, Air Soil Poll.* **14**, 81 (1980).
15. M. Sabamato, *Arch. Hydrobiol.* **62**, 1 (1966).
16. R. A. Vollenweider, "The Scientific Basis of Lake and Stream Eutrophication, with Particular Reference to Phosphorus and Nitrogen as Factors in Eutrophication," Technical Report to OECD. Paris, DAS/CSI/68.27, pp. 1–182 (1968).

17. R. A. Vollenweider and P. J. Dillon, "The Application of the Phosphorus Loading Concept to Eutrophication Research," Canada Centre for Inland Waters, NRCC No. 13690, Burlington, Ontario 1974.
18. "Industrial Waste Guide on Thermal Pollution," Federal Water Pollution Admin., 1968.
19. R.K. Linsley, M. A. Kohler, and J. L. H. Paulhus, *Hydrology for Engineers*, 2nd Ed., McGraw-Hill, New York, 1975.
20. H. F. Blaney-Criddle, and W. D. Criddle, quoted by R. D. Goodrich, *Trans. Amer. Soc. Civil Engr.* **122,** 810 (1957).

5
Deep-Well Disposal

Charles W. Sever
US Environmental Protection Agency, Retired

I. INTRODUCTION

Underground wastewater disposal and storage by well injection is under consideration today by both industries [1–5] and municipalities [2, 6–9] to help solve environmental problems. A national goal established by Congress in the Clean Water Act of 1972 (PL 92-500) was to cease all discharges of pollutants to navigable waters by 1985 [10]. The capacities of surface waters to receive effluents without violating existing water quality standards is diminishing with time. Costs of disposal or storage using conventional and advanced waste treatment techniques are increasing rapidly. These and other factors have caused both dischargers and regulatory agencies to consider subsurface injection as a mechanism for the disposal of waste fluids.

Subsurface disposal of liquid wastes into aquifers is based on the concept that such wastes can be injected through wells into confined geologic strata not having other uses, thereby providing long-term isolation of the waste material. Subsurface storage and artificial recharge involves the concept that highly treated municipal and other wastewaters are valuable and should be reused. In a nation where water deficiencies or management problems are forecast for the forseeable future, storage of treated wastewaters for reuse is destined to become a major element for consideration in water resource management [12].

Industries started injecting wastewater about 1928. However, their use of this method was negligible until emphasis on surface water pollution control prompted companies to seek out and evaluate alternatives for their wastewater discharges such as well injection.

Pollution of surface waters resulting from discharge of salt water separated from crude oil led petroleum producers to start using injection wells to return oil field brines to subsurface saline aquifers. Since its inception about 1905, more than 40,000 injection wells have been constructed for disposal of brines.

Information on disposal of treated municipal wastewater by injection is sparse. The earliest use of an injection well for municipal wastewater disposal in the United States was in 1959 at the Collier Manor Sewage Treatment Plant in Pompano Beach, Florida [9]. During 13 years of operation, the City of Pompano Beach injected about 3 billion gallons (11.4 million cubic meters) of secondary treated wastewater into a cavernous "boulder zone" through two wells 1000–1400 ft (305–427 m) deep.

Underground space is now recognized [13] as a natural resource of considerable value. A small percentage of this space, like the "boulder zone," consists of large caverns capable of receiving and transmitting extremely large volumes of wastewater for a single injection well. But most space underground consists of the area available between sand grains in the rock strata. The percentage of this space available for fluid storage and movement depends upon how much clay and silt is present and the amount and type of cementing material present [14].

The porosity of a rock is essentially its interstitial pore space. It can be expressed quantitatively as the ratio of the volume of the pore space to the total volume of the rock, and generally is stated as a percentage. Gravel or sand that is clean, uniform, and free of clay and silt will have about 30–40% of its volume available for storage space. Gravel or sand containing abundant clay and silt, cementing material, or a precipitant from injected wastes may have as little as 5–15% of its volume available. Fractures, joints, and solution channels in cemented rock formations, such as limestones, are additional types of pore space that contribute to porosity, but are difficult to measure.

Virtually all of this subsurface pore space is already occupied by natural water, either fresh or mineralized to some extent. Thus injection does not usually involve the filling of unoccupied space, but rather consists of the compression or displacement of existing fluids. Since the compressibility of water is small, creation of significant volumes of storage space through this mechanism requires a disposal strata that underlies a large geographic area.

II. BASIC WELL DESIGNS

Design of a casing program depends primarily on well depth, character of the rock sequence, fluid pressures, type of well completion, and the corrosiveness of the fluids that will contact the casing. Where fresh groundwater supplies are present, a casing string (surface casing) is usually installed to below the depth of the deepest groundwater aquifer immediately after drilling through the aquifer (Fig. 1). One or more smaller diameter casing strings are then set, with the bottom of the last string just above or through the injection horizon, the latter determination depending on whether the hole is to be completed as an open hole or gravel-packed, or is to be cased and perforated.

The annulus between the rock strata and the casing is filled with a cement grout. This is done to protect the casing from external corrosion, to increase casing strength, to prevent mixing of the waters contained in the aquifers behind the casing, and to forestall travel of the injected waste into aquifers other than the disposal horizon.

Cement should be placed behind the complete length of the surface casing and behind the entire length of the smaller diameter casing strings also, or at least for a sufficient length to provide the desired protection. It is suggested that at least 1 ½ in. of annular space be allowed for proper cementing. Casing centralizers, other equipment, and techniques such as stage cementing can give added assurance of a good seal between the strata and the casing and should be encouraged where applicable.

The majority of injection wells constructed to date have been for one of three basic purposes: injection plus external monitoring for leakage through confining layers near the well (Fig. 2A); injection plus internal monitoring for leaks in the injection tubing and casing (Fig. 2B); and injection only (Fig. 3). Numerous variations in design have been used to accomplish these three basic purposes.

Wells designed for injection plus monitoring of the confining layer generally are constructed like the well in Fig. 2A. In these type wells, the fluid chemistry and pressures outside the well in an aquifer overlying the receptor are monitored at the wellhead for changes that would indicate leakage. This monitoring is accomplished either by leaving the annulus exposed to the aquifer to be monitored as in Fig. 2A, or by some method of accessing the aquifer inside the injection well casing. In designing a well for this purpose, the engineer should remember that to obtain a sample of fluids from the monitored aquifer, fluids in the pipes must be pumped out. If these fluids are saline, then their disposal may be a prob-

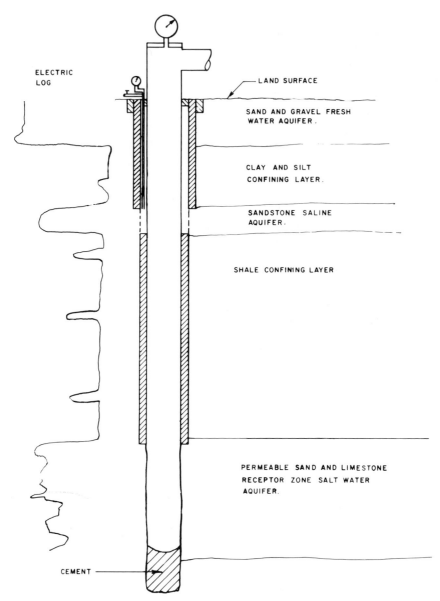

Fig. 1. Typical well construction.

lem. One variation is to fasten a small 0.25–0.5 in. pipe (usually stainless steel or neoprene) to the outside of the inner casing (in the annulus) extending from land surface to the top of the aquifer being monitored. By sampling through this "drop" pipe, the volume of fluids to be disposed is minimized. Another variation is to drill a 4–6 in. larger hole and then attach a 2–3 in. monitoring pipe outside the inner casing (in the annulus) from land surface to the aquifer being monitored.

Wells designed for self-inspection of internal leaks are constructed similar to the well in the Fig. 2B. A sampling tube is installed inside the casing to a depth near the bottom of the injection well. Either a seal is placed outside of this tubing near its bottom to prevent fluid circulation in the annulus and the annulus is pressurized, or low density fluid capable of causing back pressure, such as kerosene, is placed in the annulus. Pressure is measured at the land surface to detect fluid movement into or out of the annulus. Pressure changes in the annulus must be correlated with changes in injection rate, temperature, specific gravity of the fluids, and other factors to avoid false interpretation. In wells using this type of design, leaks in both the injection tubing and the well casing are quickly detected. Industrial waste should be injected through separate interior tubing rather than the well casing itself. This is particularly important when corrosive wastes are being injected. A packer can be set near the bottom of the tubing to prevent corrosive wastes from contacting the casing. Additional corrosion protection can be provided by filling the annular space between the casing and the tubing with oil or water containing a corrosion inhibitor.

Wells designed for injection purposes only are less expensive to construct, but lack the ability for self-monitoring described above. The well shown in Fig. 3 is typical of this type design. Separate wells must be drilled for monitoring purposes where required.

Temperature logs, cement bond logs, and other well-logging techniques can be required as a verification of the adequacy of the cementing. Cement should be pressure-tested to assure the adequacy of a seal.

Neat portland cement (no sand or gravel) is the basic material for cementing. Many additives have been developed to impart some particular quality to the cement. Additives can, for example, be selected to give increased resistance to acid, sulfates, pressure, temperature, and so forth. Other additives reduce the viscosity of the cement until it flows like water.

It is frequently desired to increase the acceptance rate of injection wells by chemical or mechanical treatment of the injection zone. Careful

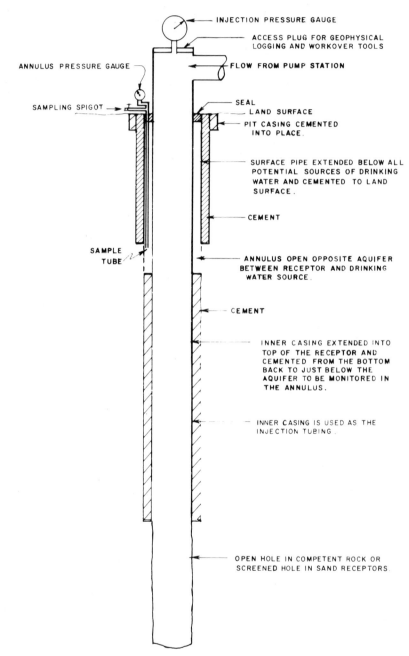

Fig. 2A. Well designed for external self monitoring.

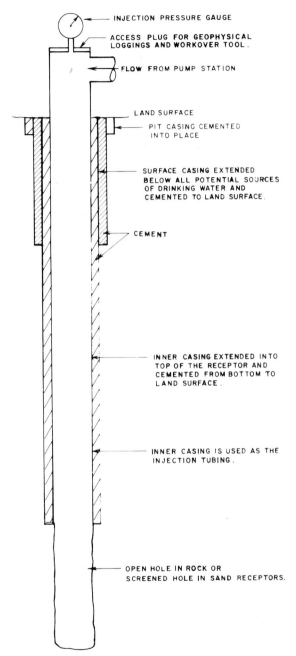

Fig. 2B. Well designed for internal self monitoring.

attention should be given to stimulation techniques, such as hydraulic fracturing, perforating, and acidizing to insure that only the desired intervals are treated and that no damage to the casing or cement occurs.

The type of well-head equipment can be a consideration in cases where the buildup of high back-pressure is a possibility. In such cases, the well head should be designed to "bleed-off" back flows into holding tanks or pits before pressures reach a hazardous level. High back-pressures can be developed by chemical reactions in the formation. For example, at Louisville, Ky. where ferric chloride solutions had been injected into dolomite and limestone, for several years an excessive buildup of carbon dioxide gas pressure caused a blowout during routine maintenance in 1980.

Surface equipment often includes holding tanks and flow lines, filters, other treatment equipment, pumps, monitoring devices, standby facilities.

Surface equipment associated with an injection well should be compatible with the waste volume and physical and chemical properties of the waste to insure that the system will operate as efficiently and continuously as possible. Experience with injection systems has revealed the difficulties that may be encountered because of improperly selected filtration equipment and corrosion of injection pumps.

Surface equipment should include well-head pressure and volume monitoring equipment, preferably of the continuous recording type. Where injection tubing is used, it is advantageous to monitor the pressure of both the fluid in the tubing and in the annulus between the tubing and the casing. Pressure monitoring of the annulus is a means of detecting tubing or packer leaks. An automatic alarm system should signal the failure of any important component of the injection system. Filters should be equipped to indicate immediately the production of an effluent with too great an amount of suspended solids.

III. EVALUATION OF A PROPOSED INJECTION WELL SITE

It would be impossible to cover all the potential problems that could develop during the construction and operation of a disposal well system. But a safe economical injection well system that will function properly can be built at most sites in geographic areas suitable for this method of disposal.

Before proceeding with an evaluation, answers to the following two questions should be obtained from the regulatory authorities.
1. What criteria will determine the degree of pretreatment that will be required before fluids can be injected?
2. What restrictions have been placed upon water quality in receptor aquifers?

Pretreatment requirements vary from state to state. One state authority may prohibit emplacement of toxic wastes underground, while another will prohibit its surface discharge and encourage injection. At least one state agency pushes for pretreatment of all wastes to drinking water standards before injection. Many state agencies and most environmental groups push for "nondegradation" of the environment. But for successful operation of the system, all wastewaters must be pretreated until the fluid to be injected is compatible with the environment in the receptor aquifer.

A common mistake made evaluating deep-well injection systems is to underestimate the degree of pretreatment required. After a well system is completed is not the time to find that after the extensive pretreatment required the wastes are in fact suitable for surface disposal. But before pretreatment can be determined, certain basic relationships between disposal operations and the physical environment must be examined. The important areas include: the receptor zones, the confinement conditions, and the subsurface hydrodynamics. Data on these subjects is available from state and federal agencies, as well as local consultants, and should be assembled and evaluated. If the results are favorable, then additional data should be collected by constructing test wells. Various potential confining layers and receptors then need to be tested. Final evaluation of data may show the disposal of wastewaters into underground formations to be an unwise solution to a disposal problem not only from the standpoint of damage to the environment, but from the standpoint of overall costs.

A. Confinement Conditions

Confining layers, although generally required by regulatory agencies, are not always essential. For example, in one of the southeastern states, wastes are being injected into the Knox Dolomite where it is about 5000 ft (1500 m) thick and intensely fractured vertically. Porosity is low, averaging about 10%. Wastes being injected are heavy, having a specific gravity of about 1.2. The heavy wastes after injection move in response to gravity downward to the base of the Knox where they are permanently stored. The salt water—fresh water boundary that now lies at about 4000 ft (1200 m) is being displaced upward at a rate of about 1 ft/yr.

Confining layers are rarely impermeable to waste movements. They just retard the movement. Most wastes are capable of slowly moving through even the denser clays. However, as movement takes place, the wastes are subjected to ion exchange, osmosis, filtration, absorption, and other forms of treatment.

Possibly the most important laboratory tests to be conducted on samples collected during drilling of the test wells are vertical permeability and ion exchange capacity on cores collected from the confining layer overlying the receptor. The vertical permeability, together with thickness of the confining bed and the anticipated differential pressure across the confining bed, should be used to predict the velocity at which wastewater will travel through the confining bed. The general premeabilities of rocks are shown in Table 1. Data generated from ion exchange and other tests may show that toxic wastes can meet drinking water use or other applicable standards after being subjected to subsurface treatment. These tests also indicate the physical and chemical changes that may take place with time during movement through the confining bed.

B. Potential Receptor Zones

A basic requirement of a receptor zone or a combination of zones is that it be capable of receiving and transmitting the volume of wastewater planned for injection. State regulatory agencies frequently place addi-

TABLE 1
Relationship of the Coefficient of Permeability to Potential of a Stratum for Use as a Receptor or as a Confining Layer

Rock type	Flow potential	Permeability range, md
Cavernous limestone	Excellent receptor	3×10^6 to 1×10^9
Gravel	Good receptor	1×10^4 to 3×10^6
Sands and sandy silts	Poor receptor or confining layer	1×10^{-2} to 1×10^4
Clay, shale	Good confining layer	1×10^{-6} to 1×10^{-2}

tional requirements on receptors based upon depth and water chemistry. For example, some state agencies prohibit injection into receptors at a depth less than 2000 ft (610 m) or where the native fluids contain less than 10,000 mg/L total dissolved solids.

Another requirement is that changes in the physical and chemical properties of the wastewater and in the receptor after injection be compatible with the goals of injection. These changes usually can be grouped as follows: changes in the wastewater induced by the environment in the receptor; changes in the wastewater caused by chemical reactions with the receptor rocks; changes in the wastewaters caused by chemical reactions with fluids in the receptor; and physical and chemical changes in the receptor resulting from reactions with the wastewater. Precipitants formed as a result of these types of changes can plug the receptor and cause the system to fail.

Knowledge of the complete chemical character of the wastewater after the pretreatment is of the foremost importance in evaluating a potential receptor. This knowledge, plus data about the physical subsurface environment available from drilling oil wells in the area or from a test well drilled at the proposed site, should enable a company to forecast the chemical stability of its waste. Knowledge of the mineralogy of the aquifer and the chemistry of interstitial fluids collected from a test well should indicate the reactions to be anticipated during injection. Laboratory tests can be performed with rock cores, formation fluids, and wastewater samples to confirm anticipated reactions.

Selm and Hulse [31] list reactions between injected and interstitial fluids that can cause the formation of plugging precipitates. These include precipitation of alkaline earth metals such as calcium, barium, strontium, and magnesium as relatively insoluble carbonates, sulfates, orthophosphates, fluorides, and hydroxides. Precipitation of other metals such as iron, aluminum, cadmium, zinc, manganese, and chromium as insoluble carbonates, bicarbonates, hydroxides, orthophosphates, and sulfides can also occur. Also, the precipitation of oxidation–reduction reaction products can occur.

Carbonate rocks generally are excellent receptors for acid wastewaters. The soluble carbonate rocks neutralize the acid and cause precipitation of many of the above metals. Where the volume of precipitant is significantly less than the volume of carbonate rock dissolved, the system will work safely. If not, then the receptor pores will generally plug and the system will fail. Undesirable effects of the reaction of acid wastes with carbonate receptors could be the evolution of carbon dioxide gas that might retard fluid movement if present in excess of its solubility.

Marine sand receptors containing clays such as montmorillonite will pass saline wastewater without change, but the clays may swell to many times their original volume when in contact with fresh water. Such swelling effectively reduces permeability and may cause well failures.

C. Subsurface Hydrodynamics

The dynamics of subsurface fluids in the receptor and overlying aquifers must be understood to the extent that the direction and rate of movement of any wastewater injected and any native fluids displaced by this waste can be estimated. Using data collected from a test well and other means, the pressure buildup in affected aquifers with time should be estimated.

A well injecting at a constant rate into an extensive confined receptor aquifer produces an area of influence that expands with time. As the formation pressure is increased, flow is radially away from the injection well, but not a steady-state flow. Theis [18] hypothesized a close analogy between groundwater flow and heat conduction, and developed the following nonequilibrium equation (Fig. 4) for determination of the coefficients of transmissibility (T) and storage (S).

$$T = \frac{114.60 Q W(u)}{s} \quad (3)$$

and

$$S = \frac{uT}{1.87 r^2/t} \quad (4)$$

where s is the pressure change at r in feet of water and $W(u)$ is the well function of u as shown in Table 2, r is the radius from the point of injection to the point of observation in feet, and t is the time since injection started in days.

The Theis method is a graphical one based on superposition of curves. $W(u)$ is plotted against u on logarithmic paper then s is plotted against r^2/t using paper of the same scale. The two plots are superimposed with the coordinate axes parallel and shifted until the position with most of the two curves matched is found as shown in Fig. 4. The coincident values of $W(u)$, s, and r^2/t are noted. By substituting these values into the above formulas, S and T for the receptor aquifer are determined. See example 3 in section IX.

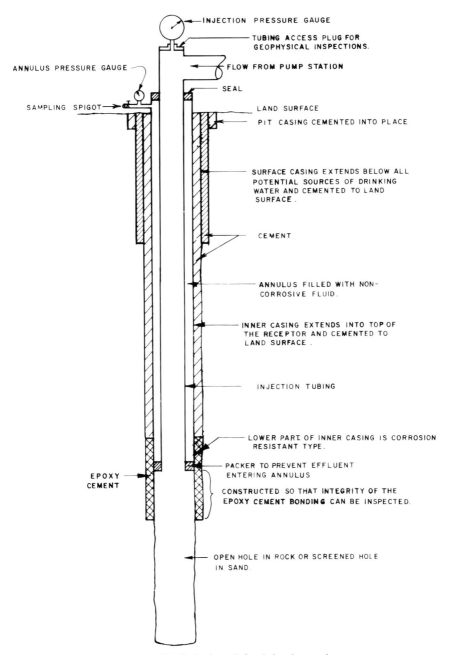

Fig. 3. Well designed for injection only.

TABLE 2
Values for $W(u)$ for Values of u^a

u	1.0	2.0	3.0	4.0	5.0	6.0	7.0	8.0	9.0
$\times 10^{-1}$	1.82	1.22	0.91	0.70	0.56	0.45	0.37	0.31	0.26
$\times 10^{-2}$	4.04	3.35	2.96	2.68	2.47	2.30	2.15	2.03	1.92
$\times 10^{-3}$	6.33	5.64	5.23	4.95	4.73	4.54	4.39	4.26	4.14
$\times 10^{-4}$	8.03	7.94	7.53	7.25	7.02	6.84	6.69	6.55	6.44
$\times 10^{-5}$	10.94	10.24	9.84	9.55	9.33	9.14	8.99	8.86	8.74
$\times 10^{-6}$	13.24	12.55	12.14	11.85	11.63	11.45	11.29	11.16	11.04
$\times 10^{-7}$	15.54	14.85	14.44	14.15	13.93	13.75	13.60	13.46	13.34
$\times 10^{-8}$	17.84	17.15	16.74	16.46	16.23	16.05	15.90	15.76	15.65
$\times 10^{-9}$	20.15	19.45	19.05	18.76	18.54	18.35	18.20	18.07	17.95
$\times 10^{-10}$	22.45	21.76	21.35	21.06	20.84	20.66	20.50	20.37	20.25
$\times 10^{-11}$	24.75	24.06	23.65	23.36	23.14	22.96	22.81	22.67	22.55
$\times 10^{-12}$	27.05	26.36	25.96	25.67	25.44	25.26	25.11	24.97	24.86
$\times 10^{-13}$	29.36	28.66	28.26	27.97	27.75	27.56	27.41	27.28	27.16
$\times 10^{-14}$	31.66	30.97	30.56	30.27	30.05	29.87	29.71	29.58	29.46
$\times 10^{-15}$	33.96	33.27	32.86	32.58	32.35	32.17	32.02	31.88	31.76

[a]After Wensel [19]

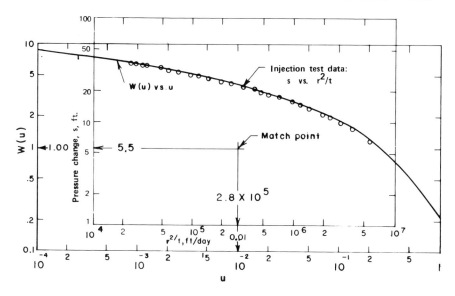

Fig. 4. Theis method of superposition for solution of the nonequilibrium equation.

IV. POTENTIAL HAZARDS—WAYS TO PREVENT, DETECT, AND CORRECT THEM

Problems with injection wells generally can be related to failures in one or more of the five areas listed below:

1. Lack of consideration of all fluid movements
2. Failure of the receptor to receive and transmit the wastes
3. Failure of the confining layer
4. Failure of an individual well either in design, construction, or operation
5. Failure because of human error.

Experience and the use of good common sense will avoid most of these problems. When injection wells are properly designed, installed, operated, and maintained, they are no different from any other good piece of equipment. If properly used, they can play an important role in removing toxic and hazardous substances from the immediate human environment.

A. Fluid Movement During Construction, Testing, and Operation of the System

Because of lack of knowledge about fluid movements, one drilling company placed salt water generated during pump testing of a receptor zone into an unlined storage pit on their site. The salt water, because of its heavy specific gravity, seeped downward into the local drinking water aquifer. It then moved laterally without mixing and some eight months later contaminated a nearby city's drinking water supply.

Each well should be designed so that local fresh water supplies are protected from contamination by a separate casing set into the top of their underlying confining layer and cemented back to land surface before the confining layer is breached during construction. Mud pits should be lined with impervious material to prevent seepage into the shallow fresh water strata. All material used during construction, such as salt, mud, acid and the like, should be stored in such a way as to assure that materials spilled from damaged containers and the like will not contaminate the fresh water supply.

Another example of a problem is the company that sought to cut costs by eliminating consulting fees. They hired a local, very capable, water-well contractor to design and construct their injection and monitoring wells. He did an excellent job of designing the injection well, four monitoring wells into the receptor aquifer, plus one monitoring well into an overlying aquifer. The only problem was that he did not understand subsurface hydrodynamics and located all of these monitor wells within 150 feet of the injection well. The waste front passed the deep monitor wells within 3 d after injection started. Four replacement monitor wells had to be drilled some distance away. Monies wasted on the first four wells (about $140,000) far exceeded the cost of hiring an experienced ground water geologist.

Making sure that the flow capacity in the test equipment is representative of the proposed permanent facilities under design is also important. An abandoned gas well being tested for conversion into a disposal well showed a capacity of only 50 GPM on gravity flow [30]. The project was almost abandoned when an engineer realized that the friction loss in the temporary piping was higher than in the permanent facilities. After the piping was changed, the injection tests showed the well would handle 450 GPM on gravity flow.

B. Failure of the Aquifer to Receive and Transmit the Injected Fluids

Failure of an aquifer generally is caused by lack of naturally developed permeability in the receiving zone or by filling of the pore space with either suspended solids from the effluent or precipitants formed by chemical or biological reactions in the receptor. Well-head injection pressure should be continuously monitored so that these problems can be detected early and failure avoided. A reduction of the ability of the receiving zone to accept and transmit wastes from any cause will increase the pressure at the well head.

Plugging of the receptor zone is by far the most common operational problem where the receptor is a sand aquifer. Most plugging problems can be avoided by one or more of the following: detailed coring to study the size and shape of pore spaces in the receptor, detailed chemical analyses of fluids and rocks in the receptor, biological cultures of both receptor fluids and wastes, analysis of pressures and temperatures in the receptor, changes expected in the wastes after injection, and proper cleanup during completion of the well.

One injector ran extensive compatibility tests in his own labs at room temperatures and pressures with no indication that a problem might exist. However, a few weeks after injection began the injection pressures started rising above the predicted pressures. New tests conducted under environmental conditions similar to those in the receptor showed that minor changes in fluid pH caused precipitants to form that partially plugged the receptor.

Another injector's effluent was found to be incompatible with the native fluids in the receptor. If the pH was raised slightly a precipitant formed; when the pH was lowered, gases were released. The solution: keep the two fluids separated by injecting a compatible buffer ahead of the effluent. The importance of compatibility tests cannot be overemphasized. These tests should be run as close to actual well conditions as possible.

C. Failure of the Confining Layer

The effectiveness of the confining layer at each site should be thoroughly investigated by monitoring changes in formation pressures and chemical water qualities during testing of the disposal wells. The most frequent cause for such failures is the creation or the opening up of vertical fractures in a receptor (Fig. 5), then propagation of these fractures by continued injection until they radiate through the confining layer. For example, a company was permitted to inject its wastes at a maximum surface pressure of 150 psi, a rather conservative pressure for most operations. But the specific gravity of the fluids was not taken into account by either the permitting agency or the company. During the operative history of this system, each time the injection pressure approached 149 psi it would miraculously (so the company officials thought) decline. Soon the pressure

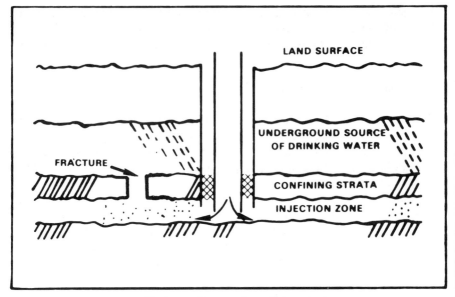

Fig. 5. Faulty or fractured confining strata.

in nearby shallow observation wells began to rise. Then wastes were detected in these observation wells and the company had to abandon its injection well system. They built surface treatment facilities at great expense. The cause of the problem was uncontrolled vertical hydraulic fracturing of the confining layers at a pressure of 149 psi. The bottom hole pressure was about 650 psi, well within the vertical fracturing gradient for the shallow unconsolidated silts and clays that made up the confining layer at this site.

Hydraulic fracturing can be avoided by adequately testing the injection system prior to injecting any effluent. Testing and operations should be planned so that operating pressures never exceed the bottom-hole pressures reached during testing.

It is almost inconceivable that an entire system would fail considering the operational history of existing wells and provided that all the precautions and testing described herein are incorporated into the evaluation of a site. However, should this occur, the disposal of effluent into the subsurface is not a strictly one-way process. If pressures begin to build up or saline water begins to increase in an aquifer above the injection zone, which would indicate leakage of a saline front moving out ahead of the waste front, the entire disposal system could be reversed by installing pumps into the disposal wells and pumping the injected fluids back up to land surface. The effluent can then be disposed via an alternate method. Much of the injected effluent can be recovered by this means. The remaining effluent would probably stop moving as pressures are reduced.

D. Failure of an Individual Well

An example of extreme well failure was the injector who went to great expense to install and cement into place fiberglass casing through which they could inject acids without installing injection tubing. However, they overlooked the fact that a cement plug had to be drilled out of the bottom of the fiberglass casing, the hole deepened and screens installed. Of course the driller did not use centralizers and his drill stem fractured the fiberglass. With time acids in the effluent ate through the cement then entered a shallow aquifer above the monitoring point. This company now has abandoned its injection well system and has drilled more than twenty monitoring wells to keep track of the movement of wastes lost through the fiberglass casing. The local regulatory agency has declared that injection wells will never be used in their state again. Inspection of this well with a caliper upon completion would have detected the fractures, which would then have been repaired and the failure avoided.

After completion of all construction, each well should be thoroughly inspected and tested using all practical geophysical and hydrogeologic methods available. Constructing each well using the best available technology followed by extensive testing and inspection prior to its use for disposal of effluents is the best practical way to avoid well failure.

Should effluents leak through a break in the casing, packers can then be installed to isolate the break. The zone between the packers can be pumped to recover the leaked fluids. Leaks of this type can usually be repaired. It is recommended that an emergency standby well be constructed at each site so that in the event of mechanical failure of some of the pumping equipment or a need develops for servicing or inspecting an injection well, the well that needs servicing can be shut down, and the emergency standby well can be put into operation. Also this emergency standby well can be utilized for observational purposes to gain additional information on changes in water quality in the aquifer during injection.

E. Failures Because of Human Error

An example of human error is an injector who noticed a sharp increase in the injection pressure. The pressure continued to rise until the maximum head pressure of the pump was reached and then the flow began to decrease. Investigation of the problem revealed that one of the cartridges in the polishing filter had been left out.

Davis and Funk [30] cited another excellent example of the importance of human error at a site where a new disposal well system was installed and shut-in pending the startup of a new plant. Compatibility tests had shown the presence of fresh water-sensitive clays in the receptor. During plant construction the transfer lines to the disposal well systems were pressure-tested with fresh water by the contractor and left full. When the well was put into service by the company the fresh water was displaced into and plugged the receptor aquifer by causing hydration and swelling of the fresh water-sensitive clays. The result was a 6000 foot deep $250,000 posthole.

Experience is the best solution to human error. However, installing automatic shutoff and alarm systems at key locations in the system will minimize the effects of human error.

V. ECONOMIC EVALUATION OF A PROPOSED INJECTION WELL SYSTEM

Costs for constructing and operating an injection well system vary tremendously from one area of the country to another. They are lower in areas with active petroleum exploration because of competitive bidding, equipment availability, and the availability of energy sources. Costs shown herein are adjusted to costs for US Department of Commerce Construction Cost Index of 153 (1981). They are broken down into general costs and indirect costs.

Figure 6 shows the range in capital costs experienced in the southeastern United States by industries and municipalities for the construction of injection well systems. An engineer should be able to take these cost data, using Engineering News Record, adjust them for inflation and other differences in the local area, add pretreatment and capital costs, and make a reasonable estimate of the total costs for disposal of a company's wastes by well injection. These costs are based upon an average well depth of 3500 ft (1067 m).

Operation and maintenance costs range from 48 cents to about $1.20/1000 gal waste water injected (1981 energy costs). This variation is caused primarily by differences in well-head injection pressures monitoring requirements. The average cost is about 78 cents/1000 gal injected.

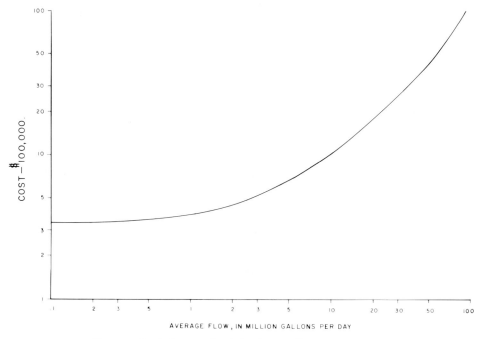

Fig. 6. Capital costs for injection well systems.

Legal fees, permitting fees, changes in insurance rates, and other indirect costs can amount to as much as 10% of construction costs and should be recognized during cost estimating.

VI. USE OF INJECTION WELLS IN WASTEWATER MANAGEMENT

Numerous pro and con considerations must be investigated before deciding upon a method of discharge for a waste fluid. For industries in certain locations, the disposal or storage of wastes in the subsurface by use of well injection may be the most environmentally acceptable practice available. The difference between storage and disposal is that storage implies the existence of a plan for recovery of the material within a reasonable time, whereas disposal implies that no recovery of the material is planned at a given site. Either operation will require essentially the same type of information prior to injection. However, the attitude of the regulatory agencies toward evaluation of a proposal to use the well injection method could be quite different for each.

A. Reuse for Engineering Purposes

The reinjection of fluids produced with hydrocarbons into oil-producing horizons to maintain oil field pressures has been practiced for years. Land subsidence in the vicinity of many oil and gas fields and in areas of substantial groundwater production is generally considered to be caused by the withdrawal of fluids. In California, for example, treated wastewaters are being injected to retard the rates of subsidence by repressurizing subsurface formations.

Along many coastal areas [15] the heavy withdrawal of potable ground water for municipal, industrial, and other uses from fresh water aquifers has caused salt water encroachment laterally within the aquifer systems. In such areas, treated wastewaters may be injected into the aquifer system to create a hydraulic barrier and hold back the encroaching salt water [16–19]. For example, since 1965 tertiary-treated sewage has been injected at Bay Park, NY, into a shallow artesian sand aquifer used for public water supply [17] to create a hydraulic barrier against salt-water encroachment.

In areas with water shortages, or where for environmental reasons other methods of discharge are not practical, it may be desirable to inject highly treated wastewaters into the subsurface for purposes of ground water storage. Wells designed for recharge purposes will see increased use as a tool for capturing and storing highly treated wastewaters such as would come from drain tiles underlying industrial or municipal land spreading disposal sites. This method should be analyzed considering engineering feasibility and health effects.

B. Injection Wells as a Part of the Treatment System

The use of injection wells for subsurface treatment of the wastewaters deserves greater consideration. The emplacement of acid wastewaters in limestone, marble, dolomite, or other carbonate formations where the acid will be neutralized through chemical reaction with the rock formations is generally more environmentally sound than surface treatment with a combination of strip mining of carbonate rocks, plus surface discharge of neutralized wastewater, plus land filling of the solids precipitated during neutralization [22, 23]. A key requirement for successful well injection is the determination that the volume of solids that may be precipitated is substantially less than the volume of rock dissolved during neutralization. An example of this operation is Kaiser's disposal well system at Mulberry, Florida, where about 150 gal/min (9.4 l/s) of fluosilicic acid (pH less than 1.0) and sodium chloride (6.5%) is being injected into a porous dolomite (Lawson Formation of Upper Cretaceous age) at a depth of about 4500 ft (1372 m). The ratio of volume of dolomite dissolved to volume of precipitant formed is about 11:7. The permeability of the receiving aquifer is increasing with time, indicating that the system is operating as planned and that the precipitant is not plugging the formation pore space.

Where odors may be a problem, where the rate of treatment is slow, or where economics are favorable, the subsurface can be used for chemical treatment and for biological treatment by anaerobic and facultative microorganisms. One of the world's largest nylon plants, Monsanto's Plant located near Pensacola, Florida, since 1963 has been disposing of its wastewaters containing nitric acid (pH 2.3) and other nitrogen compounds into the Ocala Limestone of the Eocene age. Back flushing experiments carried out by the US Geological Survey [23] in 1968 showed that the pH of the injected fluid increases rapidly in the aquifer accompanied by rapid denitrification and generation of carbon dioxide, nitrogen, methane, and other gases. Nitrate concentrations decreased from 3000 to near 0 mg/L in less than 75 min.

Still another use of the subsurface for treatment is the storage of wastewaters containing radioactive minerals that have relatively short half-lives.

C. Storage of Municipal Wastewaters for Reuse

Under certain conditions, a double benefit can be realized by injecting a good quality sewage effluent into a saline aquifer: potentially harmful viruses and bacteria that might survive the treatment process are removed from the human environment, and the effluent displaces a poorer quality (salty) groundwater, thus creating a reserve of potentially usable water in underground storage.

Several deep injection wells have been constructed in Florida, Hawaii, Louisiana, Illinois, and Texas for storage of secondary-treated sewage effluent into salt-water aquifers. Secondary- and tertiary-treated municipal wastewaters are of such good quality and in such large volumes (5–50 MGD) (220–2200 L/s) that it is much too valuable to waste in areas where water shortages are forecast for the foreseeable future. The storage of treated wastewater for future reuse is receiving increased attention in long range management planning [2, 7–9, 12, 15, 17, 24]. Expansion of this method of reuse as a tool of long-range water quality and water resource managment is being encouraged by the US Environmental Protection Agency (EPA) [25] and many state regulatory agencies as long as measures are taken to protect the public health. The method is particularly adaptable and acceptable when the planned reuse is for agricultural or other nonpotable demands.

Esmail and Kinbler [26] in their investigation of the technical feasibility of storing fresh water in saline aquifers concluded that the rate of injection and the permeability of the receiving aquifer were the two principal factors that control the recovery of the stored fresh water. Recovery is inversely proportional to the aquifer's permeability and directly proportional to the rate of injection.

D. Storage of Industrial Wastewaters

In this era of rapidly changing economics, developing technology, increasing energy costs, and demands for reuse and recycling, what is today a waste product may tomorrow become a valuable byproduct [3]. At each plant serious consideration should be given to separating streams of wastewater that contain chemicals with a potential for future reuse. These reusable chemicals could then be injected in a manner whereby they can be recovered at a future date.

E. Disposal of Toxic Wastewaters

A few regulatory agencies prohibit the disposal of toxic waste underground. Others require the best available pretreatment before injection on the premise that the safety of the method is maximized by this approach. The US Public Health Service [27] considers that only concentrated toxic wastes that cannot otherwise be satisfactorily disposed of should be considered for deep-well injection. The EPA [13] considers such disposal to be temporary until new technology becomes available, enabling more assured environmental protection.

The injection of troublesome industrial wastewaters into subsurface formations via deep wells is a relatively simple low cost disposal procedure that has attracted the attention of many manufacturing companies, particularly of the refining and chemical industries. Of the 269 known injection wells in the United States in 1973 [3], 81% were for disposal purposes by manufacturing companies. The majority, 49% or 131 wells, were for chemical and allied products, and 19% or 51 wells, were for petroleum refining wastes.

Deep-well disposal of toxic wastes has been demonstrated to be technically feasible in many areas of the country [1–5, 27]. However, ill sited and improperly designed or constructed wells can result in serious pollution problems. The effects of subsurface injection and the fate of injected wastewaters should be adequately researched to ensure protection of the integrity of the subsurface environment.

F. Disposal of Radioactive Wastes

Radioactive contaminants created by nuclear fission and other means differ from the usual industrial plant waste in their ability to emit radiation. There is no known method for neutralizing radioactivity, but radioactive isotopes decay and thus lose their activity with time.

The Halliburton Company Inc. has developed (British Patent L 054740) an improved method for the disposal of a radioactive waste by mixing the waste with cement to form a slurry then injecting it through a well. A horizontal fracture is developed hydraulically and the slurry is injected into the opening and caused to harden in place. This method, which has been used successfully since about 1967 at Oak Ridge, Tenn., is the best currently available for disposal of low and intermediate level radioactive wastes. In using this method, a conventional injection well is drilled, generally to a depth less than 1000 ft, but until an impermeable formation is transversed. The non-permeable formation is perforated by, for example, gun perforations. A fracturing fluid, which may be the waste-cement slurry, is pumped into the well under sufficient pressure (greater than 1 psi/ft of depth) to exceed the formation breakdown (fracture) pressure. The formation will fracture or part, generally in a horizontal direction at this depth. The waste-cement slurry is injected into the fracture and the fracture is sealed by allowing the cement to harden.

The same procedure then may be repeated at other depths within the well. To cause the cement to harden, the slurry may need an absorption type clay such as attapulgite. At other times, an agent such as calcium lignosulfonate that reacts chemically with the cement and retards its setting time may need to be added.

Another little used but technically sound method for the disposal of radioactive wastes (patent pending) is the injection through a well of radioactive or toxic materials into strata of low permeability at depth of 1000 to 20,000 ft, or at such a depth that the wastes are removed from the biosphere. This method involves displacement of formation connate water with liquid waste. Rocks of low permeability that are impregnated with waste become permanent storage receptors if the differential pressures across the receptor strata are maintained at a minimum. They will be retained almost indefinitely as a film held by molecular adhesion on the wall of the interstites [32–34].

Other factors also play a role in the retention of fluids in formations of low permeability. For example, the greater the amount of total interstitial surface in a rock or unconsolidated material, the greater is its specific retention. As would be expected, it is found that, as the effective diameter of the material's grain decreases, the specific retention generally increases because the total exposed surface area increases with decreasing grain size.

Capillarity is important in the retention of fluids in any granular material such as sedimentary rocks. The openings between the granules are interconnected in all directions, with the result that capillary forces act out in all directions within such materials. The moisture film around particles is held so tightly that it strongly resists any forces tending to displace it. The degree of its resistance to movement is expressed by its capillary potential, which is a measure of the force required to move this moisture from the soil.

Ion exchange is an important geochemical process. Many ions in hazardous waste products may be removed from wastewaters by means of this process. Clay minerals exhibit a marked capacity for the exchange of cations; in fact, all clay formations possess some ion exchange capacity. Clay minerals exhibiting good ion exchange are: kaolinite, halloysite, montmorillonite, illite, vermiculite, chlorite, sepiolite, attapulgite, and polygorskite. Of these, the montmorillonites are noted for the highest cation-exchange capacity, and the kaolins are noted for the most rapid rate of exchange.

The cohesive property of fluids plays an important part in their retention or movement in porous rocks. Cohesion is the ability of a fluid, or other substance, to stick together and resist separation.

The adhesive properties of a rock also play an important part in the movement or retention of a fluid in a porous rock. Adhesion is a measure of a fluid's ability to stick to the surfaces of other materials, such as to rock material in a formation.

The salty water found in deeply buried sedimentary formations generally is ancient seawater called connate water. Understanding the openings or pore spaces of the rock materials that build up on the ocean floor during geologic time and contain the connate water is the key to understanding the value of these natural spaces as permanent receptors for storage of hazardous and radioactive wastes.

G. Disposal of Municipal and Industrial Sludges

One of the more recent developments is a method [38] of dewatering, compacting, and disposing of sludges and other solid wastes from municipal and industrial waste treatment plants using the elastic rebound properties of subsurface strata to compress and dewater the waste. Hydraulic pressure is used to compress the rock and create a large opening into which the sludge is placed.

Water separated from the sludge migrates through the receptor stratum radially away from the injection well, which is the point of greatest pressure. Immediately following injection, the volume of the opening in the receptor stratum is slightly less than, but directly proportional to, the volume of waste injected. But as dewatering takes place, the volume of the opening is reduced and sludge compaction begins. During compaction the volume reduction is proportional to the amount of its suspended solids content. for example, if the sludge contained 95% water and 5% solids then the compacted volume will be about one twentieth of the injected volume. Deformation properties such as elasticity and plasticity allow the overburden to absorb the increased thickness, except that if large volumes are emplaced at shallow depth a small but measurable rise in land surface would be expected.

The end product of this method is a series of very thin (0.001 to 0.01 in.) pancakes of sludge.

VII. Protection of Usable Aquifers

Any aquifer that contains water that is both economically practical and technologically feasible to use for domestic, agricultural, industrial, or other purposes should be protected. Such aquifers are vulnerable to contamination by either the injected fluids or fluids being displaced by the injected fluids migrating into the aquifer to be protected. A variety of measures described below can be used to assure that injection well systems will not contaminate protected aquifers.

To protect groundwater, the well must be constructed so as to prevent contamination by (a) keeping injected fluids within the injection well casing and within the intended injection zone, and (b) keeping formation fluids displaced by the injected fluids from migrating into a protected aquifer. There are six major pathways [41] by which fluids may move and contaminate aquifers. The following discussion describes each pathway and summarizes a way to prevent migration through that pathway.

A. Pathway 1: Migration of Fluids Through a Faulty Injection Well Casing

The casing of a well can serve a variety of purposes. It supports the well bore to prevent collapse of the geologic formations into the hole and consequent loss of the well, it serves as the conductor of injected flui ds from the land surface to the intended injection zone, and supports other components of the well. If a well casing is defective, injected fluids may escape through the defect and enter the protected aquifer (Fig. 7). Such migration can contaminate an underground source of drinking water.

To detect migration of fluids through leaks in the casing periodic tests of the casing's integrity should be made. Several types of casing inspection tests are commercially available. For example, the down hole

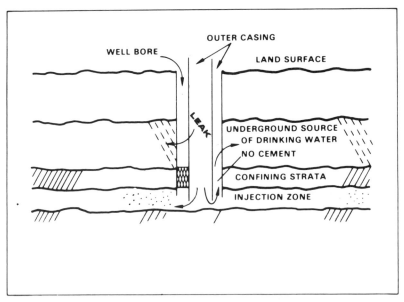

Fig. 7. Faulty well construction.

TV camera [49] can "see" what is wrong on the inside of the casing. The downhole casing inspection log called Vertilog [50] is a system for making a quantitative measurement of corrosive damage, indicating if the metal loss is internal or external, and if it is isolated or circumferential. Holes in the casing can be identified as well as casing separations.

Use of a separate tubing for injection affords protection to the casing and decreases the possibility of leakage. It isolates the casing of the well from injected fluids. By preventing this contact between casing and injected fluids, the possibility of migration of contaminants through leaks in the casing is greatly diminished. For the same reason, the use of tubing and packer also lessens the chances of corrosion of casing. By monitoring the annular space between the casing and tubing, leaks in the tubing can be detected and repaired before the casing becomes faulty. Tubing and packer offers two further advantages. It isolates the annulus (between the tubing and casing) from the injection zone, facilitating detection of any leaks in the tubing. It allows for visual inspection for deterioration of the tubing during routine maintenance. Finally, wells which inject corrosive fluids should be constructed of corrosion-resistant materials. This material is intended to prolong the operating life and continued viability of the well casing.

B. Pathway 2: Migration of Fluids Upward Through the Annulus Between the Casing and the Well Bore

A second pathway by which contaminants can enter protected aquifers is by migrating upward through the annulus between the drilled hole and the casing. Under usual injection conditions, injected fluids, upon leaving the well, enter a stratum in the injection zone that to some degree resists the entry of the fluids. This resistance results from friction and is inversely proportional to the size of the small openings in the stratum. Because fluids tend to take the path of least resistance, unless properly contained, they may travel upward through this annulus. If sufficient injection pressure exists, the fluids could migrate upward through the drilled hole into the overlying protected aquifer.

Leaks through holes in the well casing or upward fluid movement between the well's outer casing and the well bore, are illustrated in Fig. 7.

Casing should be cemented to isolate the aquifers to be protected from all underlying saline aquifers and from the injection zone. Generally two 100-ft thick cement plugs are installed. One is located immediately below the lowermost aquifer to be protected. The other is located immediately above the injection zone. The absence of leaks and fluid movement in the well bore should be confirmed periodically using geophysical logging techniques.

C. Pathway 3: Migration of Fluids from an Injection Zone Through the Confining Strata

The third way by which fluids can enter a protected aquifer is through leaks in the confining strata. Upon entry into an injection zone, fluids injected under pressure will normally travel away from the well and laterally through the receiving formation. In most cases, this occurrence is expected and gives rise to no concern, but, if the confining stratum that separates the injection zone from an overlying or underlying protected aquifer leaks significantly because it is either fractured or permeable, the injected fluids can migrate out of the receiving formation and into the protected aquifer.

For obvious reasons, there is no general well construction standard that can address this problem of migration of fluids through the confining strata.

Several steps should be taken to assure that fluids do not migrate through or around (Fig. 8) the confining strata. First, select a deep formation as a receptor. The deeper the receptor stratum selected for injection,

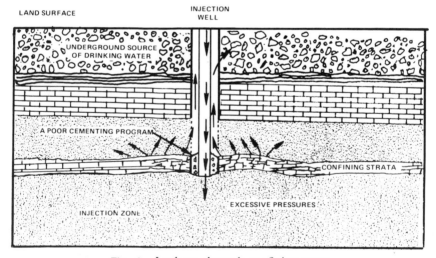

Fig. 8. Leakage through confining strata.

the greater the degree of protection. Second, place at least 200 feet of cement about the injection zone. The thicker the cement plug placed above the injection zone the larger the pathway of fluid movement before flow can enter the well bore above the plug. Third, study the confining and receptor strata. Care should be taken during drilling of the test hole to determine the permeability, thickness, and other information for the confining and injection strata, and the changes in fluid chemistry that can be expected as the fluids migrate through the confining bed. Fourth, determined fluid movement rates at various pressures. The leakage rate versus injection pressure should be evaluated prior to operation, and the injection pressure limited in order to avoid fluid movement through the confining strata into protected aquifers.

Frequently, when leakage out of the receptor stratum occurs, the adjacent aquifers (those leaked into) are permeable enough so that only limited vertical migration occurs. The equation for pressure buildup in the injection shell as a result of injecting into a zone with a leaky confining bed on one side is [42],

$$P_r = P_i + P_{DL}\left[\frac{141.2Q\mu B}{kH}\right] \quad (5)$$

where

$$P_{DL} = \text{a function of } (t_D, r/B) \quad (6)$$

$$T_D = \frac{6.33 \times 10^{-3}kT}{\phi\mu Cr^2} \quad (7)$$

$$B = \sqrt{kHh_c/k_c} \quad (8)$$

The equation for determination of pressures in the injection well where both confining beds leak is the same as given above except that:

$$B = \sqrt{\frac{kHh_c\ h'_c}{k'_ch_c + k_ch'_c}} \quad (9)$$

For examples of solutions for Eqs. (5–9), the reader is referred to the EPA publication 600/2-79-170[45].

D. Pathway 4: Vertical Migration of Fluids Through Improperly Abandoned or Improperly Completed Wells

Fluids from the pressurized area in the injection zone may be forced upward through nearby wells (Fig. 9) that penetrate the injection horizon

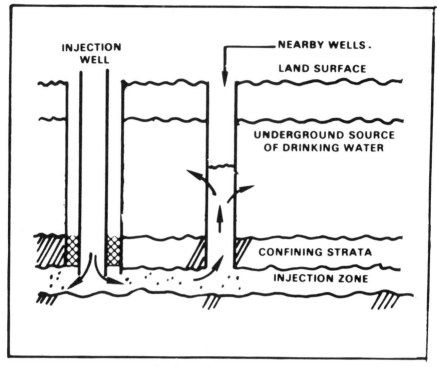

Fig. 9. Leakage through nearby wells.

within a zone around the injection well called the zone of endangering influence. The zone of endangering influence of a well may be defined as that radial distance from the well bore within which pressure increases because of injection are sufficient to cause a potential for upward migration into fresh water zones. As shown in Fig. 10, the zone of endangering influence includes all that area surrounding an injection well wherein the upward pressure in the injection zone exceeds the downward pressure of freshwater when measured using the base of fresh water as the datum. In areas where before injection the upward pressures in the injection formation already exceed the downward fresh water pressure, the zone of endangering influence is infinite or very large. In such areas an alternate fixed radius of 0.25 mi [43] was approved by EPA for the review of nearby wells.

Wells located within the area of pressure increase from an injection well should be examined to assure that they are properly completed and plugged. Corrective action should be taken where necessary to prevent

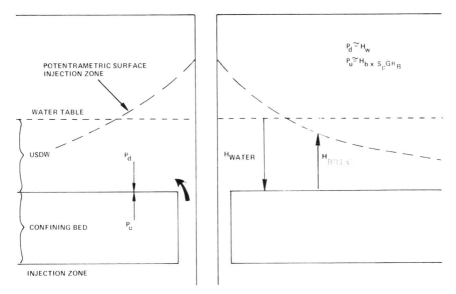

Fig. 10. Zone of endangering influence. Where $P_u > P_d$, then potential for endangerment exists.

fluids from migrating along these pathways into protected aquifers.

The key aspects influencing zone pressures during injection are: (a) the existing fluid pressures in the disposal aquifer, and (b) the pressure increases induced to effect waste fluid emplacement.

Several methods common to the practice of engineering hydrology and reservoir analysis are applicable to the solution of the zone of endangering influence (Fig. 10). These methods analyze the pressure differential that exists at the base of fresh water between increased reservoir pressures because of injection, and the pressure exerted downward by the fresh wat column.

Consider an example. Increased formation pressures cause a column of water to rise in an open hole to a level 100 ft above the base of fresh water. Thus at the base of fresh water, the formation fluid is exerting an upward pressure equal to 100 ft of hydrostatic head (of brine). However, if the fresh water aquifer contains, say, 200 ft of water, then a downward pressure of about 200 ft of head (of fresh water) is exerted at the base. Under these conditions if leakage were to occur the fresh water would "leak" downward into the brine. Only when upward pressure (P_u) is greater than the downward pressure (P_d) can there be the potential for upward movement.

The zone of endangering influence around an injection well encompasses all the area within which pressure increases due to injection are sufficient to create an upward differential pressure, measured at the base of fresh water.

This type analysis assumes "worst case" (i.e., open hole) conditions. An analysis considering friction losses through small channels, drilling mud column displacement, or seepage through beds would yield a much smaller "zone."

The pressure change in an injection zone at distance (r) caused by injection volume (Qt) may be described by the equation [47] for predicting pressure increases given below:

$$s = 162.6 \frac{Qu}{kH} \log \left(\frac{kt}{70.4\phi\mu Cr^2} \right) \qquad (10)$$

This pressure increase is additive to the existing formation pressure before injection began (P_2).

Step 1: Solve Equation (10) for any two values or r, and convert s to ft of hydrostatic head by dividing (psi) by the gradient per foot of the formation fluid (psi/ft). Add this value to P_2 (hydrostatic head of the injection zone) and plot the two values at their respective r on semilog paper. A straight line connecting the two points establishes the pressure surface of the disposal zone as it exists in space, measured in feet of head of formation brine above the top of the injection zone.

Similarly, some pressure (P_1) exists in the basal fresh water aquifer, corresponding to the weight of a column of water in a well fully penetrating that aquifer.

Step 2: Locate the stratigraphic position of the base of the lowermost fresh water aquifer on the diagram. Convert (ft of head of fresh water) to (ft of head as formation water) measured from the base of the fresh water aquifer. Draw a horizontal line to denote the pressure surface of the fresh water as it exists in space.

Step 3: The intersection (if any) of lines P_1 and ($s + P_2$) denotes the radius (r) of the "zone of endangering influence." That is, to the left of the intersection the pressure in the disposal zone is sufficient to overcome the hydrostatic pressure at the base of fresh water, and an upward potential is realized.

This method of solving the "zone of endangering influence" of an injection well represents an extremely conservative viewpoint. Also, see example 4 in Section IX.

E. Pathway 5: Lateral Migration of Fluids from Within an Injection Zone into a Protected Portion of that Strata

In most cases, the injection zone of a particular well will be physically segregated from underground sources of drinking water by impermeable material. In some instances, however, wells inject into an unprotected portion of an aquifer that is hydraulically interconnected with a protected aquifer (Fig. 11). In this event, there may be no impermeable layer or other barrier to prevent contact between contaminated fluids and underground drinking water.

Injection into unprotected portions of aquifers that contain drinking water in other areas must be done with great care. This type of injection can work if the predominant flow of the aquifer is such that injected fluids will tend to move away from, rather than toward, the protected part of the aquifer.

It is sometimes helpful to define the actual position of the waste front and its movement with time.

The minimum distance the waste will have traveled during injection may be described [46] by

$$r = \frac{v}{\pi H \phi} \qquad (11)$$

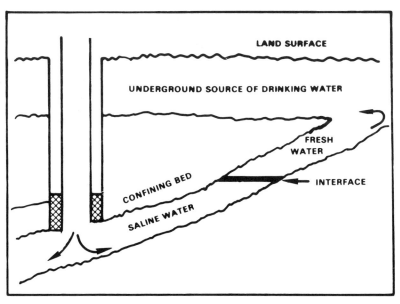

Fig. 11. Hydraulically interconnected aquifers.

A typical solution is given in Example 5 of Section IX.

In most practical situations, the minimum radial distance of travel will be exceeded because of dispersion, density segregation, and channeling through higher permeability zones.

An estimate of the influence of dispersion can be made with the equation

$$\gamma_1 = r + 2.3 \sqrt{Dr} \qquad (12)$$

This equation is obtained by solving Eq. (10.6.5) of Bear [48] for the radial distance at which the injection front has a chemical concentration of 0.2%. A dispersion coefficient of three represents a reasonable value for a sandstone aquifer (*see* Example 6 in Section IX).

It may be impossible to accurately predict the chemical character of the plume of waste 100 yr after it has slowly moved a few hundred feet in contact with subsurface minerals. However, it is important to list some of the chemical and biological reactions that will certainly occur to degrade the waste. The rates at which these reactions will occur is only speculative, however. These reactions include:

1. Dilution and dispersion
2. Biological degradation of organic compounds
3. Biochemical degradation of some species, such as nitrate, sulfate, and so on
4. Adsorption and ion exchange reactions with clay particles
5. pH neutralization
6. Precipitation and immobilization of some constituents

It is much more important to attempt to predict the direction and ultimate location of the waste front. As was pointed out in an earlier section, the disposal reservoir is confined above and below by a number of relatively impermeable, regionally persistent clay formations. Confined by these rocks, the disposal reservoir dips and thickens toward the Gulf of Mexico. In other words, the further the waste moves coastward, the deeper and more separated from fresh water it becomes. Near the coastline, the formation exists at depths exceeding 10,000 ft. At this point, the angle of dip steepens radically, and the formation dips beneath the Gulf of Mexico to depths exceeding 20,000 feet.

Therefore, over a period of millions of years, confined above and below by clay barriers and separated from fresh water by thousands of feet of rocks, a gradually decomposed waste will move at a very slow rate and remain essentially isolated in the deep subsurface.

F. Pathway 6: Direct Injection of Fluids into or Above an Underground Source of Drinking Water

The last pathway of contamination of groundwater is also the most hazardous. Direct injection of contaminated fluids into or above underground sources of drinking water presents the most immediate risk to public health. Such direct injection causes an instantaneous degradation of groundwater (Fig. 12). The injected fluids do not benefit from natural treatment processes such as filtration and ion exchange.

Many shallow injection wells, pits, septic tanks, and other similar disposal systems are used to dispose contaminants above drinking water aquifers that need to be protected. The injected fluids then percolate downward into drinking water aquifers, as illustrated in Fig. 10. EPA (1980a) proposed that wells injecting hazardous wastes directly into drinking water aquifers be banned. Drilling of new wells is to be prohibited after 1982 (EPA, 1980b).

Conversely, similar wells that inject nonhazardous fluids can be beneficial. The EPA is to conduct an assessment of this type of injection. Wells found to be causing health risks will then be phased out under this UIC program.

Casing should be installed through all aquifers to be protected and

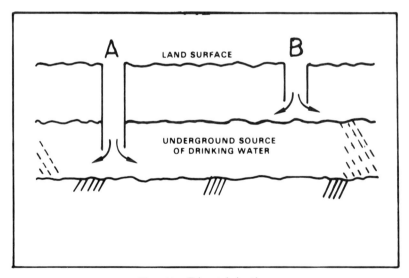

Fig. 12. Direct injection.

cemented to isolate them from exposure to fluids being injected and from all underlying saline aquifers penetrated by the well bore.

VIII. NOMENCLATURE

Symbols

A	=	area (ft^2)
B	=	leakage factor (*see* Equation 9)
β	=	formation volume factor = $\dfrac{\text{barrels of liquid at resevoir temperature and pressure}}{\text{barrels of liquid at standard temperature and pressure}}$
C	=	compressibility (psi)$^{-1}$
D	=	dispersion coefficient
H	=	reservoir thickness (ft)
h_c	=	thickness of confining layer (ft)
h_c'	=	thickness of second confining layer (ft)
I	=	hydraulic gradient (ft/ft)
k	=	average permeability (md)
k_c	=	vertical permeability of confining layer (md)
k_c'	=	vertical permeability of second confining layer (md)
L	=	leakage factor for semiconfined aquifer = $\sqrt{khh_c/k_c}$
P	=	coefficient of permeability (gal/d/ft^2)
P_{DL}	=	dimensionless pressure for semiconfined reservoirs
P_i	=	initial formation pressure (feet of water or psi)
P_1	=	hydrostatic pressure in the base of fresh water (feet of water or psi)
P_2	=	hydrostatic pressure in the injection zone (feet of water or psi)
P_r	=	reservoir pressure at radius r (feet of water or psi)
Q	=	flow or injection rate (gallons per minute or barrels per day)
r	=	radial distance from well bore (ft)
s	=	change in pressure (feet of water or psi)
S	=	coefficient of storage
t	=	time (d)
t_D	=	dimensionless time
T	=	coefficient of transmissibility (gal/d/ft)
u	=	1.87 $r^2 S/Tt$
μ	=	viscosity
V	=	$Q\,t$ = cumulative volume of waste injected (ft^3)

v = fluid velocity (ft/d)
V_1 = radial distance of dispersed travel
$W(u)$ = well function of u given in Table 1
Φ = porosity expressed as a decimal
π = 3.142

Abbreviations

BPD = barrels per day
CP = centipoise
ft = feet
gal = gallons
gpd = gallons per day
gpm = gallons per minute
md = millidarcy (1/1000 darcy)
mg = milligram
MSL = mean sea level
psi = pounds per square inch
STB = stock tank barrels, or barrels of liquid at standard temperature and pressure

GLOSSARY

Casing means a heavy pipe placed in a borehole to prevent the sides from caving, to prevent fluid loss into porous formations or to prevent entry of fluids into the well.

Confining layer means a body of impermeable material stratigraphically adjacent to one or more aquifers.

Receptor aquifer means a permeable geologic formation, group of formations or part of a formation receiving fluids through an injection well.

Safe Drinking Water Act means PL93-523 as amended by PL95-190, 42 U.S.C. 300 (f) et. seq.

Storage, coefficient of means the volume of fluid released from storage in each vertical column of the aquifer having a base of one foot square when the water surface declines one foot.

Transmissibility, coefficient of means the rate of flow of water, in gallons per day, at the prevailing water temperature through each vertical strip of the aquifer one foot wide having a height equal to the thickness of the aquifer and under a unit hydraulic gradient.

IX. Practical Examples

Example 1

The rate of leakage through a confining bed can be determined from

$$Q = PIA$$

where Q is the rate of leakage in ft³/d.

If one assumed that the permeability (P) of a typical clay confining bed averaged 0.002 gal/d/ft² (93 × 10⁻¹² L/s/cm² or 0.00027 ft³/d/ft²), that the difference in hydraulic pressure across the confining bed is 50 ft (15.24 m), and that the confining bed is 50 ft thick (hydraulic gradient [I] of 1 ft/ft), then the leakage (Q) calculated through the confining bed for a circular area (A) within 1000 ft (305 m) radius of the injection well would be:

$P = 0.00027$ ft³/d/ft²
$I = 50$ ft/50 ft $= 1$
$A = \pi(1000)^2$
$Q = (0.00027)(1)(3.142)(1000)(1000) = 848$ ft³/d

Example 2

The average velocity of fluid movement through a confining bed can be determined from:

$$V = PI/\Phi \tag{2}$$

where $P = 0.00027$ ft³/d/ft²
$I = 1$ ft/ft
$\Phi = 0.25$
Then $V = 0.00027 \times 1/0.25 = 0.001$ ft/d

At a velocity of 0.001 ft/d, the waste front would move through a 50 ft thick confining bed in 50,000 d (137 yr).

Example 3

Assume that the fluid pressure in an observation well located 500 ft (152 m) from a well injecting wastewater at a rate of 500 gal/min (32 L/s) was recorded for about a week. Values of r^2/t are computed, then plotted against changes in pressure in the observation well for different values of t. These are superimposed upon a second graph sheet that values of $W(u)$

and (u) from Table 2 had previously been plotted on at the same scale. The coincident value of the match point at $u = 0.01$ and $W(u) = 1.0$, are $r^2/t = 2.8 \times 10^5$ and $s = 5.5$ ft (Fig. 4). Solving for T and S gives:

$$T = \frac{114.60 Q W(u)}{s} = \frac{114.6(500)(1.00)}{5.5} = 10{,}420 \text{ gal/d/ft}$$

$$S = \frac{uT}{1.87 r^2/t} = \frac{(0.01)(10{,}420)}{1.87(2.8 \times 10^5)} = 0.0002$$

Having determined the T and S coefficient, long-range pressure changes with time for various distances can be forecast by rearranging the above formulas thus:

$$u = \frac{1.87 r^2 S}{Tt} \quad (13)$$

$$s = \frac{114.60 Q W(u)}{T} \quad (14)$$

For example, the pressure increase at a point of 5000 ft (1524 m) from an injection site after injecting 800 gal/min for 5 yr (1825 d) in the above described receptor would be:

$$u = \frac{1.87 r^2 S}{Tt} = \frac{(1.87)(5000)(5000)(0.0002)}{(10{,}420)(1825)} = 0.0005$$

$$s = \frac{114.6 Q W(u)}{T} = \frac{(114.6)(800)(7.02)}{10{,}420} = 62 \text{ ft}$$

Note: This method is not valid for $u > 0.02$.

Example 4

Consider the following example of how to determine that radial distance from the well bore within which pressure increases from injection are sufficient to cause an upward gradient, when considered in an open hole at the base of fresh water.

A. Specific gravity of formation water = 1.100 (0.480 psi/ft)
   (*see* Table 3)

Top of disposal zone = -3470 below mean sea level (MSL)
Observed water level = -250 below MSL
P_2 = $\overline{3220}$ ft of head (1546 psi)

TABLE 3
Density of Brines

Specific gravity of water at 60°F	Approximate total solids in parts per million, mg/L	Weight in psi of pressure per ft[a]
1.000	none	0.433
1.010	13,500	.437
1.020	27,500	.441
1.030	41,400	.445
1.040	55,400	.450
1.050	69,400	.454
1.060	83,700	.459
1.070	98,400	.463
1.080	113,200	.467
1.090	128,300	.471
1.100	143,500	.476
1.110	159,500	.480
1.120	175,800	.485
1.130	192,400	.489
1.140	210,000	.493

[a] Averages approximately 0.0043 psi/ft for each increase of 0.01 in density.

B. Specific gravity of fresh water = 1.0 (0.433 psi/ft)

Base of fresh water = −1200 below MSL
Observed water level = +40 above MSL
P_1 = 1240 ft of head (537 psi)

C. $P_r = P_2 + 162.6 \dfrac{Qu}{kH} \log\left(\dfrac{kt}{70.4\ \phi u C r^2}\right)$

where Q = 15,000 BPD
u = 1 cp
k = 1127 md
H = 300 ft
t = 7300 d (20 yr)
ϕ = 0.346
C = 6.5 × 10^{-6} psi^{-1}

DEEP-WELL DISPOSAL

Two examples for the solution of this formula for pressures at $\gamma = 10$ and $\gamma = 100$ feet in the receptor zone are given below:

Therefore,

$$P_{(r=10)} = 1398 + \frac{162.6(15{,}000)(1)}{1127(300)} \log\left(\frac{1127(7300)}{70.4(0.346)(1)(6.5 \times 10^{-6})(10)^2}\right)$$

$$P_{(r=10)} = 1398 + 63 \text{ psi} = 1461 \text{ psi}$$

$$P_{(r=100)} = 1398 + \frac{162.6(15{,}000)(1)}{1127(300)} \log\left(\frac{1127(7300)}{70.4(0.346)(1)(6.5 \times 10^{-6})(100)^2}\right)$$

$$P_{(r=100)} = 1398 + 49 = 1447 \text{ psi}$$

Example 5

The minimum distance the waste will have traveled during injection may be described [46] by

$$r = \sqrt{\frac{V}{\pi H \phi}} \tag{11}$$

Therefore for:
- $T = 5$ years
- $Q = 15{,}000$ BPD (437.5 gpm)(630,000 gpd)
- $V = 1.53 \times 10^8$ ft^3
- $\pi = 3.1417$
- $H = 300$ ft
- $\phi = 0.346$

$$r = \sqrt{\frac{1.53 \times 10^8}{326}} = 678 \text{ ft}$$

Similarly for $T = 20$ yr; $r = 1375$ ft

Example 6

In most practical situations, the minimum radial distance of travel will be exceeded because of dispersion, density segregation, and channeling through higher permeability zones.

An estimate of the influence of dispersion can be made with the equation

$$V_1 = r_1 + 2.3 \sqrt{r_1 D} \tag{12}$$

where (from Example 5)
$r_1 = 678$ ft
$D = 3$
$V_1 = 678 + 2.3 \sqrt{(3)(678)}$
$V_1 = 781$ feet

Similarly, where $r_1 = 1375$ ft,
$V_1 = 1522$ ft for the radial front of 0.2% concentrated waste.

REFERENCES

1. D. R. Rima, E. B. Chase, and B. M. Myers, "Subsurface Waste Disposal by Means of Wells—A Selected Annotated Bibliography," 1971, US Geological Survey Water-Supply Paper 2020.
2. The Environmental Protection Agency, "Compilation of Industrial and Municipal Injection Wells in the United States," October 1974.
3. D. L. Warner, "Survey of Industrial Waste Injection Wells, vols. I, II, and III," US Department of Commerce, AD756 642, June 1972.
4. E. C. Donaldson, Injection Wells and Operations Today, in *Underground Waste Management and Environmental Implications,* The American Association of Petroleum Geologists, Tulsa, Oklahoma, 1972.
5. D. L. Warner and D. H. Orcutt, Industrial Wastewater-Injection Wells in United States—Status of Use and Regulations, in *Underground, Waste Management and Artificial Recharge,* vol. 2, Amer. Assoc. Petro. Geol., Tulsa, Ok.
6. A. Sellinger and S. H. Aberbach, Artificial Recharge of Coastal-Plain Aquifer in Israel, in *Underground Waste Management and Artificial Recharge,* vol. 2, Amer. Assoc. Pet. Geol., Tulsa, Ok
7. A. Amramy, *Civil Eng.* **38** (5), 58, May 1968.
8. E. C. Donaldson and A. F. Bayayerd, "Re-Use and Subsurface Injection of Municipal Sewage Effluent-Two Case Histories," US Bureau of Mines Information Circular 8522, 1971.
9. The Environmental Protection Agency, "Ocean Outfalls and Other Methods of Treated Wastewater Disposal in Southeast Florida," Final Environmental Impact Statement, March 19, 1973.
10. The Congress of the United States of America, Public Law 92-500 known as the "Federal Water Pollution Control Act Amendments of 1972", 33 USC 1151.
11. The Environmental Protection Agency, "Environmental Impact Statement on Palm Beach County Florida Projects," 1973.
12. R. O. Bernon, "The Beneficial Uses of Zones of High Transmissivities in the Florida Subsurface for Water Storage and Waste Disposal," Florida Department of Natural Resources, Information Circular 70, 1970.
13. The Environmental Protection Agency, "Subsurface Emplacement of Fluids, Administrators Decision Statement #5," *Fed. Reg.* **39** (69), 12922–3 (Tuesday, April 9, 1974).
14. Geraghty and Miller, Inc., "The Ground Water Environment," 1973.
15. D. C. Signor, D. J. Grawitz, and W. Kamn, "Annotated Bibliography on Artificial

Recharge of Ground Water, 1955–67," US Geological Survey Water Supply Paper 1990, 1970.
16. G. Winqvist and K. Marelius, "The Design of Artificial Recharge Schemes," England Water Research Association, Vol. I, 1970.
17. J. Vocchioli, *J. New Engl. Water Works Assn.* **136,** (2), 87 (1972).
18. C. V. Theis, "The Relation Between the Lowering of the Piezometric Surface and the Rate and Duration of Discharge of a Well Using Ground Water Storage," *Trans. Am. Geophys. Union* **16,** 519 (1935).
19. L. K. Wenzel, "Methods for Determining Permeability of Water Bearing Materials with Special Reference to Discharging-Well Methods," US Geological Survey Water-Supply Paper 887, 1942.
20. C. B. Hutchinson and W. E. Wilson, "Evaluation of a Proposed Connector Well, Northeastern Deoots County, Florida," US Geological Survey Water Resources Investigations 5-74, 1974.
21. The Environmental Protection Agency, "Manual for Evaluating Public Drinking Water Supplies," 1974.
22. J. T. Barroclouth, "Waste Injection into a Deep Limestone in Northwestern Florida," in *Ground Water,* vol. 3, Number I, 1966.
23. D. Goolsby, "Geochemical Effects and Movements of Injected Industrial Waste in a Limestone Aquifer," presented at the AAPG-USGS Symposium on Underground Waste Management and Environmental Implications, Houston, Texas, 1971.
24. B. J. Schicht, Feasibility of Recharging Treated Sewage Effluent into a Deep Sandstone Aquifer, in *National Ground Water Symposium,* Environmental Protection Agency 16060GR08/71, 1971.
25. The Environmental Protection Agency, "Policy Statement on Water Reuse," July 7, 1972.
26. O. J. Esmail and O. K. Kinbler, *Water Resources Res.* **3** (3), p. 683. (1967).
27. D.L. Warner, "Deep-Well Injection in Liquid Waste," US Department of Health, Education, and Welfare, 1965.
28. M. King and D. G. Willis, *AAPG Memoir* **18,** 239, 1972.
29. Congress of the United States of America Public Law 93-523 known as the "Safe Drinking Water Act of 1974," 42 USC 300.
30. K. E. Davis and R. J. Funk, "Experience in Deep Well Disposal of Industrial Wastes," presented at the Water and Wastewater Equipment Manufacturers Association's Industrial Waste and Pollution Conference and Exposition, Detroit, Michigan, April 1–4, 1974.
31. R. P. Selm and B. T. Hulse, "Deep Well Disposal of Industrial Wastes" in *Eng. Bull. Purdue Univ.* **XLIV** (5) p. 566, 1959.
32. M. Muskat, *Flow of Homogeneous Fluids Through Porous Media,* McGraw-Hill, 1937.
33. L. J. Briggs and J. W. McLane, "The Moisture Equavilant of Soils," US Department of Agriculture, Soils Bull. 45, 1907.
34. L. A. Richards, *Physics* **1,** 318 (1931).
35. J. A. Liberman, *Civil Eng.* **25,** 44 (1955).
36. D. M. C. MacEwan, *Soc. Chin. France Bull.* **43,** 37–40 (1949).
37. D. M. C. MacEwan and Q. E. Grim, *Proc. California Div. Mines Bull.* **169,** 78 (1952).
38. C. W. Sever, "A New and Novel Method for Dewatering, Compacting and Disposing of Sludge," US Patent Office, (pending), 1983.

39. C. W. Sever, "An Improved Method for Storage of Radioactive Wastes," US Patent Office (pending), 1983.
40. D. K. Lowe and J. L. Huitt, "Propping Agent Transport in Horizontal Fractures," Society of Petroleum Engineering 40th Annual Am. Meeting, October, 1965, Paper No. 1285.
41. The Environmental Protection Agency, "A Guide to the Underground Injection Control Program," June, 1979.
42. M. S. Hantush and C. E. Jacob, *Trans. Am. Geophys. Union* **36,** 95 (1955).
43. The Environmental Protection Agency, "Part 146 Underground Injection Control Program Criteria and Standards," *Fed. Reg.* **45** (No. 123), 42500, (June 24, 1980).
44. C. S. Matthews and D. G. Russell, *Pressure Buildup and Flow Tests in Wells*, Society of Petroleum Engineers Monograph, Dallas, Tx. vol. 1, 1967, 178 pp.
45. D. L. Warner, L. F. Koederitz, A. D. Simon, and G. M. You, "Radius of Pressure Influence of Injection Wells," EPA 600/2-79-170, August 1979.
46. L. A. Browning, "Hydraulic Analysis of a Proposed Waste Injection Well, Adams County, Mississippi," EPA open files, Atlanta, Georgia, March 11, 1980.
47. L. A. Browning and C. W. Sever, "Simplified Graphic Analysis for Calculation of the 'Zone of Endangering Influence' of an Injection Well," EPA open files, Dallas, Texas, 1978.
48. J. Bear and M. Jacobs, "The Movement of Injected Water Bodies in Confined Aquifers," Underground Water Storage Study Report No. 13, Techpecon, Haifa, Israel, 1964.
49. R. Paul Coneway and Steve Coneway, "The T. V. Camera *See* What's Wrong", *The Johnson Drillers Journal,* Sept.-Oct., 1978.
50. John N. Haire and Jearald D. Haflin, "Vertilog—A Down-Hole Casing Inspection Service", 47th Annual, Calif. Regional Meeting of the Soc. of Petroleum Engineers of AIME, 1977.

6
Chemical Control of Pests and Vectors

Lenore S. Clesceri
*Department of Biology, Rensselaer Polytechnic Institute,
Troy, New York*

I. INTRODUCTION

Chemical poisons can be used in the control of pests and vectors. The term "poison" is very broad, referring to practically any chemical that can produce sickness or death if its concentration is high enough. Thus even nutrients can become poisons. For example, copper, which is highly toxic to all forms of plant life, is an essential element for plant growth at the trace level [1].

But chemical poisons, for the purpose of this chapter, will include only those agents intentionally used by humans for the control of undesirable organisms. Consequently, this discussion will omit the very large group of hazardous wastes generated by industry, government, agriculture, and various institutions such as hospitals and laboratories. Certain materials, however, fall into both categories, such as the insecticides used to control undesirable insects that are a waste from agriculture use as well.

II. PESTS AND POISONS

Among the plants and animals inhabiting the earth, there are very few taxonomic categories that do not contain some organisms capable of producing human sickness or of causing economic loss, or that simply may be classified as nuisances.

One finds bacteria, fungi, and arthropods responsible for food spoilage and economic losses through decay of natural products. There are bacteria and fungi pathogenic to humans, domestic plants, animals, and wildlife. There are pathogenic worms, as well as poisonous echinoderms, arthropods, and reptiles. One finds mollusks harboring poisonous toxins, trash fish, disease carrying arthropods, rodents and nuisance birds, and man-eating creatures of various types. There are taste-, odor-, and slime-producing microbes, allergenic plants, plants that are poisonous, and weeds occupying our lands and waters. Finally there are the viruses with no known usefulness for humanity.

To combat this array of undesirable organisms, an arsenal of chemical poisons has been developed that can control the various pests. Some of these agents are simple inorganic compounds whereas others are organic compounds of relatively complex structure.

It is generally agreed that materials used to control, destroy, or mitigate undesirable organisms (other than bacteria, rickettsia, and viruses) are called pesticides, whereas disinfectants, antibiotics, and virucidal agents are used to control the microbes. Yet in many cases pesticides indirectly control the microbe through control of the vector as, for example, in the battle against the tick that harbors the rickettsia known to induce Rocky Mountain Spotted fever.

Among the pesticides are chemical poisons that are rodenticides, molluscicides, nematocides, miticides, insecticides, fungicides, algicides, and herbicides (weed and brush killers, defoliants, and desiccants). Plant growth regulators and insect sex attractants are also normally considered among the pesticides.

Federal legislation in the United States strictly controls the use of pesticides through the Federal Environmental Pesticide Control Act of 1972. This act requires the classification of all pesticides according to whether they may be used generally or are restricted. A pesticide is classified as restricted if it can be hazardous to human health or the environment. These restricted pesticides can be applied only by people certified by the Environmental Protection Agency (EPA) as competent in using and handling pesticides.

EPA has developed a five volume registry of pesticides [2] containing uses of all registered products. The registry specifies dosage, mode of application, and limitations along with a classification for restricted or general usage. The registry does not recommend one pesticide over another and thus is most useful to persons with some prior knowledge. Chemical and generic names are used, but not trade names. For trade name information, one is referred to the *Pesticide Handbook* [3], which is an annual publication containing information on trade names, generic names, and the composition of commercial pesticide formulations. The series, *Advances in Pest Control Research* [4] is an excellent reference on the fundamental aspects of pest control.

Federal regulations also require US Department of Agriculture registration of any pesticide shipped in interstate commerce. The Department of Health and Human Services has established tolerance levels for residues resulting from pesticide treatment of agricultural products. Pesticides added to processed foods are considered additives and are subject to regulations of the Federal Food, Drug and Cosmetic Act.

Only registered pesticides can be marketed in the United States. Stringent registration requirements limit the proliferation of new pesticides since it has been estimated that an average of 4 million dollars and six years of development and testing are required to market a new pesticide under the regulation.

For any pesticide, one must check federal, state, and local regulations regarding its use in chemical poisoning programs. The chemical examples that are given in this chapter do not imply that current usage is permitted. This would be dangerous since the listing of chemicals that can be used in vector and pest control is changing rapidly as new information becomes available.

III. CONTROL OF ORGANISMS PATHOGENIC TO HUMANS

Some of the microorganisms are among the organisms that are a health problem. These organisms include certain bacteria, protozoa, fungi, and algae. Viruses and rickettsia are also in this group since methods used to control them are similar to those used for bacterial control. Recently, however, much concern has been expressed over the inadequacy of detection methods to quantitatively determine infectious viruses. Researchers recognize that water deemed safe by the coliform index may be virologically unsafe. Disinfection of water by chlorination is the most common choice for the control of viruses, rickettsia, and infectious bacteria such as the enterics and the various spirochetes responsible for leptospirosis. Chlorination also is often used to control other water-borne pathogens such as the protozoa *Endameba histolytica* as well.

Certain water-borne flatworms and round worms that infect humans are also controlled by disinfection techniques. More commonly, however, since foods are common vehicles of infection, proper cooking measures can control the spread of these parasitic worms. A summary of food and water-borne diseases, along with their etiological agents, prevention, and control is given by Salvato [5].

Microorganisms that produce respiratory diseases are usually spread via the air or by direct contact with an infected individual. These organisms are best controlled through personal cleanliness of the infected individual. Other measures include disinfecting aerosols and disinfection of surfaces upon which the pathogens fall.

Disinfection is likewise used to control fungi that produce various forms of skin disease (dermatophytosis). People most commonly contact skin diseases of fungal origin where conditions include moisture, warmth, and other people, such as swimming pools and shower rooms. Dermatophytosis is controlled at the source by measures capable of killing the fungal spores. The simplest of these is the drying of the infected area, which destroys both the vegetative fungi and the fungal spores. Chemical means normally used are either disinfection with sodium hypochlorite (500 mg/L available chlorine) or scrubbing the facility frequently with a hot solution of strong detergent followed by a rinse with a disinfectant [5].

Health hazards originating from algae arise via their endotoxins, some of which are highly poisonous to humans. Most noteworthy are shellfish contaminated by the algae *Gonyaulax*. The control of algae will be discussed later in this chapter.

IV. CHEMICAL CONTROL OF VECTOR ORGANISMS

A. Vertebrates

The rat is one of the most destructive vectors of disease and human parasites. Because of growth and proliferation, rats annually cause millions of dollars of damage to stored foods and materials in the United States alone.

Some of the diseases carried by the rat are bubonic plague, caused by the bacterium *Pasteurella pestis,* transmitted to humans by fleas, and leptospirosis, which is transmitted by direct bite. Leptospiral jaundice also is transmitted to humans and, more frequently, to the domestic dog.

Rat poisons are toxic against other rodents as well, such as mice, rabbits, and squirrels. Rats migrate in the absence of food, water, or shelter; they are also confined to a certain population size by existing conditions. It is obvious then that a rat poisoning program without accompanying cleanup procedures is a strictly temporary relief since survivors or newcomers will eventually multiply to again fill the vacant capacity of the previously infested environment.

For some purposes fumigation is the best method to exterminate rats. The most effective fumigant, in that it is the most rapidly and generally lethal of all highly volatile substances, is hydrogen cyanide. It is extremely lethal to all aerobic life since it acts by blocking respiratory metabolism. It is a classic example of a very potent, but essentially acute, poison in that there are no long-term, low-level accumulative effects.

Among the other fumigants that may be used are sulfur dioxide and methyl bromide.

More commonly, the placing of poisoned baits is employed in rat poisoning programs. Care must be taken that the poison selected is rapidly detoxified within the rat or that the poisoned animals are immediately removed to avoid transmission to other animals.

Anticoagulants such as Warfarin are the preferred rat poisons for most baiting purposes, but continuous feedings are required for 3–10 d, causing high labor costs for large programs. These compounds act by causing capillaries to break down, resulting in internal bleeding, and by interfering with normal blood coagulation.

Among the more rapid-acting rodenticides are sodium fluoroacetate, fluoracetamide, and α-naphthyl thiourea (antu). The fluoro compounds are highly toxic to other vertebrates as well as to rats and are banned in many localities.

Some other diseases transmitted to humans by vertebrate vectors are anthrax, caused by the bacterium *Bacillus anthracis* and carried by livestock; psittacosis, caused by a virus and carried by birds; and rabies also viral caused and carried by dogs, cats, livestock, and many wild animals.

B. Arthropods

More people have died from malaria than the combined deaths of World War I and II. The protozoan parasite is carried by mosquitoes of various species in the genus *Anopheles*. In addition, mosquitoes transmit dengue fever, encephalitis, filariasis, Rift Valley fever, and yellow fever among others of less importance. A summary of mosquito-borne diseases along with etiologic agent, reservoir, mode of transmission, incubation period, and control is given by Salvato [5]. A listing of some human diseases transmitted by arthropods is given in Table 1.

TABLE 1
Some Diseases Transmitted by Arthropods[a]

Disease	Causative agent	Vector	Reservoir
Cholera[b]	Bacteria, *Vibrio cholera*	Housefly, *Musca domestica*	Humans
Dengue[b]	Virus	Yellow fever mosquito, *Aedes aegypti*	Humans
Dysentery, amebic[b]	Protozoan, *Endamoeba histolytica*	Housefly, *M. domestica*	Humans
Dysentery, bacillary[b]	Bacteria, *Shigella dysenteriae* and other species	Housefly, *M. domestica*	Humans
Encephalitis (St. Louis, Western, and Eastern)	Virus	Mosquitoes, *Culex tarsalis*, and others	Birds and mammals
Filariasis[b]	Worm, *Wuchereria bancrofti* and possible *malayi*	Mosquitoes, *Culex, Aedes, Anopheles*, and *Mansonia*	Humans
Malaria	Protozoan, *Plasmodium vivax, P. falciparum, P. malariae*	Mosquitoes, *Anopheles*	Humans
Plague	Bacteria, *Pasteurella pestis*	Oriental rat flea, *Xenopsylla cheopis* and other fleas	Rats and wild animals
Tularemia	Bacteria, *Pasteurella tularensis*	Deerfly, *Chrysops*. Hard ticks, *Dermacentor*	Rabbits and other wild animals
Yellow fever[b]	Virus	Yellow fever mosquito, *Aedes aegypti*	Humans and monkeys

[a]Adapted from Pratt [6].
[b]Not found in the United States.

The insecticide qualities of DDT were observed in 1939. In 1942 the United States used the compound to control the massive malaria outbreaks sustained by the military. It became available commercially in 1945. Malaria has largely been controlled in most parts of the world through the use of DDT, which was the first synthetic organic insecticide of significance. Since then, vast numbers of synthetic organic insecticides have been developed. Many have been subsequently banned due to their effects on human health and the environment.

The insecticides can be most conveniently classified into organic and inorganic types.

1. Organic insecticides.
    a. Chlorinated hydrocarbons such as DDT, lindane, methoxychlor, and dieldrin.
    b. Organophosphorus insecticides such as parathion, malathion, and diazinon.
    c. Carbamate insecticides such as carbaryl (Sevin) and propoxur (Baygon).
    d. Natural substances such as nicotine and rotenone.
2. Inorganic insecticides.
    a. Arsenicals such as Paris green, lead arsenate, and arsenic trioxide.
    b. Sulfur and its compounds.
    c. Others such as compounds of fluorine and mercury, and hydrogen cyanide.

One can also classify insecticides according to usage.

1. Stomach poisons. There are poisons absorbed through the alimentary tract which are usually applied before an insect feeds. Most of the stomach poisons are inorganic chemicals and familiar examples are the arsenicals and sodium fluorosilicate.
2. Contact poisons. These are poisons which kill by direct contact with some part of the insect's body. (a) The contact may be made by direct application of sprays, dusts, or aerosols. An example of this type is parathion. (b) The insecticide may be applied to a surface on which the insect passes (residual treatment). The chlorinated hydrocarbon compounds are insecticides of this type. (c) The insecticide may be applied in a closed space as a fumigant. Among these are hydrogen cyanide, methyl bromide, and paradichlorobenzene.
3. Systemic poisons. These compounds are taken up by the plant or animal at the point of application whereby they are transported to another part of the body. Some of the organophosphorus insecticides are systemic poisons.

Most of the organic insecticides act as nerve poisons. They induce continuous nerve action through inhibition of cholinesterase, producing tremors, convulsions and, eventually, death. Since the biochemistry of pest insects is no different from that of beneficial insects (e.g., honeybee or preying mantis), and very similar to that of higher animals, including humans, these chemicals must be used very cautiously.

In contrast to the nerve poisoning role of nerve poisons, the chlorinated hydrocarbon insecticides interfere with the organisms production of enzymes resulting in a variety of abnormalitites such as the thinning of eggshells and subsequent unsuccessful development to maturity, and the elevation and depression of hormone levels producing erratic behavior.

Many arthropods besides the mosquito are disease vectors for humans. Houseflies carry disease microorganisms and parasitic worms. They spread such diseases as typhoid fever, paratyphoid fever, bacillary and amebic dysentery, cholera, typhus, anthrax, gastroenteritis, conjunctivitis, trachoma, and various viral diseases. Wood ticks and dog ticks infected with ricksettsia carry Rocky Mountain Spotted fever, Q fever, and typhus. They are also vectors for tuleremia caused by the bacterium *Pasteurella tularensis*.

The list of pesticides suitable for these vector arthropods is long. The nonpersistence of the organophosphorus and carbamate insecticides (1–12 wk) have made these the poisons of choice in most instances. And yet, when dealing with a disease problem, the use of more persistent insecticides such as the chlorinated hydrocarbons (2–20 yr has been justified [7]). Recommendations for the use of specific agents can be found in various sources [5, 7, 8, 9, 10], but interested persons should consult the EPA registry (2) for current regulations regarding the use of any of the pesticides since these are rapidly changing.

In mosquito control carbaryl has been used extensively for control of the adult, and malathion both for adult larvae in outdoors spraying. Dichlorvos (chlorinated organophorphorus insecticide) and malathion have been widely used in outdoor spraying for fly control.

V. CHEMICAL CONTROL OF ORGANISMS DESTRUCTIVE OR PATHOGENIC TO PLANTS

Because humans can chemically supplement the effects of climate, weather, and natural predators on plant pests, the ability to produce large agricultural yields has been greatly increased.

Of all plant parasites, fungi are the most numerous [11]. The taxonomic groups of fungi with the largest number of plant pathogens are the fungi imperfecti and the ascomycetes. These organisms infect plants by (1) killing tissue prior to their spread through the secretion of pectic enzymes producing soft rot and eventual death or (2) feeding directly from the living plant cell without necessarily killing the plant tissue.

Agricultural fungicides are applied to soil, seed, propagating material, and growing plants for combating fungal infections. The method of application and choice of fungicide is specific for the particular plant [10, 12]. Agriculturalists have used dithiocarbamate fungicides and inorganic sulfur compound preparations widely and effectively to control pathogenic fungi.

Substantial losses both during the growing season and in storage result from the action of insects on crops and foodstuffs. Early synthetic insecticides were inorganic (Paris green, silicofluoride, thiocyanates). The organic synthetic insecticide era has produced a formidable array of chemicals with insecticidal (and often homicidal) properties. Much of the search has been directed toward better insect selectivity, safety to higher animals, and desirable levels of persistence.

The problem of insect resistance has stimulated the development of new insecticides. Nearly all of the established insecticides have induced resistance in some insects. The mechanism of insect resistance may be morphological, behavioral, or biochemical. The selective advantage of some adaptation in structure, behavior, or metabolism in members of a population produces resistance in certain insects. Very commonly the successful mutant possesses an enzyme mechanism for detoxifying an organochemical poison, thereby necessitating the use of a different class of compounds. This problem of resistance does not exist as much with the inorganic pesticides since the possibility for chemical modification by the organism is much less among the inorganic poisons.

A general rule worth observing is that two classes of organic pesticides may not be used simultaneously. Such use encourages the disastrous appearance of a double mutant resistant to both agents. The agents should be used sequentially instead, if needed. Thus, mutants resistant to the first agent will be killed by the second. Further mutation is unlikely to occur without a loss of the initial resistance.

The use of an insecticide like dichlorvos is not ecologically wise according to this rule. Dichlorvos contains the insecticidal moiety of both the organophosphorus and the organochlorine insecticides making it doubly lethal, but also conducive to the selection of a double mutant resistant to both classes of insecticides.

Unfortunately, all the pests that we wish to control are more adaptable than humans to a changing environmental chemistry. Thus, although pesticide resistance in insects is common, humans are unlikely to acquire a similar resistance.

The guidelines for control of insects that are destructive or pathogenic to specific plants are beyond the scope of this book and are found in numerous references [2, 4, 8, 10]. Other organisms that destroy plants that must be considered are the nematodes and molluscs. For the most part, nematodes, which are small, unsegmented worms, dwell in the soil and feed on plant roots. Some, however, invade bulbs, stems, or leaves. In so doing, they produce damage to the plant themselves and also open the way for further damage from fungi. Methyl isothiocyanate and formaldehyde are commonly used nematicides.

Slugs and snails are especially damaging in a wet spring. They damage many leaf and root vegetables and produce considerable economic loss. The most effective chemical against them has been a polymeric form of acetaldehyde called meta. Upon contact with meta, desiccation occurs from excessive loss of water from slime secretion.

VI. CHEMICAL CONTROL OF NUISANCE ORGANISMS

A. Terrestrial Plants

Attention also should be directed to a large category of miscellaneous terrestrial plants that compete with the plants man has chosen to cultivate. Chemical weedkillers or herbicides for the control of these plants are widely used in agriculture, for lawns, and along roadways and railways.

Herbicides are called pre-emergents when they are sprayed before the weed plant appears. The germinating seed absorbs the herbicide until a lethal concentration occurs within the young plant. Foliage application or post-emergent application is made directly on the plant.

Herbicides may be selective or nonselective, i.e., they may have a biochemical specificity that permits selective killing of weeds and sparing of cultivated plants, or they may be broad-spectrum nonselective herbicides for total kill programs.

The herbicides can be further classified on the basis of hormonal or nonhormonal action. The phenoxyacetic acids (2,4-dichlorophenoxyacetic acid and related compounds) are the oldest and most widely used of the hormonal herbicides [12]. They produce their effect by competing with the action of natural hormone, resulting in lethally abnormal growth. The excessive root thickening produced in treated plants is a typical effect of 2,4-dichlorophenoxyacetic acid (2,4-D) poisoning. The phenoxyacetic acids have been the herbicides of choice for broad leafed weed control in grass and cereal crops.

Among other hormone herbicides available, pichloram has been found to be extremely effective in killing trees [12] and is probably the most persistent and most active of the herbicides.

In contrast with most other pesticides, the hormonal herbicides have a very low toxicity to animals because of their specific effect on plant metabolism. This is also apparently true for the nonhormone substituted ureas and triazines that act by inhibiting the photolysis step in photosynthesis. Of this group atrazine and simazine have been widely used for massive weed killing programs, and are unusually nontoxic toward corn crops.

Among nuisance plants, one must mention again the fungi, but this time as agents that product tremendous damage to wood and wood products. In general, damage can be lessened through various treatments. Polymerizing chemicals applied to lumber reduce porosity and water adsorption and mimic the lignification process in nature. Heartwood, which is more widely lignified, is more resistant to fungal attack than sapwood. Lumber can also be treated by pressurized or brush application of various fungicides. Most preservatives are applied in an oil base to facilitate penetration and prevent warping. For outdoor purposes, creosote, which is a crude mixture of aromatic hydrocarbons, is extensively used. Pentachlorophenol and the copper naphthenates have also been widely used as wood preservatives.

B. Aquatic Plants

Aquatic plants are valuable as food and shelter for aquatic animals. Floating or emergent vegetation is rarely desirable, however, when the body of water is used for recreation. Excessive algal growths severely restrict the use of the water for drinking, cooling, or recreation. They compete with the aquatic fauna for oxygen during low light intensity and, upon death, products of their decomposition produce objectionable tastes and odors in the water.

Thus it is frequently desirable to chemically rid the water of its flora. It is only as a last resort that chemical control of aquatic weeds is employed, however, and then only after approval by the proper local, state, or federal agency. Weed killers such as 2,4-D, applied as pellets or granules, are effective for rooted plants, whereas spray application is effective for floating weeds. Copper sulfate should not be used for aquatic weeds since the concentration required to destroy the vegetation will most certainly kill any fish in the vicinity [5].

Copper sulfate is, however, commonly used for the control of algae. In general, the use of about 2.5 lb. of copper sulfate per million gallons of water at 2–4 wk intervals between April and October will prevent the growth of algae in the temperate zone. Salvato [5] gives copper sulfate dosages for specific organisms as well as dosages found to produce fish kills. He notes the range of sensitivities for various fish. For example, whereas 1.2 lb of $CuSO_4$/million gallons of water is toxic to trout, it takes 16.6 lb in that volume to produce a toxic effect in black bass.

$CuSO_4$ is usually applied in a grid pattern with parallel lines about 25 ft apart. Application techniques vary and one can spray a $CuSO_4$ solution (0.5–1%), blow the pulverized salt, or meter out crystals of the salt through porous containers towed behind a boat.

It should be pointed out that a chemical control program is best done with complete knowledge of the system being treated. The chemistry of the water and its physical characteristics will contribute to the success or failure of the program. For example, we know that $CuSO_4$ is rapidly inactivated in alkaline hard water by precipitation as the basic carbonate, and that fish kill can occur as readily from asphyxiation as from toxicity following a $CuSO_4$ treatment. Decomposition of algae with the concomitant consumption of oxygen may exhaust the oxygen supply producing such a massive fish kill, especially when the water temperature is high. Thus for an effective program, investigation is highly warranted before chemical application.

C. Arthropods

The mention of mosquitoes and houseflies both as nuisances and disease vectors is well-justified and certainly few will contest this double entry.

More than an annoyance, termites are highly destructive and impose serious restraints in building construction. Redwood and certain species of cedar are quite resistant to termites, whereas pine and fir are readily attacked under certain conditions. There are both above ground and below ground types of termites. Below ground termites are very sensitive to light and moisture and, consequently, protection is best achieved through lighting and ventilating subterranean areas.

Above ground termites may be of the damp wood or dry wood destroying type. The dry wood type, along with other nuisance insects (such as roaches, fleas, bedbugs, and ants), are quite susceptible to sorptive dusts such as silica gels. These materials remove the lipoid protective layer covering the insect's body and cause a rapid loss of water and death. Both types are controlled with pentachlorophenol, as well as a wide range of other insecticides, including lindane, chlordane, and dieldrin.

Ants can be a serious problem, but have the beneficial effect of controlling termites by entering their tunnels and attacking them. Where ants are to be controlled, chlordane has been most effective, but malathion or diazinon have also been useful.

Pediculosis or infestation with lice can be controlled by dusting with insecticidal powders. Lindane or malathion have been commonly the active ingredients in these powders.

Blackflies are blood-sucking insects that breed in flowing streams. Larvae are found attached to rocks and vegetation in the streams. Certain species in Mexico, South America, and Africa carry disease, and they occur as annoying pests in other parts of the world. DDT has been effective against them, as has methoxychlor.

Clothes moths and carpet beetles do extensive damage to woolen or wool containing materials. Chlordane, lindane, and dieldrin are effective but dangerous, especially inside the home. Preferable control is through exposure to sunlight and storage in tight containers with naphthalene, paradichlorobenzene, or camphor.

D. Vertebrates

The use of poisons to regulate pest populations of vertebrates is rarely popular. With the exception of the killing of rats, many people oppose chemical measures (or any measure) to control the higher animals. These animal pests, however, produce considerable damage and can wipe out whole crops of grains or fruits, interfere with planting by feeding on the sown seed, destroy trees through bark stripping, and damage lawns by burrowing.

Easiest to control are the burrowing animals. Gassing the burrows with hydrogen cyanide or phosphine is effective, as is baiting the burrows with poisoned earthworms.

Overpopulation of certain fish species that have no value for angling or fish production may result in a decision to completely remove all fish. This can be done with several fish toxicants such as rotenone, antimycin

A, and toxaphene. A certain amount of success results from partial and selective poisoning [14].

A chemical approach that seems acceptable is the use of repellents. The ultimate effect is to force the animal to alternative foods or to starve it. Populations become regulated according to the size of the food supply. In making part of the food supply unavailable through the use of repellents, competition for the available food becomes more intense, resulting in successful and unsuccessful competitors.

The effect of repellents is most often too transient for effective control. Since higher animals have a learning capacity, they quickly find out if a repellent is only distasteful and not harmful. Under such conditions, protection is short if the animal is hungry. For a critical period, however, such as seed germination, the use of repellents can be highly effective. Such an example is the coating of corn seed with liquid coal tar to repel crows.

Another chemical approach is the use of nonlethal narcotics, such as chloralose [9]. This can be used to dust grain, which will produce sleepiness in birds shortly after consumption. Selective control of pest birds can be made by dusting large seeds so that small song birds cannot eat them. The undesirable drugged birds can then be destroyed, allowing the desirable ones to recover in a safe place with no ill effect.

The major difficulty with chemical approaches for vertebrate pest control is that the enormous testing work necessary for the use of the chemical is usually not justified unless a large market exists. Testing with respect to effect on nontarget organisms, biodegradation, lethal dose, and so on is expensive and many of the vertebrate pest problems, although very serious, are also very limited in scope.

VII. POLLUTION FROM CHEMICAL POISONS

A. Zone of Influence

The effect of the chemical poison on target organisms, as well as on possible fringe organisms, must be examined upon selection of chemical poisoning as a control measure [15].

Tolerance figures are arrived at, for the most part, by studying the influence of the poisoning agent on small samples of organisms under laboratory conditions for relatively short durations. The objective of these studies is the evaluation of the survival or death of the organism as expressed in LC_{50} values. These are a statistical estimate of the concentration of the poison required to kill 50% of the population under study. Such values are usually reported for a given length of exposure. Sometimes this information is reported as survival time for a given percentage of organisms at a given concentration of lethal agent. The point is, however, that these data do not tell us much about survival rates in nature since the level of the chemical agent causing death depends on the organism's physiology and total environment. Variation in these factors modify the laboratory-based LC_{50} value. In addition, to kill or not to kill is only part of the role played by the poisonous agent. The effect on growth and development for nontarget and target organisms that survive is a role of longer duration and far greater complexity that remains to be unraveled for most chemical poisons.

For aquatic life, toxicity of an agent is commonly reported as a TLm96 value. These aquatic toxicity ratings are defined as the 96-h static or continuous flow concentration producing a toxic effect. For aquatic life, toxicity of an agent generally decreases with increasing salinity, turbidity, or organic content of the water. These effects appear to be the result of decreasing the concentration of the toxic agent either by decreasing its aqueous solubility or by adsorption to particulate matter. In general, the toxic effect of the pesticide is proportional to temperature, reflecting the increased metabolic activity of organisms to a certain maximum level as temperature is elevated.

A sampling of the comparative susceptibility of fish families to some insecticides is given in Table 2. A compendium of toxicity values for selected aquatic organisms to various pesticides is found in Liptak [17]. A more extensive listing is the National Institute of Occupational Safety and Health Registry [18].

Chemical poisons have a wide sphere of influence and thus must be used with caution and concern for the entire influenced zone. Obviously the rule must be to use the least toxic agent and the minimum quantity for the particular job. Consideration should be given to the most sensitive stage of the pest's development in order to permit this rule to be most effective. In addition, the state of the application equipment and weather conditions must be given consideration before commencing operation.

Although a discrete area may be slated for a spraying or other poisoning operation, the nature of the soil, rainfall, and presence of migratory animals, all contribute to the size of the zone of influence.

TABLE 2
Susceptibility of Fish Families to Insecticides by Static Test (TLm96)[a]

	Organochlorines, ppb			Organophosphorus compounds, ppb		Carbamates, ppm
	DDT	Lindane	Toxaphene	Methyl parathion	Malathion	Carbaryl
Salmonidae						
Rainbow trout	7	27	11	2.8	0.2	4.3
Brown trout	2	2	3	4.7	0.2	2.0
Coho salmon	4	41	8	5.3	0.1	0.8
Percidae						
Yellow perch	9	68	12	3.1	0.3	0.7
Centrarchidae						
Red-ear sunfish	5	83	12	5.2	0.2	11.2
Bluegill sunfish	8	68	18	5.7	0.1	6.8
Largemouth bass	2	32	2	5.2	0.3	6.4
Ictaluridae						
Channel catfish	16	44	13	5.7	9.0	15.8
Black bullhead	5	64	5	6.6	12.9	20.0
Cyprinidae						
Goldfish	12	131	14	9.0	10.7	13.2
Fathead minnow	19	87	14	8.9	8.6	14.6
Carp	10	90	4	7.1	6.6	5.3

[a]Adapted from Brown [16].

B. Biological Magnification

It must also be recognized that immediate influence does not tell the whole story since biological magnification can eliminate organisms that were not influenced by the initial contact with the agent. Thus, acceptable water concentrations of pesticides such as the organochlorine insecticides have been found to result in concentrations in fish-eating birds that far exceed levels acceptable for human consumption.

This magnification occurs when a substance is not excreted as rapidly as it is ingested. The concentrated dose is then passed on to its preda-

tor in the food chain, which in turn concentrates the substance further.
These are the effects that are so difficult to access since an actual lethal effect may not appear until several or many years after exposure to the poison.

C. Toxic Effects on Human Health

Although beyond the scope of this chapter, one must express concern about the possible human health problems that can arise from the application of chemical poisons to the environment in pest and vector control.

Toxic effects on human health may be acute, chronic, carcinogenic, mutagenic, or teratogenic. Acute effects imply that the effect is elicited rather quickly (within hours to a few days) following exposure. The acute effect may range from a mild headache and discomfort to death. Chronic effects occur following long exposure to a chemical. Chronic effects are exemplified by chemically induced liver cirrhosis and ulcers. Carcinogenic and mutagenic effects involve the alteration in the information content (DNA) of a cell from that seen in the normal growth of that cell. Teratogenic effects involve the toxic effects a chemical may have on the developing embryo.

Among the chemicals that humans handle, organophosphorus insecticides rank first among occupational poisonings. High risk groups include pesticide manufacturers and spray operators. Some estimated fatal dosages of various organophosphorus insecticides to humans are given in Table 3.

Apart from occupational hazards, it is difficult to prove that any chemical in the environment is a carcinogen at environmental concentrations. There are also little human data available on the mutagenic or teratogenic effects of pesticides, although this is an area of intensive study.

TABLE 3
Organophosphate Toxicity to Humans

Compound	Estimated fatal adult dose, g
Malathion	60
Diazinon	25
Methylparathion	0.15
Parathion	0.02

VIII. ALTERNATIVES TO CHEMICAL POISONING

There is no general prescription that can apply to every situation involving the need for pest control. There are, however, some general rules that can be applied as outlined in a study of pest control strategies for the future made under the aegis of the National Research Council [19].

- Learn to tolerate a certain level of the pest since most pests are very adaptable to host, environment, and the control scheme, making them very difficult to eradicate completely.
- Take direct action against pests by suppressing reproduction.
- Enhance the destruction of pests by drawing them to natural predators or making them susceptible to parasites.

A. Prevention Programs

Although there are many alternatives to the use of chemical poisons for the control of pests, most of them do not provide the immediate relief produced by chemical control. They are, nonetheless, much more desirable as means for pest control. When used in a preventative program, the effect is a stable, controlled environment without the hidden costs of chemical poisoning. Many of the alternatives rest in the realm of biological control of one type or another. Some of these attempts have backfired as the result of insufficient study before employing them. Others, however, are tried and true measures. Some of these alternatives are:

1. Sanitation

Community and local sanitation to eliminate food, water, and harborage for vertebrate pests (rodents, pigeons, and so on). This is the simplest and most effective means for controlling such animals. A similar policy applies to the control of wildlife, although this is much more complicated since many food alternatives exist outside of the urban area. Thus, removing a single food source or even a class of food sources will in all probability exert some control over the particular wildlife species, but not to the extent that may be desired.

2. Proper Construction

Avoiding the contact of wooden structural members with the ground and elimination of buried wood will discourage the development of termite colonies. Tight construction prevents the use of the building as harborage for pests. Treatment of wooden members with preservatives extends the life of the structure.

3. Elimination of Breeding Places

This is the most desirable way of controlling most arthropod pests. The housefly thrives on any kind of waste (rotting vegetation, manure, sewage sludge, and so on). Proper sanitation, installation of garbage grinders in kitchen sinks, and use of fly traps are all measures that help in controlling flies without insecticides.

The elimination of standing water, controlled fluctuation of water levels to alternatively flood and dry breeding areas, and the stocking of ponds with top-feeding fish that consume larvae are effective means for combating the mosquito where it breeds.

Burning brush and long grasses that harbor mosquitoes, chiggers, and so on is another alternative to chemical poisoning.

4. Multiple Crops on Agricultural Land

Insect infestation that frequently occurs with monoculture is controlled by multiple plantings. Alternating rows of susceptible with nonsusceptible crops serves to confine the pests and prevent the widespread infestation that can occur when an unlimited supply of food is contiguous to the initial infection.

5. Natural Predators

The use of natural predators that consume developing or adult stages of pests is ideal if the predator itself does not become a pest. Ladybugs are among the most effective destroyers of a variety of plant eating insects. The preying mantis lives exclusively at the expense of other insects. Yellow jackets feed on soft-bodied insects. Adult dragonflies feed on mosquitoes, whereas their young, naiads, prey on the developing stages of the mosquito and other insects in the waters below.

6. Sterile Mating

Some success has been met in combating undesirable organisms by capturing, sterilizing (usually with radiation), and releasing sterile mates into the environment. The relative effectiveness of this program, of course, depends upon how many fruitful matings can be prevented.

7. Competition

Enhancement of the growth of a competitor is frequently possible. For example, one can replace weeds with a desirable ground cover by seeding among the weeds. The cutting of weeds, especially when seeded with grass, will accelerate succession to a more stable vegetation.

8. Use of Attractants

Many insects locate food, mating partners, and sites to lay eggs by being able to respond to chemical stimuli emanating from these sources. Traps permeated with these materials (pheromones) may physically restrain the insect by means of a sticky surface or by drawing it through a small orifice, by killing the insect outright by electric shock, or poisoning it with insecticides. Both natural and synthetic attractants are used. Light is also commonly used as an attractant for certain insects.

9. Physical Removal of Aquatic Weeds

Preferable to the use of chemical poisons is the physical removal of aquatic weeds. This is feasible for the rooted macrophytes through weed cutting procedures. Weed cutting and removal of the plants is better than removal by dredging, since it leaves the sediment less disturbed to continue its role in the decomposition of sedimented organic material.

10. Use of Hormones

Hormonal pesticides can be used to selectively disturb growth, development, and reproduction. When applied in excessive quantities or at abnormal times in a life cycle, they disrupt a wide range of body functions. Dormancy-breaking chemicals produce germination before a suitable environment is available, resulting in seedling death. In a like manner, insect diapause has been terminated prematurely through the use of juvenile hormones.

11. Cold, Heat, Dehydration, and Radiation

For certain pest insects and microorganisms the use of cold, heat, dehydration, or radiation are effective means of protection.

12. Use of Pathogenic Microorganisms

Effective control of pests through the use of specific pathogens has been demonstrated for a variety of insects. A particularly dramatic example is the decimation of the European pine sawfly that produced tremendous damage to pine plantations and nurseries in southeastern Canada and the northeastern United States in 1949. Infection of the insects with a virus imported from Sweden quickly brought the insects under control. Only two pathogens are federally registered for use in insect control programs. These are *Bacillus popilliae* and *Bacillus thuringiensis* for the control of Japanese beetles and various caterpillars, respectively. Several other pathogens await proof of safety and reliability and are currently being tested.

B. Conclusion

Within the limitations provided for by state and federal regulation chemical poisons can be used in pest and vector control. Coupled with preventative programs, chemical control of pests and vectors protects human health and the economy. The use of any control measure is bound to produce side effects as well as the particularly desired effects. Therefore, the impact of the side effects must be carefully considered by a knowledgable authority before commencing a control program. State Health Department, State Cooperative extension programs, the Office of Pesticide Programs of the US Environmental Protection Agency, US Department of Agriculture, universities and consulting firms can provide guidance to the selection of suitable control measures.

REFERENCES

1. J. Levitt, *Responses of Plants to Environmental Stresses*, Academic Press, New York, 1972, p. 542.
2. *Compendium of Registered Pesticides*, Vol. 1-5, Technical Services Division, Office of Pesticides Programs (updated continually).
3. D. E. H. Frear, *Pesticide Handbook*, College Science Publishers, State College, Pennsylvania (annual).
4. R. L. Metcalf, *Advances in Pest Control Research*, Interscience, New York, 1957–present.
5. J. A. Salvato, Jr., *Environmental Engineering and Sanitation*, 2nd Ed., Wiley-Interscience, New York, 1972.
6. H. D. Pratt, *Mosquitoes of Public Health Importance*, Center for Disease Control, US Public Health Science, Department of Health, Education and Welfare, 1960.
7. *Public Health Pesticides*, Center for Disease Control, US Public Health Service, Department of Health, Education, and Welfare, 1970.
8. E. R. de Ong, *Chemistry and Uses of Pesticides*, 2nd ed., Reinhold, New York, 1956.

9. G. S. Hartley and T. F. West, *Chemicals for Pest Control*, Pergamon, New York, 1969.
10. E. R. de Ong, *Chemical and Natural Control of Pests*, Reinhold, New York, 1960.
11. E. Moore-Landecker, *Fundamentals of the Fungi*, Prentice-Hall, Englewood Cliffs, New Jersey, 1972.
12. R. J. Lukens, *Chemistry of Fungicidal Action*, Springer-Verlag, New York, 1971, p. 14.
13. C. L. Hamner and H. B. Tukey, *Science* **100,** 154 (1944).
14. G. W. Bennett, *Management of Lakes and Ponds*, Van Nostrand Reinhold, New York, 1970.
15. D. Pimentel, *Ecological Effects of Pesticides on Non-target Species*, Executive Office of the President, Office of Science and Technology, June, 1971.
16. A. Brown, *Ecology of Pesticides*, Wiley, New York, 1978, p. 171.
17. F. L. Mayer, Jr., in *Environmental Engineers' Handbook*, B. G. Liptak, ed., Vol. I, Chilton, Radnor, Pennsylvania, 1974, pp. 408–410.
18. *Registry of Toxic Effects of Chemical Substances*, National Institute of Occupational Safety and Health, 1979.
19. G. L. McNew, in *Pest Control Strategies for the Future*, National Academy of Sciences, Washington, DC, 1972, pp. 119–153.

7
Management of Radioactive Wastes

Donald B. Aulenbach
Department of Chemical and Environmental Engineering

Robert M. Ryan
Division of Radiation and Nuclear Safety, Rensselaer Polytechnic Institute, Troy, New York

I. INTRODUCTION [1]

A. Historical

Radioactivity has always been with us. However, humankind was not aware of it until recently, primarily because there were no means to measure it. In 1896 Henri Becquerel first discovered that penetrating radiation was emitted spontaneously from a uranium compound. This phenomenon was given the name radioactivity by Pierre and Marie Curie. The Curies were successful in isolating polonium and radium from uranium ore in 1898. Within the next ten years, three distinct types of radiation were identified and named alpha, beta, and gamma radiation [2, 3].

Only a few months before Becquerel's discovery of radioactivity, Wilhelm Roentgen observed the emission of a penetrating radiation from a cathode ray tube. He named this X-radiation, standing for an unknown quantity, and X-rays are thus electromagnetic radiations similar to visible light, but with greater energy and penetrating power. They are similar to gamma rays emitted by radioactive materials. Alpha and beta radiation consists of electrically charged particles emitted from the nucleus of disintegrating radioactive atoms. The alpha particles are identical with helium atoms devoid of their orbital electrons. The beta particles are electrons.

One of the most interesting observations from Roentgen's discovery of the X-rays was their penetrating power, particularly the ability to penetrate human flesh and produce a shadow image of bones on a photographic plate. Continued use, however, revealed some harmful effects such as loss of hair and painful peculiar burns to the skin and flesh. Numerous early experimentors eventually developed malignant tumors and died of cancer.

Some doctors in the early 1900s, apparently thinking that the therapeutic agent in some of the natural mineral spring waters of the popular health spas of Europe was the radioactivity, prescribed radiation, from other sources both internal and external, for the treatment of various human aliments, particularly arthritis.

As the consequences of the excessive use of radiation were observed, precautions began to be taken to avoid any unnecessary exposure of humans to radioactivity. Guidelines were proposed to prevent exposure to harmful levels of radiation and radioactivity. As time passed and more information was obtained, these guidelines were revised to further prevent undue harm. Some of these guidelines are shown in Table 1. It may be seen that as information is obtained, the levels of radioactivity

TABLE 1
Suggested Maximum Permissible Radiation Levels

Title and Author	Effective year	Rems, rads, or roentgens per		
		Year	Week	Day
Safe intensity, Rollins	1902	—	—	10
Tolerance dose, Mutscheller	1925	50	—	0.2
US Advisory Committee on X-ray and Radium Protection	1936	25	—	0.1
ICRP (Handbook 52)	1950	15	—	—
NCRP (Handbook 69)	1961	5 (Age-18)	3/13 wk	—
US Code of Federal Regulations, Title 10, Chap. 1, Part 20	Current	5	1¼/13 wk	—

recommended as safe for humans have been continually reduced. Further, the most recent guidelines take into consideration the specific effects of various radiations on different parts of the body. These more complicated calculations have been developed in the interest of complete safety to the population.

B. Effects of Radioactivity on Matter [4]

Radiation may be divided into two broad classes: (1) electromagnetic radiation and (2) corpuscular or particulate radiation. Characterization of the electromagnetic radiation, which consists of photons, X-, or gamma-rays, is made on the basis of energy. The corpuscular radiation may be classified as: (1) heavy charged particles (alpha particles, deuterons, and photons and the recoil nuclei from which these come); (2) light charged particles (beta-particles and positrons); (3) neutrons, and (4) other subatomic particles. In addition to the type of radiation, other factors controlling the effects of radiation are the density and the atomic number of the material absorbing the radiation.

The mechanism of interaction between the radiations and the absorber is the transfer of energy from the radiations to the absorbing material. This is accompanied by ionization and excitation. When any of the orbital electrons are removed from an atom, the atom becomes ionized. The electron removed has a negative charge and the remaining portion of the atom has a positive charge. This pair of particles is termed an ion pair. These ions formed can react chemically with other matter, move in electric fields, recombine and emit light energy, or serve as condensation nuclei.

The interactions of radiation with matter can occur with both elastic or inelastic collisions. During elastic scattering, no ionization or excitation is produced in the atom with which the collision occurs. The usual elastic collision results in a deflection of the radiation with little effect on the much heavier absorber atom. Inelastic collisions result in a change in internal energy in one or both of the reactants, ionization or excitation, and a change in the sum of the kinetic energies of translation.

The number of ion pairs formed by ionizing radiation per unit of pathlength through which the radioactive material passes in the absorber is called the specific ionization. The energy loss per unit length of path, linear energy transfer (LET), is a direct function of the mass and charge, and an inverse function of the velocity of the radiation.

The heavy charged particles travel through the absorbing medium in nearly straight lines and the energy is lost almost entirely through inelastic collisions. The Coulomb field of the radiation interacting with that of the electrons of the reacting atoms results in ionization or excitation of the atom. In a few instances, nuclear transmutation and nuclear scattering may be additional methods of energy dissipation. Alpha particles have a high specific ionization primarily because of their high charge and large mass.

Electrons that originate from the nucleus of the atoms are normally called beta particles. Those that are of orbital origin or that are formed in pair production are more frequently called photoelectrons, recoil electrons, or secondary radiation. The interactions with matter are the same regardless of the source. The electron interacts with matter repeatedly until it is slowed down enough to combine with a positron, an ionized atom or an un-ionized atom. In doing this, the beta particle may follow a tortuous path in the absorber. Beta particles and positrons from the same source may have a spectrum of energies rather than a single energy such as normal alpha and gamma particles have. The electron may lose its energy by (1) resonance absorption; (2) collision with nuclei or electrons; (3) excitation of atoms; (4) radiation production and, (5) the electrodisintegration of nuclei. When passing through matter, beta particles will undergo inelastic collisions. These collisions may result in ionization if an electron is removed from the orbit of the absorber or the reacting atom is excited to a higher energy state. If an electron is removed, it becomes another beta radiation, frequently called a delta-ray. Three types of radiative interactions with beta particles include (1) bremsstrahlung, (2) annihilation and, (3) Cerenkov radiations. Bremsstrahlung or braking radiations are produced from a negative change in the velocity vector produced by interaction of a beta particle with the nucleus. Annihilation radiation results from the combination of an electron and positron that have lost most of their kinetic energy. Cerenkov radiation is produced if a particle is accelerated beyond the velocity of light in the medium. When the particles decelerate, energy is released as a proton in the visible blue light region.

X- and gamma-rays may dissipate their energy by (1) photoelectric effect, (2) Compton effect and, (3) pair production. However, for low energy photons and high Z number absorbers, Rayleigh scattering may also be significant. In the photoelectric effect, an electron is knocked out of orbit by absorbing all of the energy of the incident photon. Any excess energy is transmitted to the photoelectron as kinetic energy minus the electron's binding energy. In this process the absorbing atom is ionized and the photoelectron dissipates its energy like any other beta particle. In the Compton effect an electron may be knocked from orbit without utilizing all of the energy of the photon. This results in a Compton or recoil electron and a photon of lower energy than the incident photon. Photons with energy greater than 1.02 MeV may interact with an atomic nucleus and produce a pair of electrons, one being positively charged, called a positron. This is called pair production. Any energy above that required to form the pair is shared as kinetic energy of the two particles formed. Rayleigh scattering occurs when a photon interacts with an atom and is deflected through a small angle, but does not excite or ionize the atom, and essentially loses no energy.

Neutrons are intimately involved in nuclear reactions. They are normally classified as thermal (approximately 0.025 eV), slow (less than 100 eV), and fast (greater than 0.1 MeV). Some authors additionally break these down into epithermal (0.1–100 eV) and intermediate (100–100,000 eV). Very high energy neutrons (greater than approximately 20 MeV) may cause the literal explosion of an absorbing nucleus in a process called spallation or star formation. Normal high energy neutrons will not be absorbed by interaction with the nucleus of an absorbing material. Ultimately these neutrons are slowed down by elastic and inelastic collisions. Some resonance capture may take place during the slowing down to thermal neutrons. Here they can cause fission in ^{235}U, ^{239}U, and ^{233}U. This is the basis for the fission process on which nuclear reactors are designed.

C. Effects of Radioactivity on Humans [1]

The radiations considered in this chapter are called ionizing radiations because they are energetic enough to disrupt some of the atoms or molecules with which they come in contct into positively and negatively charged particles called ion pairs. The radiations under consideration include X-ray, alpha-, beta-, and gamma-radiation, and several other minor radiations. Their sources include not only anthropogenic radioactivity, but also some natural radiation, including cosmic radiation and terrestrial radiation from certain naturally occurring radioactive materials in the earth. Because of the extremely low levels of exposure to natural radiation (about 130 mrem/yr) for the "average" person in the United States, it is difficult to establish any positive detrimental effects on humankind from this amount of radiation. It is only when people are exposed to many times this amount of radiation that a positive correlation between radiation and the effects can be established. In many instances, high radiation doses are administered to patients for diagnostic or therapeutic medical reasons with full knowledge of the potential effects of the radiation; however, it is concluded that the benefits in this case may outweigh the possible harm, particularly to an older person. The onset of cancer, one of the effects of high levels of radiation, normally takes many years, and may not occur in the patient during the relatively short period of his or her remaining lifetime.

The effects of radiation on humans are considered at two different levels of dose and time. The first is that of the relatively large dose delivered to the human body over a period of seconds to hours. This is called an acute radiation exposure. These normally cause clinically observable symptoms called somatic effects. The acute radiation syndrome complex for various levels of whole body exposed is listed in Table 2 [1, 5]. It may be noted that without medical attention an acute whole body dose of about 300 rems would be expected to result in mortality in approximately 50% of the cases exposed within 30 d. With relation to human exposure, both

TABLE 2
Probable Effects of Acute Whole-Body Radiation Doses [1,5]

Acute dose, rem	Probable Clinical Effect
0–75	No effects apparent. Chromosome abberations and temporary depression in white blood cell levels found in some individuals.
75–200	Vomiting in 5–50% of exposed individuals within a few hours, with fatigue and loss of appetite. Moderate blood changes. Recovery within few weeks for most symptoms.
200–600	For doses of 300 rem or more, all exposed individuals will exhibit vomiting within 2 h or less. Severe blood changes, with hemorrhage and increased suseptibility to infection, particularly at higher doses. Loss of hair after 2 wk for doses over 300 rem. Recovery within 1 month to 1 yr for most individuals exposed at lower end of range; only 20% survive at upper end of range.
600–1000	Vomiting within 1 h, severe blood changes, hemorrhage, infection, and loss of hair. From 80 to 100% of exposed individuals will succumb within 2 mo; those who survive will be convalescent over a long period.

the severity and the frequency of appearance of biological effects are dose-dependent. Also the effects of a given exposure vary widely from individual to individual and it is impossible to predict precisely how any one individual will respond. The symptom complex is listed for the whole body dose. Higher doses delivered to more remote parts of the body such as the hands, feet, arm, or leg would not result in similar symptom complexes. Also, if the total dose is distributed over a longer period of time, the somatic effects are lessened. There appears to be an ability of the human body to repair damage caused by ionizing radiation. Part of this results from the ongoing process of repair and replacement occurring in all parts of the body, providing the damage imparted does not exceed the capacity of the repair mechanism.

Other effects that appear with relatively high dose rates that either do not cause any serious somatic effects or that may cause somatic effects that have been overcome are the so-called late effects or long-term effects of exposure to radiation. Many of these occur months or years after the exposure has taken place. These include cataract formation, leukemia, and cancer. The mechanisms for these late effects are not precisely or completely known, but could be caused by damaged cells that manage somehow to reproduce and then multiply the damage until there are sufficient numbers of defective cells to cause the clinical symptoms to appear.

The second range of radiation exposure of concern is low doses over an extended period of time, referred to as chronic exposure. During a normal life span, an "average" individual in the United States accumulates a total radiation dose of about 10–20 rems to the whole body from natural background radiation, routine medical X-rays, and the small amounts of anthropogenic radioactivity present in the environment. Persons working in the radiation industry are normally allowed to accumulate a lifetime exposure of about 200 rems. Since only a relatively small portion of the total population is engaged in the nuclear industry, this is considered one of the acceptable risks in working in this industry. It must be pointed out that this is not an average exposure rate, but a maximum recommended rate to any individual. In all cases, effort should be made to reduce the total exposure to as low as reasonably achievable (ALARA). The accumulation of radiation dose for occupational workers ranges from 0.2 to 5 rem/yr, which is considered a low rate. It is postulated that natural repair mechanisms can overcome, at least in part, any biological effects of these low radiation doses. However, there is some speculation that there may be no level of exposure below which there is certainty that there are no effects on humans. These speculations have been reached by extrapolating known research results far beyond the limits of the original measurements. Since there is presently little way of proving or disproving this point, the matter is subject to a great deal of controversy. In any event, the exposure of humans to radiation should be kept ALARA at all times.

The concept of possible effects at even the lowest levels of radiation is based upon the genetic effects or mutations that do not appear in the exposed individual, but rather in subsequent generations [6]. Some persons are of the opinion that the same may be true of certain somatic effects. The possibility of inducing mutations with various doses of ionizing radiation has been investigated in many organisms from bacteria to mice. Radiation doses as low as 500–1000 mrem/d administered continuously have given rise to observable genetic and somatic effects in small mammals. However, these radiation levels are many times higher than the average radiation dose added to the human population by the operation of nuclear facilities. Thus, we still have no positive conclusion about the possible genetic effects of continued exposure to extremely low levels of radioactivity.

It may be concluded that radiation does effect humans and therefore every effort should be made to limit exposure to ALARA.

D. Energy Relations

Demands for energy have been constantly increasing in both the United States and throughout the rest of the world. In particular, demands for electrical energy have been increasing at an even more rapid rate because of its desirability as a clean source of energy. Recent studies of the electrical energy demands for the United States indicate that they will most likely continue to increase at a rate of 3.6% per year [7]. (In 1976 the rate of increase was 5.5%.) This rate represents a projected doubling of electrical power production capacity approximately every 20 yr. Critics of this projected rate suggest that we are now becoming conservation minded and will reduce our electrical energy demands. However, most likely the conservation measures will merely result in temporarily cutting back demands to a minimum, after which they will rise again because of increased uses for electricity. Even at a more conservative estimate of increasing at 2.2%/yr because of conservation efforts and price increases, this results in a doubling time of about 30 yr. In any event it may be seen that demands for electrical energy will continue to rise in the future.

Other sources of energy presently available include natural gas, oil, and coal. It is almost assured that natural gas and oil will not play as significant a role in the future as these fossil sources have in the past [8]. Natural gas is presently in very short supply. Present predictions of oil reservoirs indicate a capacity for approximately 20 yr, with possible predictions of extending this to 30 yr based upon anticipated findings of new oil sources. At the present time, the United States imports in the order of 42% of its oil from foreign nations. This is very undesirable both from the standpoint of the balance of payments and from the undesirability of depending upon foreign nations for a large portion of our energy needs. Furthermore, oil has a much more important role in the transportation industry, and for petrochemicals, pharmaceuticals, and fertilizers. Thus, it is undesirable to utilize oil as a source of heat for generating electricity.

The United States has the world's largest reserves of coal, predicted to be sufficient to supply our demands for the next 300 yr. However, there are problems associated with the use of coal as a source of energy. One of the greatest problems is that of mining the coal or removing it from the ground. There are studies underway to utilize the energy from coal without even removing it from the ground. Removal today encompasses the concept of strip mining. Here the overburden is removed, exposing the coal, which is in turn removed by large digging devices. Provisions must be made for handling the overburden and returning it to the mined area after the mining operation has been completed. Additional coal may be located at greater depths within the earth, and it is not practical to remove these by conventional strip mining techniques. Furthermore, studies have shown that discharges of materials into the air from a 1000 MW(e) coal fired power plant may result in an estimated 50–60 fatalites per year. Furthermore, coal-induced air pollutants may cause discomfort to parts of the population susceptible to lung disorders such as emphysema.

Numerous other energy sources have been suggested to provide the electrical power we desire [9]. These alternatives include solar power, wind energy, geothermal power, combustion of organic matter including algae, terestrial plants, and trash, ocean thermal gradients, and osmotic pressure generation between fresh and salt waters. Each one of these alternatives has advantages and disadvantages. One of the greatest disadvantages of the majority of them is the fact that insufficient work has been conducted to determine the feasibility and costs of long-term use of these energy sources. Of these alternatives, solar, wind, and geothermal seem to have the greatest promise. Solar power is very desirable since it is clean and present in most locations at certain times. It is feasible for use in heating homes and can be used to supplement existing heating systems. For the production of electrical power, large areas of solar cells or disc-shaped collectors would be required to focus energy to produce electricity either directly or indirectly from the production of steam. An estimate has been made that about 36 mi^2 (93 km^2) of area would be needed for each 1000 MW(e) power plant. Wind has been promoted as being feasible. Unfortunately, little operating experience has been acquired with the few windmills that have already been constructed. Furthermore, in countries, particularly in Europe, where windmills have been in operation, they are not popular with the public. It is estimated that to generate approximately 1.0–1.5 MW(e) would require a windmill almost 300 ft (100 m) tall with blades 200 ft (70 m) in diameter. Therefore, to provide 1000 MW(e) electrical power, approximately 700–1000 windmills would be required. To replace the estimated 1985 nuclear capacity of 300,000 MW(e) with windmills would require in the order of 200,000 to 300,000 windmills. Geothermal power is being developed in a few remote areas in the world. This source, however, still suffers from problems of noise and air pollution because of the escaping of steam and other gases brought up with the steam.

The hydroelectric power generating sites in the United States have been developed to near capacity. Furthermore, there is great reluctance among environmentalists and people in general to further dam wild rivers to provide reservoirs for hydroelectric power generation.

Thus, it may be seen that for the near future, our commitment is to the utilization of nuclear fuel as an energy source for generating electricity. In the longer range, other alternative energy sources may surpass nuclear energy. Every effort to achieve this must be made.

The case for the utilization of uranium for generating electricity is based upon proven reliability, proven availability and proven safety. It was fortunate (or unfortunate?) that in the 1940–1950 era a decision was made to consider nuclear power as an energy alternative. Simultaneous to this decision for commercial power Admiral Hyman Rickover made the decision to build a nuclear navy and chose pressurized water reactors over sodium-cooled (Sea Wolf) reactors. Today we have a navy with nuclear submarines, crusiers, frigates, and aircraft carriers. The decision has been made to have nuclear power as a significant source of energy, and we must live with this decision until some alternate source of energy can be developed satisfactorily. The initial decision was made during the time of Truman's presidency and has been supported by all presidents since that time. Thus, the commitment to nuclear energy has been made and consideration must be made as to how to handle all problems from nuclear power generation and to minimize the hazard to the public, particularly from radioactive wastes.

Figure 1 [10] suggests a possible energy flow pattern for 1980. The

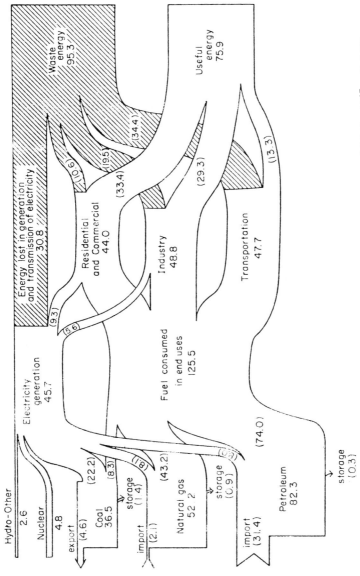

Fig. 1. Production and consumption of energy in the United States, 1975 (in 10^{17} cal) [41].

units are given in millions of barrels of oil per day equivalent. In order to convert to other sources the following factors may be used:

$$
\begin{aligned}
1 \text{ Barrel crude oil} &= 5,800,000 \text{ BTU} \\
1 \text{ ft}^3 \text{ natural gas} &= 1,000 \text{ BTU} \\
1 \text{ Ton coal} &= 26,000,000 \text{ BTU} \\
1 \text{ kW-h} &= 3,412 \text{ BTU}
\end{aligned}
$$

The energy flow pattern depicts (a) supply and demand; (b) form of use; (c) end uses; and (d) efficiency. This reference was based on a predicted nuclear (uranium) power production by 1980 of 300,000 MW(e). This level was not reached by 1980, but may be reached by approximately 2000.

To achieve energy independence, maintain a viable economy, and provide jobs for the future will require four efforts: (1) Conservation of all sources of energy must be practiced. This can reduce the rate of increase of energy demands, but it is not feasible to reduce it to zero. (2) Coal production must be increased. At the present the United States is depending upon its least abundant resources, oil and natural gas, for its prime sources of energy, principally because of the convenience of the use of these resources. The most abundant resource, coal, with reserves amounting to three times the energy contained in the Middle East oil reserves, supplies only 18% of our energy needs. (3) Alternate energy sources must be studied and developed. Hydropower, geothermal power, and the burning of trash, and even the wind can help in the short term (less than 10 yr), but in anticipated amounts of less than 10% of our total electrical energy needs. It appears that solar energy sources may become significant in the period beyond the next 10 yr. (4) Domestic resources of known uranium reserves indicate there is sufficient fuel for the projected 30 yr plant life of all the light water-cooled nuclear power plants predicted to be built through the late 2020s. This is on the order of 300 plants. If the next generation of nuclear power plant, the breeder reactor, is developed, there is almost an infinite energy supply from uranium as shown in Fig. 2 [11].

E. Magnitude of the Problem

Problems of radioactive wastes are not exclusively limited to the United States. Nuclear matters are international in character and regardless of the future United States approach to nuclear power, radioactive wastes will

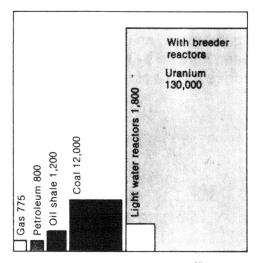

Fig. 2. US available energy in quads (1 quad=10^{10} BTU) shown graphically [source: A National Plan for Energy Research, Development & Demonstration, ERDA-48 (1975)].

be around for a long time. Even if the United States withdrew from nuclear power production in the near future, there would still be nuclear wastes around with their problems of major regulatory effort because of irreversible commitments to nuclear power already made by most of the developed countries in the world [12]. Even the oil-rich Middle East nations concur that light-water reactors offer the only plausible route towards energy independence. Iran concedes that oil is a wasting asset and is planning to build a nuclear capability in that country.

In Europe, 15 nations have joined an association called "Foratom" that is investigating the problems of nuclear waste disposal. Norway is the only European nation that will not have any nuclear power capacity by 1985. However, it does maintain a nuclear research center. Conversely, Portugal has no nuclear research center, but will be producing significant amounts of nuclear power by 1985.

The first European nation to install nuclear power was the United Kingdom, which now has 7500 MW(e) of nuclear power on line and expects to double that figure in the next 3 yr. The French Atomic Energy Commission has indicated that 35,000 MW(e) of nuclear power has been authorized by 1984, which will provide 20% of the total energy consumption of France and nearly 75% of the total electricity consumed by France in 1985. This capacity will require the reprocessing of 900 tons of reactor fuels in 1985 and 1650 tons in 1990. West Germany has a present capacity of 4000 MW(e) and projects 45,000 MW(e) by 1985. That growth, as in all other nuclear nations will produce a major waste disposal problem [12]. Some coordination of international problems in nuclear matters has been conducted by the International Atomic Energy Agency (IAEA), which has headquarters in Vienna. This organization already is involved in some aspects of the waste disposal problem and has a program aimed at standardizing the surface and underground storage of nuclear wastes.

In the United States, projected waste generation has been calculated assuming that the installed generating capacity will rise from 70,000 MW(e) in 1980 to 625,000 MW(e) by the year 2000. It is anticipated that most of the electricity will be generated by light-water reactors with a small amount from fast breeder reactors. Commercial liquid-metal-cooled fast breeder reactors are expected to be on stream by 1993 and will constitute 10% of the installed capacity by the year 2000. The year 1985 appears to be the time when the United States expects to begin development of the first national site for the permanent disposal of high-level radioactive wastes.

Various estimates of the amounts of radioactive wastes from the nuclear power industry have been made. Blomeke [13] made some calculations of the amounts of wastes anticipated on the basis of the installed nuclear electrical capacity rising from 6000 MW(e) in 1970 to 940,000 MW(e) in 2000. His predictions are summarized in Table 3. It was assumed that the high-level wastes from fuel reprocessing would be stored

TABLE 3
Summary of Fuel Cycle Wastes Projected for the Year 2000 [13]

	Accumulated		Number of annual shipments	Probable disposition
	Volume, 10^3 ft^3	Activity MCi		
High-level solidified	510	270,000 (β)	590	Salt (900 acres used)
Alpha solid	25,000	420 (Pu)	2500	Salt (50 million ft^3)
Cladding hulls	874	40 (Pu)	990	Salt (210 acres used)
IRL-alpha solid	5300	20 (Pu)	11,600	Salt (11 million ft^3)
Nonalpha solid	150,000	—	30,500	Burial (3000 acres used)
Tritiated water	32,000	3.7 (^{3}H)	3750	Deep wells
Noble gas	25[a]	190 (^{85}Kr)	170	Surface vaults

[a]Pressurized at 2200 psi.

as liquid for 5 yr, after which they would be solidified. On this basis there would be 26 million gallons of high level wastes in tank storage by the end of 2000 and 510,000 ft^3 of solidified high level wastes accumulated. About 1000 MW of thermal power and 270,000 MCi of beta activity will be associated with these high level wastes. In order to ship the solidified wastes to a federal repository there would have to be 600 shipments per year with about 14 vehicles being in transit at any time. Approximately 900 acres of burial area would be required for the accumulated wastes through the year 2000. Alpha wastes containing plutonium will probably also be shipped to a salt mine repository. By 2000, a total estimated annual generation of 2.5 million ft^3 of alpha wastes is predicted. This would be compacted by a factor of 10. By the year 2000 25 million ft^3 of this waste containing more than 50 tons of actinide elements will have been accumulated. Approximately 900,000 ft^3 of compacted cladding hulls will have been generated by the year 2000. These hulls will contain more than 3 tons of plutonium and will require 1000 shipments per year utilizing 200 acres of repository by the year 2000. It is estimated that approximately 2 ft^3 and approximately 9 ft^3 of compacted hulls will be generated per ton of LWR and LMFBR fuel reprocessed, respectively. Over 5 million ft^3 of IRL-alpha solid wastes containing 1600 kg of plutonium will have been accumulated by the year 2000. The volume of compacted non-alpha solid wastes is expected to approach 15 million ft^3 per year by 2000, requiring more than 30,000 truck shipments per year to burial sites that would require about 300 acres of land per year. Each 1000 MW(e) power plant generates from 2000 to 4000 ft^3 of non-alpha solid wastes per year. For noble gases it is assumed that krypton and xenon would be removed together from the off-gas streams at nuclear power plants and reprocessing plants, compressed in standard gas cylinders, and shipped to a repository for long term containment. By the year 2000, we should have accumulated almost 14,000 such cylinders containing 1500 MCi of ^{85}Kr. Assuming the internal recycling of all aqueous streams at power plants, a minimum of about 40,000 gal/yr of excess water containing about 5 MCi/mL of tritium will be generated per 1000 MW(e) LWR. The tritium in this waste will approximate 4 MCi in the total volume of 270 mgal by the year 2000. If natural salt formations are used for disposal of all wastes containing known or detectable amounts of plutonium, over 1100 ac of salt mine area will be required for high-level and cladding wastes, and in addition, about 61 m ft^3 of space will have to be mined in salt for the alpha and IRL-alpha solid wastes. An additional 3000 acres of burial site for lower-level radioactive wastes would also be required.

Belter [14] indicated that between 1960 and 1963 the AEC carried out an interim land burial program where over 7 million ft^3 of solid radio active wastes were disposed of by land burial at Oak Ridge Tennessee and at the NRTS in Idaho. Between 1963 and 1970, the AEC buried approximately 1.6 million ft^3/yr of solid wastes, whereas commercial waste burial increased to about 0.75 million ft^3/yr up to 1970. The volume of alpha waste generated in AEC facilities over the next 30 yr is anticipated to decrease because of program changes at waste generating sites, improved technology and economics for scrap recovery, more efficient packaging and handling of solid wastes, and the cost of special handling and long-term storage.

Although disposal at sea is being discontinued, during 1967 off the coast of Portugal and 1969 off the coast of the United Kingdom, sea disposal accounted for 60,000 containers with about 30,000 Ci of beta/gamma activity at a water depth of about 5000 m. Japan has also carried out several extensive sea disposal operations off its coast.

An estimate was made [14] of the amount of tritium produced from operating nuclear power reactors in the United States during the period 1967 through 1970. BWRs have released between 3 and 50 Ci tritium/yr and PWRs have released 100–7400 Ci/yr, both with plants in the size of 200–500 MW(e). For projected plants in the 1000 MW(e) range, BWRs produced in the range of 100 Ci/yr and PWRs less than 1000 Ci tritium/yr. In addition it is estimated that approximately 1–2 × 10^3 ft^3 of solid waste materials will be generated per 1000 MW(e) nuclear power plant. Considering all of the low level wastes from all atomic energy applications, it was estimated that 6 million ft^3 of low level solid wastes will require disposal by 1980.

There are many other estimates of the amounts of radioactive wastes produced in the total nuclear power industry. In the United States, because of precautions and antigonisms, the number of nuclear power plants actually being completed is less than had been predicted in previous years. This is resulting primarily in only a delay in the ultimate completion of the projected nuclear power capacity. Regardless of the number of plants in operation and the estimated amounts of wastes generated for each plant, it is sufficient to say that large amounts of radioactive wastes exist today and will continue to be generated in the future.

II. SOURCES OF RADIOACTIVE WASTES

A. Nuclear Fuel Cycle

At the present time there are many pressures to expand the nuclear industry. These include demand for electricity, potential shortages of certain fossil fuels, and the effects of present methods of producing electricity upon health and the environment. The growth of the nuclear power industry will depend upon technology development, establishment of its health and environmental costs, and public acceptance.

The extent and the environmental impact of the total nuclear fuel cycle have been estimated for mid-1972 and projected for the estimated installed nuclear power generation capacity for 1979–1980 [15]. It was estimated that in mid-1972 approximately 10,000–12,000 MW(e) of nuclear electric generation capacity was in existence in the United States. The projected installed nuclear electric generation capacity for 1980 was 140,000–150,000 MW(e). The nuclear industry demands and the number of plants required for each process in the nuclear fuel cycle in 1972 and projected to 1980 are shown in Table 4. It may be seen that through 1980 there is an adequate number of uranium mines and mills, isotopic enrichment plants, and fuel fabrication plants; however, additional plants will be needed for UF_6 production and fuel reprocessing.

1. Mining

Mining is the process involving the removal of the relatively concentrated uranium ore ($\sim 2\%$) from the ground for further use in the nuclear fuel industry. Essentially all of the uranium produced is obtained from either open pits or underground mining. About one-half of the total ore is produced in 29 open pit mines. About 70% of the ore is mined in New Mexico and Wyoming, with an additional 17% in Colorado and Utah [15].

The mining of uranium has not been found to cause measurable increases in environmental radioactivity outside the immediate vicinity of the mines. The primary problem in underground mines is the radon-222 (^{222}Rn) and its short-lived daughters. Physical attachment of the daughters to airborne dust particles is probably the most significant problem. This is of primary concern to the miners themselves [16]. However, this is not unique for uranium mines, since radon occurs in most all deep mines [17]. Minor problems of transfer of radioactivity to the environment occur in the water that comes into contact with the mine. In underground mines, the water may be from underground streams that flow through the mines and emerge at some point to the surface water. In the open pit mines rain will fall into the mine and pass either through the ground or over the surface, ultimately reaching some source of water supply. In addition, some of the materials removed from the mine, whether the non-uranium materials in an underground mine or the overburden from the open pit mine, may contain low concentrations of uranium. Where these tailings are stored on the surface in the area of the mine,

TABLE 4 [15]
Total Nuclear Fuel Cycle Industry

Types of plants	1972			1980 (est.)	
	Plant size, thousands of MT	Annual requirement, thousands of MT	No. of average plants required to meet power demand	Annual requirement, thousands of MT	No. of average plants required to meet power demand
Uranium mines, ore	250–750	4500	10 (220)[a]	17,000	30–40
Uranium mills, U_3O_8	0.5–1.1	9	12 (20)[a]	34	40–45
UF_6 production, U	5–15	8	1+ (2)[a]	34[b]	3 low enriched plants and 3–4 natural plants
Isotopic enrichment, SWU	7.5	5	1 (3)[a]		3
Fuel fabrication, U	0.3–1.2	1.2	3+ (10)[a]	5.3	8–10
Fuel reprocessing, U	0.3–1.5	0.2	1 (2)[a]	3	3–4

[a]Number in parenthesis is number of plants actually available.
[b]Including recycle of recovered U from reprocessing operations. Nuclear power generation basis: 10–12,000 MWe 1972; 140–150,000 MW(e) 1980.

rainfall may fall upon these tailings, dissolving some of the radioactive materials and again transporting them to adjacent water courses.

A summary of the environmental considerations for an open pit uranium mine of a 4.8×10^5 MT/yr capacity is shown in Table 5.

2. Milling

In the milling operation the uranium is extracted from the ore and concentrated as a semirefined U_3O_8 product called yellowcake that is the feed material for the production of uranium hexafluoride (UF_6). Both mechanical and chemical processes are involved in the milling operation. The ore is first crushed and ground, after which it is treated with a sulfuric acid or a sodium carbonate leaching solution to extract the uranium. The leach liquors are purified and concentrated by ion exchange or salt extraction and the uranium is removed by chemical precipitation, with the solid product calcined, pulverized, and drummed for shipment as yellowcake [15].

After removal of the uranium from the ore, the material remaining is called the tailings. These are stored around the plant and any water that is involved in the process or that flows over the tailings, as, for example, rainwater, is collected in ponds. In general, because of the location of these milling operations, evaporation is responsible for removal of the majority of this water. In some instances, this water may infiltrate to the ground water or run off in surface water during periods of heavy rainfall. It is the goal of the milling operation to maintain all the liquids at these plant sites. The tailings may contain significant amounts of radium-226 (^{226}Ra). This is rather insoluble in water, but does dissolve slowly and enters the biosphere especially in water and aquatic biota. At one time, tailings were being incorporated into construction materials for public use, but this practice has now been discontinued.

A summary of the environmental considerations for a uranium mill processing 1000 MT/yr of U_3O_8 is shown in Table 6.

3. Chemical Purification

The U_3O_8 concentrate extracted from the ore must be converted to the volatile compound uranium hexafluoride (UF_6) for enrichment by the gaseous diffusion process. Two processes are used for the UF_6 production. The hydrofluor process consists of continuous successive reduction hydrofluorination and fluorination of the ore concentrates followed by fractional distillation of the crude UF_6 to obtain a pure product. The sec

TABLE 5
Summary of Environmental Considerations
for Open Pit Uranium Mine—4.8×10^5 MT/yr [15]

Natural resource use	
Land (acres)	
Temporarily committed	290
Permanently committed	10
Overburden moved (MT $\times 10^{-6}$)	14
Water (gallons $\times 10^{-6}$)	
Discharged to ground	650
Fossil Fuel	
Electrical energy (MW-h $\times 10^{-3}$)	1.3
Equivalent coal (MT $\times 10^{-3}$)	0.5
Effluents	
Chemical, MT	
Gases	19
SO_x	5
NO_x	0.05
Hydrocarbons	0.1
CO	

ond method employs a wet chemical solvent extraction step at the head end of the process to prepare a high purity uranium feed prior to the reduction hydrofluorination and fluorination step. Roughly equivalent quantities of UF_6 are produced by each method. The hydrofluoride process produces effluents primarily in the gaseous and solid states, whereas the wet solvent extraction releases mostly liquid effluents. A model UF_6 production plant has been assumed based upon half of its output by each process. The environmental considerations from a UF_6 production plant producing 5000 MTU/yr are shown in Table 7.

Several process off-gases are generated in the preparation of UF_6 from yellowcake. Most of these are combustion products. Some are volatilized solids and gases involved during calcining and fluorination. Several off-gas treatments are applied to minimize the airborne effluents released to the environment. Fluorine and oxides of nitrogen are more significant sources of potential adverse environmental impact [15].

There are two major liquid effluent streams associated with UF_6 production. The liquid effluent from the raffinate stream from the solvent extraction process is not released to the environment, but is held indefinitely

TABLE 6
Summary of Environmental Considerations
Uranium Mill—1000 MT/yr U_3O_8 [15]

Natural resource use	
Land (acres)	
Temporarily committed	50
Permanently committed (limited use)	250
Water, gal/yr × 10^{-6}	
Discharged to air	3600
Fossil Fuel	
Electrical energy, MW-h/yr × 10^{-3}	15
Equivalent coal, MT/yr × 10^{-3}	3
Natural gas, scf/yr × 10^{-6}	380
Effluents	
Chemical, MT/yr	
Gases	
SO_x	200
NO_x, 40% from natural gas use for process heat	87
Hydrocarbons	5
CO	1.7
Liquids, × 10^{-3}	
Tailing solutions	1300
Solids, × 10^{-3}	
Tailings	500
Radiological, Ci/yr	
Gases, including airborne particulates	
Rn-222	460
Ra-226	0.1
Th-230	0.1
U-natural	0.2
Liquids	
U and daughters	11
Solids	
U and daughters	6700
Thermal, BTU/yr/× 10^{-9}	390

TABLE 7
Summary of Environmental Considerations Uranium Hexafluoride Production Plant— 5000 MTU/yr [15]

Natural resource use	
Land (acres)	
Temporarily committed	1400
Permanently committed	10
Water, gal/yr × 10^{-6}	
Discharged to air	100
Discharged to water bodies	1100
Total	1200
Fossil Fuel	
Electrical energy, MW-h/yr × 10^{-3}	58
Equivalent coal, MT/yr × 10^{-3}	21
Natural gas, scf/yr × 10^{-6}	850
Effluents	
Chemical, MT/yr	
Gases	
SO_x	800
NO_x, 25% from natural gas for process heat	280
Hydrocarbons	2.2
CO	5.5
F	3.0
Solids	1000
Radiological, Ci/yr	
Gases	
Uranium	0.4
Liquids	
Ra-226	0.7
Th-230	7.4
Uranium	0.7
Solids (buried)	
Other high level	7.7
Thermal, BTU/yr × 10^{-9}	830

in a sealed pond. The second stream is made up of mostly cooling water and dilute scrubber solutions that represent the bulk of the water use.

4. Enrichment

Isotopic enrichment of the uranium is necessary to provide fuel for a light-water moderated nuclear reactor. The concentration of ^{235}U in natural uranium is about 0.7% and the enriched uranium content must be in the order of 2–4%. Enrichment is accomplished by the different rates of diffusion of the various isotopes of uranium in the gaseous state (UF_6). About 1700 concentration stages are needed to produce 4% ^{235}U enriched UF_6. There are presently three facilities operating in the United States to conduct this gaseous diffusion enrichment process. A summary of the environmental considerations for a uranium enrichment complex is shown in Table 8 [15]. Other methods are proposed such as centrifuge and laser separation, but none are operational in the United States at this time.

Large amounts of electrical energy are required in order to enrich the fuel. (An estimate has been made that a nuclear power station at 80% load factor produces about 22 times as much energy as is consumed to produce its annual fuel requirement [15].) The primary source of environmental impact associated with the enrichment of uranium is related to the gaseous effluent from the coal fired stations used at present to generate the required electric power. Waste gas emissions, including particulates, of approximately 6600 MT are associated with the production of the annual fuel requirement. This also results in the discharge of large amounts of heat to the environment. Small quantities of airborne fluorine are released from diffusion plants. Measurements in unrestricted areas indicate the concentrations are well below the range for which deleterious effects have been observed. In addition, oxides of nitrogen and sulfur are released at the diffusion plant. Conservative estimates of the off-site concentrations of these contaminants yield levels that are at or slightly below EPA standards. Furthermore, the total quantity of these effluents is insignificant in comparison with the combustion products generated by the supporting electric power plant.

5. Fabrication of Fuel Elements

For the fabrication of the fuel elements, the UF_6 which has been enriched in the ^{235}U isotope to 2–4% is converted to UO_2 that is then formed into pellets and sintered to achieve the desired density. The finished fuel pellets are loaded into zircaloy or stainless steel tubes fitted with tops that are welded in place. The completed fuel rods are then assembled as appropriate in the core of the reactor. The summary of environmental considerations for a 900 MTU/yr fabrication plant are summarized in Table 9 [15].

The most significant effluents from the standpoint of the potential environmental impact are chemical. Nearly all the airborne chemical effluents are from the combustion of the fossil fuels used to produce the electricity to operate the fabrication plant. The only significant airborne chemical effluent from the process is fluorine. The most significant chemical species in the liquid effluents are nitrogen compounds that are generated from the use of ammonium hydroxide for the production of UO_2 powder and from the use of nitric acid in the scrap recovery operation.

In general, fuel fabrication is carried out in such a manner as not to increase levels of radioactivity in the environment [16].

6. Reactor Operation

The two most common types of reactors in operation are the boiling water reactor (BWR) and the pressurized water reactor (PWR). From the standpoint of potential generation of radioactive wastes, the two types of reactors are similar. The major difference between the two systems is that in the BWR the water that is heated in the reactor core is the same steam that passes through the turbine generating the electricity. In a PWR there is a dual system with an intermediate heat exchanger and steam generator. The water that is heated in the reactor core is used to heat a secondary cycle of steam that in turn drives the turbine.

The total amount of radioactivity within an operating nuclear power plant depends upon the power level and the time it has been in operation. There is a buildup in radioactivity because of the fission products with an increase in the order of 1.7×10^{10} curies between refueling operations that occur about once a year. When the reactor is shut down the generation of radioactivity ceases and the level decreases. However, large amounts of heat are still produced by this residual radioactivity. The initial activity of radionuclides in the nuclear reactor core at the time of shutdown is shown in Table 10 [18]. This is the same amount of radioactivity that is considered potentially available for release in the event of a hypothetical accident. A typical radioactivity inventory for a 1000 MW(e) nuclear power reactor is shown in Table 11 [18].

In general there are two types of potential wastes that are generated in a nuclear power reactor. These are the normal fission products from the fuel system and the activation products from impurities in the air or the

TABLE 8
Summary of Environmental Considerations
for Uranium Enrichment Complex [15]

Natural resource	1972 10.5 MT SWU	1980 18 MT SWU
Land, ac		
Temporarily committed	1500	1500
Permanently committed	Nil	Nil
Water, gal/yr × 10^{-6}		
Discharged to air	8100	12,000
Discharged to water bodies[a]	10^6	1.8×10^{6c}
Fossil Fuels		
Electrical energy, MW-h/yr × 10^{-3}	28,500	41,000
Equivalent coal, MT × 10^{-3}	10,000	10,000[c]
Effluents		
Chemical, MT/yr		
Gases		
SO_x	400,000	400,000
NO_x	100,000	100,000
Hydrocarbons	1000	1000
CO	2500	2500
Particulates	100,000	100,000
F	45	45
Liquids		
Ca^+	490	880
Cl	740	1130
Na^+	740	1130
SO_4^{2-}	490	880
Fe	36	65
NO_3	240	450
Radiological, Ci/yr		
Gases		
Uranium	0.2	0.3
Liquids		
Uranium	1.8	3.1
Thermal, BTU/yr × 10^{-9b}	290,000	420,000

[a]Power plant cooling.
[b]67% at power plants.
[c]About 1/3 of electrical power from nuclear stations in 1980.

TABLE 9
Summary of Environmental Considerations
For 900 MTU/yr Fuel Fabrication Plant [15]

Natural resource use	
Land, acres	
Temporarily committed	100
Permanently committed	Nil
Water, gal $\times$ 10^{-6}	
Discharged to water	135
Fossil fuel	
Electrical energy, MW-h $\times$ 10^{-3}	44
Equivalent coal, MT $\times$ 10^{-3}	16
Natural gas, scf $\times$ 10^{-6}	94
Effluents	
Chemical (MT)	
Gases	
SO_x	600
NO_x	160
Hydrocarbons	1.6
CO	4
F	0.1
Liquids	
N as NH_3	220
N as NO_3	140
F	10
Solids	
CaF_2	680
Radiological, Ci	
Gases	
Uranium	0.005
Liquids	
Uranium	0.5
Th-234	0.3
Solids (buried)	
Uranium	0.2
Thermal, BTU $\times$ 10^{-9}	230

water used in either the cooling process or the heat exchange system. The principal radionuclides present in reactor effluents are ^{3}H, ^{58}Co, ^{60}Co, ^{85}Kr, ^{89}Sr, ^{131}I, ^{131}Xe, ^{135}Xe, ^{134}Cs, ^{137}Cs, and ^{140}Ba. Gaseous and volatile nuclides such as ^{85}Kr, ^{131}Xe, and ^{133}Xe contribute to external gamma dose as a result of immersion or being surrounded by the gas. The other radionuclides contribute to the dose externally by surface deposition and internally via the food chain and inhalation.

During the operation of the reactor, the fission products are retained within the fuel pellets encapsulated in the stainless steel or zircaloy cladding. Some of the more volatile fission products diffuse out of the fuel pellets and normally occupy the annular space between the fuel and the cladding, referred to as the "gap region." During reactor operation cladding defects can result from mechanical or thermal stresses, corrosion, and other causes, thus allowing the escape of small amounts of radioactivity into the primary coolant. Some reactors have experienced maximum activity levels in the coolant equivalent to leakage from 0.5% of the fuel rods. The fission products that are released to the primary coolant include the noble gases, primarily isotopes of krypton and xenon. These are normally separated from the coolant and enter the plant off-gas system, whereas the remaining fission products become dissolved or suspended in the coolant and are usually removed by the liquid coolant purification system. Neutron activation can create radioactive isotopes from traces of air dissolved in the gaseous coolant and impurities and corrosion products in the liquid coolant system. Most of the gases are radioisotopes of nitrogen and oxygen that have half-lives of less than 30 s. The dissolved solids and particulates in the liquid coolant are controlled in the plant's liquid treatment system.

In addition to the radioactive liquids and gases generated in a nuclear power plant, tritium is also a potential problem. Tritium is generated from fission, from neutron activation, from neutron capture, and by neutron activation of the boron and lithium used to control reactivity and the chemistry of the cooling water system. In power reactors, about one fission in 10,000 produces a third fission fraction which is tritium [19]. Most of the fission products and tritium remain in the zircaloy clad fuel rod, but some may be released at the time of fuel reprocessing. One of the main problems with tritium is that it normally forms the oxide that is water. Thus, it becomes an integral part of the water from which it is difficult to separate. So far, tritium from the nuclear industry is a small portion of the existing environmental inventory. It could constitute a significant future fraction if all of it is released to the environment. Meth-

TABLE 10
Initial Activity of Radionuclides in the Nuclear Reactor Core at the Time of Shutdown [18]

Radionuclide	Radioactive inventory source, Ci × 10^{-8}	Half-life, d
Cobalt-58	0.0078	71.0
Cobalt-60	0.0029	1920
Krypton-85	0.0056	3950
Krypton-85m	0.24	0.183
Krypton-87	0.47	0.0528
Krypton-88	0.68	0.117
Rubidium-86	0.00026	18.7
Strontium-89	0.94	52.1
Strontium-90	0.037	11,030
Strontium-91	1.1	0.403
Yttrium-90	0.039	2.67
Yttrium-91	1.2	59.0
Zirconium-95	1.5	65.2
Zirconium-97	1.5	0.71
Niobium-95	1.5	35.0
Molybdenum-99	1.6	2.8
Technetium-99m	1.4	0.25
Ruthenium-103	1.1	39.5
Ruthenium-105	0.72	0.185
Ruthenium-106	0.25	366
Rhodium-105	0.49	1.50
Tellurium-127	0.059	0.391
Tellurium-127m	0.011	109
Tellurium-129	0.31	0.048
Tellurium-129m	0.053	0.340
Tellurium-131m	0.13	1.25
Tellurium-132	1.2	3.25
Antimony-127	0.061	3.88
Antimony-129	0.33	0.179
Iodine-131	0.85	8.05
Iodine-132	1.2	0.0958
Iodine-133	1.7	0.875
Iodine-134	1.9	0.0366
Iodine-135	1.5	0.280
Xenon-133	1.7	5.28

(continued)

TABLE 10 (continued)

Radionuclide	Radioactive inventory source, Ci × 10^{-8}	Half-life, d
Xenon-135	0.34	0.384
Cesium-134	0.075	750
Cesium-136	0.030	13.0
Cesium-137	0.047	11,000
Barium-140	1.6	12.8
Lanthanum-140	1.6	1.67
Cerium-141	1.5	32.3
Cerium-143	1.3	1.38
Cerium-144	0.85	284
Praseodymium-143	1.3	13.7
Neodymium-147	0.60	11.1
Neptunium-239	16.4	2.35
Plutonium-238	0.00057	32,500
Plutonium-239	0.00021	8.9 × 10^6
Plutonium-240	0.00021	2.4 × 10^6
Plutonium-241	0.034	5,350
Americium-241	0.000017	1.5 × 10^{-5}
Curium-242	0.0050	163
Curium-244	0.00023	6,630

ods are presently being devised to handle this radioactive material before it becomes a hazard to the general public.

Liquid metal fast breeder reactors (LMFBR) may additionally produce wastes containing ^{239}Pu, ^{238}Pu, ^{241}Pu, ^{24}Na, and ^{22}Na. Normal release rates are expected to be low but the implications of handling large amounts of radioactivity and of accidents will have to be taken into account [16].

7. Spent Fuel Reprocessing

After about 3 yr service in a reactor the ^{235}U fraction is decreased and the fission products in the nuclear fuel buildup, reducing the effectiveness of the fuel. Under normal operations, the partially spent fuel rods are removed from the reactor core and stored in the on-site spent-fuel storage area approximately 90 d or longer under water. After this decay time, the fuel rods are shipped to a reprocessing plant for separation of the fission products from the usable fuel. In the process the zircaloy cladding is removed by dissolving in ammonium chloride solution or the

TABLE 11
Typical Radioactivity Inventory for a 1000 MW(e) Nuclear Power Reactor [18]

Location	Total inventory, Ci			Fraction of core inventory		
	Fuel	Gap	Total	Fuel	Gap	Total
Core[a]	8.0×10^9	1.4×10^8	8.1×10^9	9.8×10^{-1}	1.8×10^{-2}	1
Spent fuel Storage pool (max.)[b]	1.3×10^9	1.3×10^7	1.3×10^9	1.6×10^{-1}	1.6×10^{-3}	1.6×10^{-1}
Spent fuel Storage pool (avg.)[c]	3.6×10^8	3.8×10^6	3.6×10^8	4.5×10^{-2}	4.8×10^{-4}	4.5×10^{-2}
Shipping cask[d]	2.2×10^7	3.1×10^5	2.2×10^7	2.7×10^{-3}	3.8×10^{-5}	2.7×10^{-3}
Refueling[e]	2.2×10^7	2×10^5	2.2×10^7	2.7×10^{-3}	2.5×10^{-5}	2.7×10^{-3}
Waste gas storage tank	—	—	9.3×10^4	—	—	1.2×10^{-5}
Liquid waste storage tank	—	—	9.5×10^1	—	—	1.2×10^{-8}

[a]Core inventory based on activity 1/2 h after shutdown.
[b]Inventory of 2/3 core loading; 1/3 core with 3-d decay and 1/3 core with 150-d decay.
[c]Inventory of 1/2 core loading; 1/6 core with 150-d decay and 1/3 core with 60-d decay.
[d]Inventory based on 7 PWR or 17 BWR fuel assemblies with 150-d decay.
[e]Inventory for one fuel assembly with 3-d decay.

stainless steel cladding is removed in a sulfuric acid bath. The fuel reprocessing methods vary depending upon the material from which the fuel is fabricated, but all fuel reprocessing plants use some form of the Purex Process, a solvent extraction system, using tributyl phosphate (TBP) diluted with kerosene. The fuel elements are prepared for processing by shearing into small links and removing the cladding. The fuel is then dissolved in nitric acid, following which a series of successive TBP extraction and stripping steps results in separating the original components into the transuranic elements, uranium, plutonium, and fission products. High concentrations and large amounts of radioactive wastes remain in the fission product fraction [20]. A summary of environmental considerations of a 900 MTU/yr fuel reprocessing plant is shown in Table 12 [15].

A fuel reprocessing plant may release airborne effluents containing ^{85}Kr, tritium, and minute quantities of radioactive iodine, other fission products, and transuranic particulates that might pass through the off-gas treatment and filtration system. High-level liquid wastes are produced that contain nearly all of the nonvolatile fission products. These are normally stored rather than discharged. Of the three fuel reprocessing facilities that have been in operation, only the Nuclear Fuel Service plant released radionuclides into the liquid effluents [15].

In a more recent study [21], the impacts of reprocessing and waste management per reference reactor per year (RRY) are summarized in Table 13 [21]. A reference reactor per year is considered to be a 1000 MW(e) reactor assumed to be operating at 80% of its maximum capacity for 1 yr. This is equivalent to the AFR (annual fuel requirements) as used in WASH-1248 [15]. It must be pointed out that these values are annual requirements as opposed to the previous tables that were total values over the lifetime of the installation described.

8. Dose Commitment

A summary has been prepared of the amount of radioactivity that would be released to individuals in the environment from the various processes in the nuclear fuel cycle [21]. This summary is shown in Table 14. This tabulation is a summary of numerous references as listed in the table. It considers the radioactivity per individual from a normally operating nuclear fuel system, and also includes the dose potential from decommissioning a no longer used reactor and from potential sabotage. In general, it may be seen that the overall impact upon individuals in the environment from radioactivity from the nuclear fuel cycle is being kept to a minimum.

TABLE 12
Summary of Environmental Considerations
of 900 MTU/yr Fuel Reprocessing Plant [15]

Natural Resource Use	
Land (acres)	
Temporarily committed	2000
Permanently committed	100
Water, gal $\times$ 10^{-6}	
Discharged to air	100
Discharged to water	150
Fossil fuel	
Electrical energy, MW-h $\times$ 10^{-3}	12
Equivalent coal, MT $\times$ 10^{-3}	4
Effluents	
Chemical, MT	
Gases	
SO_x	160
NO_x	185
Hydrocarbons	0.5
CO	1
F	14
Liquids	
Na^+	140
Cl^-	0.5
SO_4^{2-}	14
NO_3^- (as N)	5
Radiological, Ci	
Gases (including entrainment)	
Tritium $\times$ 10^{-3}	405
Kr-85 $\times$ 10^{-3}	9000
I-129	6×10^{-2}
I-131	6×10^{-1}
Fission products	26
Transuranics	1×10
Liquids	
Tritium $\times$ 10^{-3}	65
Ru-106	100
Thermal, BTU $\times$ 10^{-9}	1600

B. Research, Development, and Commercial Applications

Radionuclides have experienced considerable use in research and development and in many commercial applications. These include tracer measurement, luminous compounds for use on dials, static eliminators, level gages, flow gages, thickness gages, and numerous other applications. In many cases, the radioactive source is sealed and therefore does not present a problem to the environment under normal use. However, there is a potential for rupture of the sealed containment and specifically for ^{226}Ra there is the problem of the production of radon gas that may escape from the container. The classic case of ingestion of radionuclides comes from the radium dial painters who licked their brushes in order to get a finer point. The incidence of cancer among these workers was far greater than the average. In general, tracer studies present only a low potential hazard to the environment because the radionuclides can be detected at extremely low levels, even though they may be dispersed over a relatively large area or volume in their tracer use.

Another problem involved with this type of use of radioactive materials is the potential hazard in shipping and handling radioactive materials. Proper containment is required and limits in the spacing and/or number of packages per shipment are defined. However, there is always the potential for an accident involving the carrier vehicle. Thus, shipping of radioactive materials always has the potential of creating a radioactive hazard.

C. Medical

The medical profession makes use of large quantities of both sealed radioactive sources and radiopharmaceuticals. Although the largest use is for diagnostic purposes, various amounts of radioactive materials are also used for therapeutic purposes. Radioactive tracers have been used to evaluate body cavity sizes, blood flow, and flow of material through the digestive system. They also have found considerable use in locating tumors and cancers within the body. Most of the radioactive materials ingested are passed from the body in the total excreta that may be considered to include the exhaled air. In large hospitals and medical research facilities the total amount of radioactive wastes generated may be a significant problem.

As with the research and development uses of radionuclides, the medical uses present a potential hazard in the shipping and handling of

TABLE 13
Summary of Impacts of Reprocessing and Waste Management per RRY [21]

Waste resource use	Reprocessing[d]		Waste management[a]		Transportation[c]		Fuel cycle totals[a,b]	
	(NUREG—0116)	(WASH-1248)	(NUREG-0116)	(WASH-1248)	(NUREG-0116)	(WASH-1248)	(NUREG-0116)	(WASH-1248)
Land, ac								
Temporarily committed	32[b]	3.9	3.2	—	—	—	94	63
Undisturbed area	28.5	3.7	2.9	—	—	—	73	45
Disturbed area	3.5	0.2	0.28	—	—	—	22	18
Permanently committed	0.12	0.03	2.6[c]	0.2	—	—	7.1	4.6
Overburden moved, millions of MT	0.1	—	0.0015	—	—	—	2.8	2.7
Water, millions of gal								
Discharged to air	6.6	4.0	0.38	0.13	—	—	159	156
Discharged to water bodies	54.8	6.0	0.051	0.13	—	—	11,090	11,040
Discharged to ground	—	—	0.96	—	—	—	124	123
Total water	61.4	10.0	1.4	0.26	—	—	11,373	11,319
Fossil fuel								
Electrical energy, thousand MW-h	4.0	0.45	0.62	0.0077	—	—	321	317
Equivalent coal, thousand MT	1.5	0.16	0.22	0.003	0.016	—	117	115

Natural gas, million scf	28.6	—	3.3	—	—	124	92
Chemical, MT				*Effluents*			
Gases, MT							
SO_x	5.4	6.2	0.030	0.045	—	4400	4400
NO_x	21.9	7.1	0.031	0.62	2.6	1190	1177
Hydrocarbons	0.5	0.02	0.02	0.062	—	14	13.5
CO	0.5	0.04	0.007	0.38	—	29.6	28.7
Particulates	0.6	1.6	0.02	0.022	—	1154	1156
Other Gases							
F^-	0.05	0.11	—	—	—	0.67	0.72
HCl	6E-4	—	0.013	—	—	0.14	—
Liquids							
SO_4^{2-}	<0.02	0.4	—	—	—	9.9	10.3
NO_3^-	—	0.9	—	—	—	25.8	26.7
Fluoride	—	—	—	—	—	12.9	12.9
Ca^{2+}	—	—	—	—	—	5.4	5.4
Cl^-	0.09	0.2	—	—	—	8.5	8.6
Na^+	<0.02	5.3	—	—	—	12.1	16.9
NH_3	—	—	—	—	—	10.0	16.9
Tailings solutions (thousands)	—	—	—	—	—	240	240

(continued)

TABLE 13 (continued)

Waste resource use	Reprocessing[d] (NUREG—0116)	Reprocessing[d] (WASH-1248)	Waste management[d] (NUREG-0116)	Waste management[d] (WASH-1248)	Transportation[c] (NUREG-0116)	Transportation[c] (WASH-1248)	Fuel cycle totals[a,b] (NUREG-0116)	Fuel cycle totals[a,b] (WASH-1248)
Fe	—	—	—	—	—	—	0.4	0.4
Solids	—	—	0.42	—	—	—	91,000	91,000
Radiological, Ci Gases, including entrainment								
Rn-222	—	—	0.0071	—	—	—	74.5	74.5
Ra-226	—	—	5.3E-7	—	—	—	0.02	0.02
Th-230	—	—	5.3E-7	—	—	—	0.02	0.02
Uranium	0.000039	—	7.9E-6	—	—	—	0.034	0.032
Tritium (thousands)	18.1	16.7	14g	—	—	—	18.1	16.7
Kr-85 (thousands)	400	350	290g	—	—	—	400	350
I-129	0.3	0.0024	1.3g	—	—	—	1.3	0.0024
I-131	0.83	0.024	—	—	—	—	0.83	0.024
Fission products	0.18	1.0	0.003	—	—	—	0.021	1.0
Transuranics	0.023	0.004	0.0014	—	—	—	0.024	0.004
C-14	24	—	19g	—	—	—	24	—
Liquids								
Uranium and daughters	—	—	5.4E-6	—	—	—	2.1	2.1
Fission and activation products	—	—	5.9E-6	—	—	—	5.9E-6	0.0034
Ra-226	—	—	—	—	—	—	0.0034	0.0034

Th-230	—	—	—	—	0.0015	0.0015
Th-234	—	—	—	—	0.01	0.01
Tritium (thousands)	—	2.5	—	—	—	2.5
Ru-106	—	0.15	—	—	—	0.15
Solids (buried onsite)[e]						
Other than high level (shallow)	0.52	—	4,700	—	5,300	601
TRU and HLW (deep)	—	—	1.1E+7	—	1.1E+7	—
Thermal, billions of BTU	75.5	61	88	1.0	0.014	0.03	3,462	3,360

[a] Maximized for either of the two cycles: U-only and no recycle.
[b] Including columns A–E of Table S-3A of WASH-1248.
[c] For wastes only.
[d] Differences between NUREG-0116 and WASH-1248 estimates of reprocessing impacts are attributable to the use of a new model plant for this Supplement.
[e] Disposal included here did not appear in WASH-1248.
[f] Not released to the environment.
[g] Major radionuclide releases in the Waste Management column are attributable to the disposal of spent fuel (no-recycle option) and the conservative assumption of complete release of gaseous nuclides in the geologic repository.
[h] The contributions to temporarily committed land are not prorated over 30 yr, since the complete temporary impact accrues regardless of whether the plant services one reactor for one year or 57 reactors for 30 yr.

TABLE 14
Dose Commitments Summary [21]: Normal Operations

	Dose	Basis	NUREG-0116 Section	Reference
Reprocessing	330 person-rem/RRY	US Population	4.1	GESMO, Tables IV E-8, E-9, and E-12
	22 person-rem/RRY	Occupational		
High-level waste				
Tank storage	(see reprocessing)			
Solidification	10 person-rem	Total annual dose to the population within 50 mi of the facility	4.2.2 and Table 4.2	NUREG-0082, Section V-B
Interim storage at FRP	(See reprocessing)			
Transportation	0.011 person-rem	20 Transport workers		GESMO, Chapter IV, Section G
	0.046 person-rem	General public: 20 on-lookers	4.9	
	0.041 person-rem	General public: 7.6×10^5 residents	Table 4.35	
Interim storage at RSSF	$< 5 \times 10^{-3}$ person-rem/yr	Population within a 50-mi radius in year 2005.	4.2.5	Letter to W. P. Bishop from J. R. LaRiviere, Oct. 1, 1976, Attachment 4.
Disposal	3.6×10^{-7} rem/RRY-individual	50-yr dose commitments (HLW plus TRU)	4.4 and Table 4.18	GESMO, p. IV H-46, and Tables IV H-18 and H-19

	9.0×10^{-4} person-rem/ RRY Population		

Transuranic wastes

Treatment

Combustible	0.6 mr/year	At a reprocessing plant	4.3.2.4	BNFP, FES, Reprint 1, p. V-15.
Noncombustible	(see reprocessing)			
Interim storage				
Transportation	1.25 person-rem	For 80 transport workers	4.9	GESMO, Chapter VI, Section G
	0.12 person-rem	General public: 80 on-lookers	Table 4.35	
	0.11 person-rem	General public: 2.7×10^6 residents		
Disposal	3.6×10^{-7} rem/RRY-individual	50-yr dose commitments	4.4	GESMO, p. IV H-46 and Tables IV H-18 and H-19
	9.0×10^{-4} person-rem/ RRY-population	(TRU plus HLW)	Table 4.18	

(*continued*)

TABLE 14 (*continued*)

	Dose	Basis	NUREG-0116 Section	Reference
Long-term risks—disposal				
All HLW accumulated to the year 2000 (commercial)		Table 4.19 50-yr accumulated dose; released in year 2100	4.4.2	BNWL-1927
		Table 4.20 50-yr accumulated dose; released in year 2000		
		Table 4.21 50-yr accumulated dose; released in year 102,000		
		Table 4.22 50-yr accumulated dose; released in yr 1,002,000		
		Table 4.23 50-yr dose commitments for meteorite		ORNL-TM 4639
		Table 4.24 impact at 1000 and 100,000 years after the year 2000		
Spent fuel as a waste				
Water basin storage	110 person-rem, worldwide—^{85}Kr cumulative 1975–2000		Table 4.25	GESMO, Chapter IV, Table IV K-1
	1.2×10^4 person-rem occupational dose, cumulative 1975 to 2000			
Disposal	260 person-rem	Population dose normalized to model reactor	Table 4.28	GESMO, Page IV E-30 (see paragraph 4.6.3.4)

Low-level wastes:	1% of established guidelines—burning an entire drum of waste.	4.7.3.4	ERDA-1537
Transportation	For 75 transport workers	4.9	GESMO, Chapter IV, Sec. G.
	0.61 person-rem		
	General public: 70 onlookers	Table 4.35	
	0.14 person-rem		
	General public: 2.4×10^6 residents		
	0.13 person-rem		
Decommissioning	Occupational exposure normalized to model LWR fuel requirements	Table 4.31	"Operational Health Physics During Dismantling of the Elk River Reactor," D. McConnon
	37 person-rem		
Sabotage			
Spent fuel storage	20 mrem-whole body at site boundary	4.10.3	NEDM-20682, "Sabotage Analysis for Fuel Storage at Morris"

the radioactive materials. Also there is a greater potential for the radioactive contamination of technicians and nontechnical persons in a medical facility. Special precautions must be taken in handling radiopharmaceutical liquids that are normally used for ingestion by the patient. Also, special procedures must be taken for handling the bodily wastes of persons taking radioactive materials internally.

It must be mentioned that the greatest source of radiation in the medical profession is in the diagnostic and therapeutic use of X-rays. However, since these do not produce radioactive wastes under normal operation, they are not considered in this report.

III. TRANSPORT MECHANISMS

Great efforts are made to confine all radioactive materials and prevent their release to the environment. However, in some instances there are planned releases of small quantities of radioactive materials to the environment. Also accidents may occur in which radioactive materials gain access to the environment. The pathways by which these radioactive materials ultimately reach humans are defined as transport mechanisms. Although the mechanism is usually considered to terminate in humans, there may be instances in which the radioactivity more adversely affects an intermediate before it reaches the ultimate target. Usually, however, this also has an indirect effect upon humans. Thus, the transport mechanism systems discussed here will always terminate in humans.

A. Air

Radioactivity in the air may be in the form of gases, particulates, and aerosols. The particulates can be further broken down into those that remain suspended in the atmosphere and those that tend to fall out or precipitate under conditions of low turbulence. However, some of the dusts that may have settled out under conditions of low wind velocity may be resuspended under conditions of greater wind velocity.

From Fig. 3 it may be seen that there are numerous pathways for radioactivity to travel through the atmosphere, all of which ultimately reach the human target [22].

The most direct pathway is by direct radiation from the radioactive materials in the air surrounding the individual. This creates a whole-body exposure to external radiation. The next most direct route is through inha-

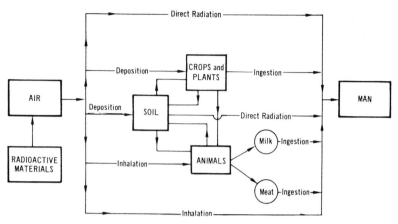

Fig. 3. Pathways between radioactive materials released to the atmosphere and man [22].

lation of the radioactive materials in the air. Some of this material is taken up by the body and then becomes internal radiation. For the basis of calculations, it is normally assumed that particles greater than 10 μm are intercepted in the nasal passages, and do not reach the lungs. Unless specific data are available, it may be assumed that of the soluble material entering the nasal system, 25% is retained in the lower respiratory tract and is absorbed in the blood, 50% is intercepted in the upper respiratory tract and is swallowed, thereby entering into the gastrointestinal system, and 25% is exhaled. For insoluble materials, equivalent numbers are 12, 63, and 25%, respectively.

Another route involves desposition onto the leaves of crops and plants. Whereas some of this merely remains on the surface of the leaves, where it may be removed by washing or rainfall, other portions of the radioactive materials may be taken up into the plant material. If these plants are then eaten, particularly if they are not thoroughly washed before consumption, there may be a direct transfer of radioactivity to humans through the gastrointestinal system, resulting in an internal radiation dose. As an alternative to this cycle, animals may eat the crop that has been contaminated with radiation. From this point there are numerous routes that the radioactive materials can follow. The most direct route is by incorporation in milk, assuming that the plant is consumed by a cow or similar animal. If, on the other hand, the animal is consumed for its meat, any radioactivity transferred to the flesh of the animal may be ingested by humans, thereby causing internal radiation to the individual. The animals may also inhale the radioactive materials, ultimately transporting them in a similar manner to the milk or meat. Another alternate route involves the animal waste materials that are deposited onto the soil. Here they may be taken up by the plants and crops, which again may be consumed directly by humans or recycled through the animal. Some materials on the soil may also be taken up completely by the animal in another form. Thus, where the crop upon which the radioactive material has been deposited is consumed as forage by animals, there may be a significant recycling through the system, with humans ingesting radioactive materials through the milk and/or through the meat. With the potential for concentration of specific radionuclides in various plants and animals and/or various portions of these plants and animals, as will be described in a later section, it may be seen that there is the potential for a buildup of radioactive materials in portions of this system. Since human beings are at the top of the food chain, they are affected by all of the intermediate steps and the concentrations that may occur along the way.

B. Water

Radioactive materials may gain access to the aqueous environment through several mechanisms. The radioactive materials may be present in the water as soluble and suspended radioactive materials in a flowing stream. However, suspended materials are normally removed prior to discharge to a stream. Greater quantities of radioactive particulates would be carried in a stream at higher flow velocities. Radioactive materials may be deposited onto the soil through direct deposition. Precipitation on the surface of the soil may dissolve some of the radioactive materials that have been deposited onto the soil, and depending upon the velocity of the runoff may also carry away certain particulate radioactive materials. Some of the precipitation may gain access to the groundwater. Radioactive materials present in the ground, including improperly buried radioactive materials, may be leached from the soil system. A third mechanism for entry of radioactive materials into the aqueous environment would be by incorporation into precipitation. Rainfall is an effective scrubber of radioactive materials from the air. Snowfall has been noted to be particularly high in radioactive materials that may have been present in the atmosphere at the time of formation of the snow.

Some pathways between radioactive materials released to the ground and surface waters and humans are shown in Fig. 4 [22]. The most direct pathway would be by immersion in water containing the radioactivity,

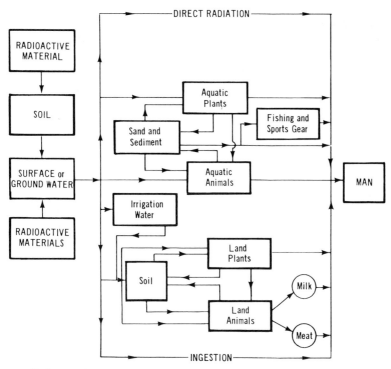

Fig. 4. Pathways between radioactive materials released to ground and surface water (including oceans) and man [22].

such as during swimming. This would be considered as external radiation. Another source of external radiation would be from fishing, and sporting gear that had been immersed into the body of water containing the radioactivity. To an extent a boat hull could also be considered to be an external source of radiation particularly where the boat is removed from the water at frequent intervals. All of the other transport mechanisms to humans involve ingestion of the radioactivity in some form and may be interpreted as internal radiation.

Radioactivity in the water may be taken up by aquatic plants. Of particular note here is the potential for these plants to concentrate the radioactivity in their cell structure. Some larger aquatic plants are consumed directly by humans. Others, particularly the microscopic plants (algae), are consumed by aquatic animals. Here, there is a chain of events involving first the microscopic animals (zooplankton) that in turn are consumed by small fish that in turn are consumed by larger fish that finally are the ones most desirable for human consumption. Each step along this way can result in a concentration of the radioactive materials either in the entire organism or a specific portion of the organism. When both aquatic plants and animals die, they normally sink to the bottom of the body of water where their nutrients and potential radioactivity may be released back into the aquatic environment for reuse by both aquatic plants and aquatic animals. Thus, there is a potential for multiple concentration of the radioactive materials present.

Radioactive materials in the water reach the soil either through direct precipitation or through application by irrigation. Interactions similar to those for the fallout of radioactive air contaminants onto the soil may occur. Frequently, irrigated corps are consumed directly by humans; however, grazing fields are also irrigated, potentially resulting in the uptake of radioactivity in the forage crops. From here the radioactive materials may be consumed and recycled through the land animals, to the soil and back to the plants. Consumption of the milk and meat of these animals can result in the ingestion of significant amounts of radioactive materials.

It may be seen that monitoring radioactivity transport is not a simple matter because of the many alternate pathways which are available.

C. Concentration

Many biological systems are capable of concentrating radioactive materials in their cell materials or in particular portions or specific organs of the larger organism. It must be kept in mind that the concentration of radioactive materials is similar to that for the concentration of the stable material of the same chemical species. Thus, the concentration of radioactive materials is considered to be the same as the concentration of heavy metals. As a matter of fact, the concentration of heavy metals is often determined by the use of radioactive tracers of those specific metals. Whether or not the concentration of radionuclides in the biological system is harmful to humans depends upon whether the location where the radioactive materials is concentrated is the part consumed by them. For example, radioactivity concentrated in bones, hair, skin, scales, or shells would not present a hazard because these portion of the organisms are not directly consumed. On the other hand, concentration in the flesh or in an organ that is consumed, such as the liver, the kidneys, and so on, could result in a considerably greater concentration of radioactive materials. Table 15 [23] lists some concentration factors for certain metals. It may be seen that some of the concentration factors are very significant, potentially allowing for the concentrating of radioactive materials from levels in the environment that would be well below acceptable tolerance levels to lev-

TABLE 15
Typical Metal Concentration Factors of Selected Aquatic Organisms [23]

$$\frac{\mu g \text{ metal/g organic}}{\mu g \text{ metal/g } H_2O}$$

Metal	Marine organisms					Fresh water organisms		
	Phytoplankton[a]	Zooplankton[a]	Macrophytes[b]	Mollusks[c]	Fish[d]	Macrophytes[e]	Mollusks[f]	Fish[g]
Ti	2700	—	—	—	—	—	—	10
Cr	7800	—	2880	21,800	—	—	267	23
Mn	3800	3900	—	2300	373	1450	—	190
Fe	28,300	114,600	—	14,400	—	3642	—	85
Ni	570	560	1050	4000	235	—	650	90
Co	—	—	—	—	50	1367	300	60
Cu	2800	1800	2890	3800	127	158	1500	228
Zn	5500	8800	7000	27,300	533	318	2258	

[a]Martin and Knauer (1973).
[b]Hagerhall (1973).
[c]Pringle et al. (1968).
[d]Goldberg (1962).
[e]Merlini et al. (1970).
[f]Mathis and Cummings (1973).

els that could be potentially harmful to the individual consuming these concentrated radioactive materials.

Some means of calculation of the uptake of radioactive materials into various portions of the environment have been determined. Many of the equations are complicated and require the use of a computer for solution. Also many of the factors contained in the equations are not absolute numbers, so estimates or assumptions have to be made in numerous cases. As a typical example of such a calculation, the daily uptake of radioactive materials by a cow is shown in the following example [24]:

The daily intake of the cow (I_d) depends heavily on the particular dairy management practice employed and the season of the year; thus, an exact description of the factors affecting this term leads to a more complicated equation for (I_d):

$$I_d = Q_f \left[\frac{R}{Y_f} \left(\frac{1 - e^{-\lambda_E t_f}}{\lambda_E} \right) \left(\frac{D_{fm} + \epsilon_{fm}\omega_{fm}}{30} \right) + \frac{B_f}{P} \left(F_{fA} + F_{fW} \right) \right]$$

$$+ Q_{gd} e^{-\lambda_g t_{\theta_g}} \left\{ \frac{RT_g}{Y_g} = \left[\sum_{m=a}^{h} \left(D_{gm} + \epsilon_{gm}\omega_{gm} \right) e^{-\lambda_E t_\beta} \right] + \frac{B_g}{P} \left(F_{gA} + F_{gW} \right) \right\}$$

$$+ Q_{sd} e^{-\lambda_s t_{\theta_s}} \left\{ \frac{RT_s}{Y_s} = \left[\sum_{m=a}^{h} \left(D_{sm} + \epsilon_{sm}\omega_{sm} \right) e^{-\lambda_E t_\beta} \right] + \frac{B_s}{P} \left(F_{sA} + F_{sW} \right) \right\}$$

$$+ G_d A_m N_d + L_{dm}\omega_m$$

where

A_m Radionuclide concentration in air during month "m", pCi/L
a First month of crop contamination
B_f Forage crop concentration factor, pCi/kg per pCi/kg soil*
B_g Grain concentration factor, pCi/kg per pCi/kg soil*
B_s Stored feed concentration factor, pCi/kg per pCi/kg soil*
C_d Radionuclide concentration in milk, pCi/L
D_{fm} Deposition of forage crop during month "m", pCi/m²
D_{gm} Deposition of grain crop during month "m", pCi/m²
D_{sm} Deposition on stored feed crop during month "m", pCi/m²
F_{fA} Radionuclide concentration in plowlayer of forage crop from air deposition, pCi/m²
F_{fW} Radionuclide concentration in plowlayer of forage crop from water deposition, pCi/m²
F_{gA} Radionuclide concentration in plowlayer of grain crop from air deposition, pCi/m²

F_{gW} Radionuclide concentration in plowlayer of grain crop from water deposition, pCi/m^2
F_{sA} Radionuclide concentration in plowlayer of stored feed from air deposition, pCi/m^2
F_{sW} Radionuclide concentration in plowlayer of stored feed from water deposition, pCi/m^2
G_d Branching rate of dairy cow, m^3/d
h Month of crop harvest
I_d Daily ingestion of radionuclide by dairy cow, pCi
L_d Daily ingestion of drinking water by dairy cow, L
m Month
N_d Radionuclide inhalation retention factor
P 2.24 × 10^2 kg soil/m^2 plowlayer*
Q_f Daily fresh forage ingestion by dairy cow, kg
Q_{gd} Daily grain ingestion by dairy cow, kg
Q_{sd} Daily stored feed intake by dairy cow, kg
R Deposition retention factor
S_d Coefficient of transfer of radionuclide from diet to milk, pCi/L per pCi/d
T_g Grain crop translocation factor
T_s Stored feed translocation factor
t_f Time between successive removals of forage crop, d
t_β Time between deposition and harvest (h-a), d
$t_{\theta g}$ Time between harvest and consumption of grain, d
$t_{\theta s}$ Time between harvest and consumption of stored feed, d
Y_f Forage crop yield, kg/m^2*
Y_g Grain crop yield, kg/m^2*
Y_s Stored feed crop yield, kg/m^2*
ϵ_{fm} Irrigation rate for forage crop during month "m", L/m^2
ϵ_{gm} Irrigation rate for grain crop during month "m", L/m^2
ϵ_{sm} Irrigation rate for stored feed during month "m", L/m^2
ω_{fm} Radionuclide concentration during month "m", pCi/L, in irrigation water applied to forage
ω_{gm} Radionuclide concentration during month "m", pCi/L, in irrigation water applied to grain
ω_{sm} Radionuclide concentration during month "m", pCi/L, in irrigation water applied to stored feed crop
λ_E Effective decay constant, d
λ_r Radioactive decay constant, d

*Plants are measured in kilograms of fresh weight whereas soils are measured in kilograms of dry weight.

It is important to remember that the air deposition and water and soil concentrations to be used with the above equation must be those concentrations estimated for the area where the food was produced, not where it was consumed (if the two are different). In the instance of animal feed, however, it was assumed that all three types—forage, stored feed, and grain—were grown in the same area where they were consumed by the animals.

The values of S_d used in the milk pathway equation are tabulated in Table 16 [24].

Similar equations are available for other intake parameters in different organisms and for human food consumption [25]. Space does not allow inclusion of all of these equations here. It may be seen that there can be considerable concentration of radioactive materials in the transport from the source of the radioactivity to man.

IV. WASTE MANAGEMENT

A. Principles of Treatment

The nuclear wastes that must be either disposed of or recycled as part of the overall nuclear industry may be broken down into 11 categories as follows: (1) high-level solidified wastes; (2) cladding hulls; (3) noble gases; (4) iodine; (5) light-water reactor tritium (water); (6) fuel pellet tritium (solid); (7) carbon-14; (8) low-level transuranium isotopes; (9)

TABLE 16
Coefficients of Transfer of Radionuclides from Diet to Milk [24]
Transfer Coefficient (S_d), pCi/L Milk per pCi/d Intake

H-3	2.0 E-2[a]	Ni-63	6.7 E-3	I-129	1.0 E-2
N-13	2.2 E-2	Cu-64	1.4 E-2	I-131	1.0 E-2
C-14	1.5 E-2	Zn-65	3.0 E-2[b]	I-132	1.0 E-2
Na-22	5.0 E-2	Sr-89	1.0 E-3[b]	I-133	1.0 E-2
Na-24	5.0 E-2	Sr-90	1.0 E-3[b]	I-135	1.0 E-2
Cr-51	2.2 E-3	Zr-95	5.0 E-6	Cs-134	5.0 E-3
Mn-54	2.5 E-4	Nb-95	2.5 E-3	Cs-137	5.0 E-3
Fe-55	1.2 E-3	Mo-99	7.5 E-3	Ba-140	6.0 E-4
Fe-59	1.2 E-3	Ru-103	1.0 E-6	La-140	5.0 E-6
Co-58	1.0 E-3	Ru-106	1.0 E-6	Ce-141	2.0 E-5
Co-60	1.0 E-3	Te-132	1.0 E-3	Ce-144	2.0 E-5

[a]2.0 E-2 means 0.02.

intermediate-level transuranium isotopes; (10) nontransuranium isotopes; and (11) ore tailings [12].

Just as in any other industry, proper management of the wastes generated by the atomic energy industry must be followed. However, in the case of the atomic energy industry, the wastes present a particular problem in that the materials, in addition to being objectionable from the normal standpoint, are also radioactive and may present a hazard to the environment from the standpoint of their radioactivity. Extra precautions must be taken in handling the radioactive materials. Methods that may have been sufficient for the removal of the stable waste substances may have to be upgraded in order to bring the levels of radioactivity below maximum acceptable limits.

It must be pointed out that there is no way to change the radioactivity except through its normal natural decay scheme or by induced transmutation. Thus, the radioactivity is always somewhere. The only things that can be done are to change its form and/or its state, and then separate it using the best means known. Effective separation of all of the radioactive materials is essential. Only the smallest traces may be allowed to remain in the original medium: air, water, or land. Even biological treatment is merely a means of changing the form of the radioactive material, incorporating it into cell material. However, the radioactivity remains in the cell and for effective removal of the radioactivity the entire organism must be removed from the system before it dies and resolublizes the radioactive material that was removed during the life of the organism.

Another point that must be made is the fact that any means of removal or separation of radioactive materials is just as efficient as the similar removal of nonradioactive substances. Under normal waste treatment operation, complete efficiency is not required. For many radioactive materials better than 99% removal of the radioactive substances must be achieved. The efficiency of the system is not a function of the radioactivity, but of the molecule and its physical or chemical form, and the removal is the same as for stable substances of the same type.

One caution is the potential problem caused by concentration of the radioactive materials, particularly in the sediment or sludges of a treatment facility. Whereas it is difficult to concentrate the radioactive material to a point where it would cause a chain reaction, it is possible that concentration in sludges may reach levels that would kill the organisms because of the radioactivity, thus preventing the normal biodegradation of the major portion of the sludge. Such concentrations may also be harmful to workers in the area and appropriate precautions should be taken.

Although there may be many specific processes involved in waste treatment, the overall principles may be categorized into three main processes. In each process, numerous specific types of treatment may be used to accomplish the ultimate goal of adequate waste management with least impact upon the environment. These three principles are described as follows.

1. Concentrate and Contain

The first principle of waste management is the concept of concentrate and contain. This implies the concentration of the radioactive material and separating it from the main body of materials such as air, water, or soil. The concentrated radioactive material must then be placed in an appropriate container and stored for a long time. This procedure is of most value for high-level wastes or concentrations of radioactivity in the media that are relatively high. Such wastes originate primarily in the fuel reprocessing system. High-level wastes include those having relatively high radiation energies and those that have intermediate half lives and are present in high concentration. An effort is made to separate these radioactive materials from the nonradioactive materials and then these high level wastes are placed in a container for storage for an appropriate period of time. In some instances, this storage may be on the order of hundreds or thousands of years. Maintaining the integrity of a waste container on a disposal site over many generations presents some potential problems. There are even two philosophies of containment. One is to remove the concentrated materials to a remote location where they will never interfere with present or future life. The other philosophy is to store them where they may be monitored and where access will allow them to be used, modified, changed, or moved to a site that would be more appropriate in the future. Consideration must also be given to the heat produced by such concentrated radioactive wastes, since the end result in all radiation decay schemes is the production of heat. Some of the liquid wastes stored at the Hanford site may self-boil for hundreds of years. Thus, special precautions must be taken to monitor these areas for a long period of time.

2. Dilute and Discharge

The second major method for handling radioactive materials is the principle of dilute and discharge. This method implies that with sufficient dilution the radioactive materials will be dispersed in the environment and will not raise the level in the environment to detectable or harmful levels. Space does not permit description of the methods for determining what levels may be harmful. Suffice it to observe from regulations (10 CFR 20) what concentrations will be acceptable. The dilute and discharge principle is normally applied to substances that are present in low concentrations, which have low energies of radiation, and which have either extremely short or extremely long half-lives. Two concepts are involved in this principle. One is that extremely low levels of radioactivity will not be harmful and that such levels can be achieved by dilution. The other is that by applying this method to radioactive materials that have relatively short half-lives, by the time the radioactivity returns to the human environment, the radioactivity may have decayed to levels well below any potentially harmful levels. This method of disposal is particularly applicable to the low level wastes from the nuclear industry and the laboratory and research facilities that normally use only very small amounts of radioactive materials for the studies involved. Low level wastes from the normal operation of nuclear reactors also come under this category.

3. Delay and Decay

The concept of delay and decay may be considered as a separate principle or a modification of the principle of dilute and discharge. It implies the presence of short-lived radionuclides whose activity may be decreased to extremely low levels by a reasonable time or delay prior to discharge into the environment. After adequate decay to low levels, dilution and discharge is the normal method for ultimate disposal of such radioactive wastes. Provisions must be made for the holding of the radioactive materials for a sufficient amount of time to allow the decay to low levels of radiation to take place. Depending upon the half-life, this may be as short as a few hours or as long as a few weeks or even a few months. With proper dispersion into the environment, further decay will take place before this radioactive material reaches the human environment.

B. Plan for Waste Management

There are numerous options available for the specific type of treatment to accomplish the principles of waste management. These depend upon the physical state of the material (solid, liquid, or gas), the amount of radioactivity involved, the type of radioactivity (alpha, beta, gamma, neutrons), the energy of the radiation involved, the half-life of the specific radionuclide, the concentration of radioactive material in the total mass to be disposed of, and the total volume of waste material to be handled. A general schematic plan has been proposed to portray the general techniques for handling radioactive wastes [21]. This is based upon the three fuel cycles that are possible with LWRs: (1) no recycling (with no reprocessing); (2) recycling of uranium only, and (3) recycling of both uranium and plutonium. For the two recycle options, the environmental impacts from reprocessing and waste management are similar with the exception that plutonium is an additional waste in the uranium only recycle option. The wastes from the nuclear fuel cycle are grouped into six major classes, as shown in Fig. 5. Four of these arise from the two recycle systems: (1) high-level wastes (HLW); (2) transuranium-contaminated wastes (TRU); (3) nontransuranium-contaminated wastes (often called low-level wastes, LLW); (4) contaminated facilities and large equipment. In addition the uranium fuel cycles give rise to (5) spent fuel from the no-recycle option and (6) unused plutonium from the uranium-recycle option. It may be seen that with the exception of the disposal of facilities and large equipment, the other five treatment processes are similar in that they involve some type of treatment and/or packaging, with transport to some ultimate disposal site. The various types of treatment involved are discussed in the next section.

C. Methods of Treatment

1. Physical

Physical removal of the radioactive material from its carrier medium is normally applicable to particulate radioactive materials. These may range in size from slightly greater than colloidal particles to large bulky products and even building demolition materials. The latter large materials may be easily handled and may be removed from the immediate environment by such simple means as trucking or other similar means of transportation. Smaller particles may require other means of separation. Included in this category are materials that have been rendered insoluble by chemical means, as will be described later. In general, the environments considered are air and water.

The basic techniques for the separation of the radioactive particulates and the medium may be divided into two main categories: (1) removal of the radioactive material from the medium, and (2) removal of the medium from the radioactive material.

Removal of the radioactive material from the medium by physical

MANAGEMENT OF RADIOACTIVE WASTES 337

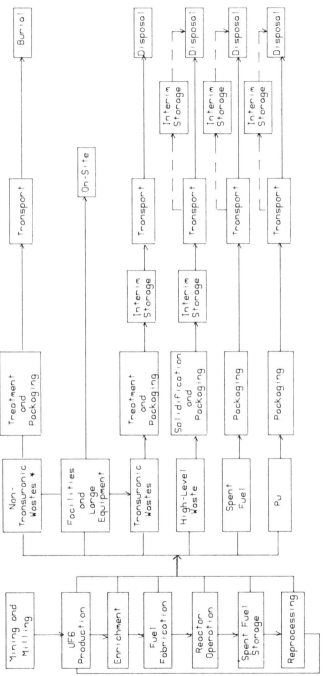

Fig. 5. Waste management flow diagram [21] *Low-Level Wastes.

means normally implies the presence of the radioactivity in particulate form. The most common technique involves sedimentation with or without prior coagulation and flocculation. Settling is a function of the differential size and density of the particulate matter with relationship to its medium, air or water. Under quiesent conditions the particles will settle out and will accumulate on the bottom of the settling chamber. From here they must be removed for further treatment and/or ultimate disposal.

A second method for removal of the particulate radioactive material is filtration. Depending upon the size of the particles to be removed, various porosity filters may be employed. For rather large particles or only rough filtration, sand beds are frequently used. For medium sized particles, beds containing fine sand or anthrafilt may be employed. Other filter media, such as cloth, paper, or fine mesh screens, may also be employed for this separation. For very fine particulates, filtration through a membrane can be used. Membranes are readily available on the market with a 0.22 μm effective pore size. Such filters are capable of removing particles slightly smaller than the effective pore size. This suggests that particles of colloidal size can be removed by membrane filtration.

A third method for separating the particulates from the medium is the use of a centrifuge. In this case, the particles are moved to the walls of the centrifuge because of their greater density with respect to their supporting medium. Means must be provided to remove the separated radioactive materials and also to allow the medium, now free of radioactive particulates, to pass into the environment.

A fourth technique for removal of particulates from the medium is the use of foam separation. Here the wastewater is subjected to pressure and supersaturation with air. When the pressure is released, fine bubbles are created that are trapped on the particulates, floating them to the surface. A skimmer will remove the floated particulates.

A second means of physical separation is the removal of the medium from the radioactive material. This generally involves the principle of distillation in which the medium, usually water, is boiled from the mixture and the radioactive materials remain in the boiling flask. This method is effective for even soluble materials, and therefore has application where the other physical methods may not be effective.

2. Chemical

Chemical treatment of radioactive wastes usually involves changing the physical form of the radioactive material so that it may be handled in another way. In general, this involves the addition of some chemical to convert soluble materials to an insoluble form that can be separated from the medium by precipitation, filtration, and so on. Various chemical precipitation techniques are available, not all of which are applicable for all radioactive materials. Therefore, an individual study must be made for the most effective chemical removal technique for each specific radionuclide. Some of the processes involved include the conversion of the radioactive materials to insoluble hydroxides, sulfides, and salts of heavy metals. Precipitation as sulfides is generally effective with the metals that may be in solution. Depending upon the metal involved, precipitation as a salt may be more effective.

When the radioactive materials are present in an aqueous medium in very small size particles, coprecipitation with iron, manganese, or aluminum salts is often effective. By adjusting the concentration of the metal salt added, finding the pH of the isoelectric point of the specific substance to be removed, and using polyelectrolytes to achieve a more precise isoelectric neutralization of the charge on the colloidal particles, a system may be devised that can be very effective in removing the finely divided and/or colloidal radioactive particles.

Combustion or incineration is considered a chemical process by which the radioactive materials may be separated from the nonradioactive materials. This assumes that the radioactive materials are in the inorganic form, whereas the organic matter is nonradioactive. In general, combustion involves the conversion of organic matter to CO_2 and water, with the inorganic material remaining in ash that should contain the bulk of the radioactive material. Precaution should be taken to prevent the fly ash from escaping from the system. Controls such as scrubbers may be installed in the effluent stack from an incinerator. Other devices to prevent fly ash from escaping from the system are bag houses and electrostatic precipitators.

The normal procedure for the ultimate separation of radioactive materials from the medium involves chemical precipitation, followed by some means of physical separation as described previously. Thus, chemical methods of separation usually involve physical methods for actual separation of the particulates formed in the chemical process.

3. Physical-Chemical Systems

There are some processes that are difficult to categorize into physical or chemical processes and therefore may be considered as physical-chemical systems.

Ion exchange is considered one of these physical-chemical processes that is very effective in the removal of radioactive materials, particularly of soluble materials in an aqueous environment. By selection of the appropriate ion exchange resin for the radioactive materials to be removed from the system, effective removal of soluble radioactive materials can be achieved. Materials having the highest valence are more readily removed. The radioactive materials are transferred to the ion exchange medium, where they may be concentrated for further handling and disposal.

Activated carbon is effective in removing organic material from both liquid and gaseous systems. The materials to be removed are absorbed on the large surface area of the activated carbon and the purified air or water is allowed to escape the system. The activated carbon containing the concentrated radioactive materials must then be disposed of in a satisfactory way. Provisions may be made for regenerating the activated carbon and removal of the radioactive materials to another form that will still require ultimate disposal.

Reverse osmosis is a system involving placing of a liquid under pressure and forcing the water through a semipermeable membrane that will restrict the passage of nonwater molecules through it. Thus the water is separated from the radioactive materials in this process. This is effective where the concentration of the radioactive materials is not extremely high in the water. It results in concentration of the radioactivity in a small amount of water and allows for any further treatment of this concentrated solution prior to ultimate disposal.

Solvent extraction is an effective means of concentrating radioactive materials. This again is most applicable in an aqueous environment. It precludes finding a solvent in which the radioactive material is more soluble than in the water. Thus an investigation must be made into the proper solvent and the proper conditions for separating the aqueous and the solvent phases. By the use of a different solvent or different conditions such as pH, the radioactive materials removed in the solvents may be returned to the aqueous phase for ultimate handling and disposal.

4. Biological

Various biological methods are available for treatment of sewage and industrial wastes. The same principles may be applied to the treatment of wastes containing radioactive materials. The effectiveness of the separation is a function of the form or phase in which the radioactive material is found in the liquid wastes. Several precautions must be taken in considering biological treatment. The principle involved is normally the uptake of the pollutant material, in this case the radioactive material, into the biological cell. In order to achieve effective removal of the radioactive material, the cells must be removed from the aqueous system. Most frequently this involves a sedimentation step, but for removal or separation of radioactivity from the aqueous phase, the efficiency of most common sedimentation processes is not sufficient for the satisfactory removal of radioactive material. Furthermore, the biological organisms must be separated from the aqueous system in a relatively short period of time. For example, a sewage oxidation pond normally involves the growth of algae in a pond with a retention time in the order of 30 d. Certain algae and other organisms that may have taken up the radioactivity may have a life span of less than 30 d. Therefore, these organisms will die, settle to the bottom, and rupture or lyse, thus releasing the cell materials back into the aqueous environment. If they have taken up radioactive material, these radioactive materials likewise will be returned to the aqueous environment, essentially resulting in little to no overall reduction in radioactivity. Another potential problem, depending upon the level of the radioactivity in the waste to be treated, may be the buildup of radioactivity in the sludges to levels high enough to interfere with the life of the biological systems present. This could result in no treatment of any type by the system.

The most common types of biological treatment include the trickling filter, the activated sludge process with its various modifications, oxidation ponds, and contact stabilization. The trickling filter process involves contact of the liquid containing organic materials with stones or other media on which the organisms are able to attach. The organisms remove the nutrient from the liquid and convert this into energy and cell growth. Radioactive materials, both soluble and insoluble, may be taken up according to the needs of the organisms attached to the filter medium. The biological films on the filter medium are periodically or continuously sloughed off and must be separated in a settling tank following the trickling filter.

D. Ultimate Disposal

In general the means of treatment of radioactive wastes involve basically the change of form or state of the radioactive materials followed by separation of the radioactive materials from the stable materials. This still leaves the radioactive material to be disposed of in some manner that will not be harmful to the environment. In general, these wastes are now in a concentrated form, so that a smaller volume of materials needs to be handled, but at the same time results in radioactive wastes that are difficult to handle and that will require monitoring for a long period of time. This section will deal with some of the methods used for ultimate disposal of the concentrated radioactive wastes.

1. Liquid Storage

Some provisions have to be made for storage of high radiation level liquid wastes. It is assumed that no simple method has been devised for further concentrating the radioactive materials in the wastes, although in some instances sludges have formed over the many years that liquids have been accumulated and stored. High-level liquid wastes are stored in underground tanks at the Hanford Works in Richland, Washington, at the Savannah River Plant near Aiken, South Carolina, and at the National Reactor Testing Station in Idaho. A typical storage tank is shown in Fig. 6. Originally tanks were made of mild steel, but some of these have been found to leak after about 25 yr of service. Newer tanks are made of higher alloy steel and are coated for corrosion resistance. One of the problems encountered with these wastes is that they may self-boil for a period of hundreds of years. Also extensive monitoring systems must be devised to identify leaks as soon as they occur and preparations must be made to remove the contents of a leaking tank into a spare tank. Thusfar, in about 25 yr of operation of tanks of this type, no serious contamination of groundwater has occurred even due to leaks.

Maintaining storage facilities for liquid radioactive wastes for periods of hundreds or even thousands of years presents many problems, both technological and political. It is difficult to conceive of the implications involved in storage and commitment of space over such long periods of time. Therefore, efforts are underway to solidify as much as possible the liquid radioactive wastes that are presently in storage. Appendix F of AEC Regulation 10 CFR50 requires that the inventory of high-level liquid wastes at fuel reprocessing plants be limited to that produced in the prior 5 yr and that it be converted to a solid form and transferred to a federal repository within 10 yr of its separation from the irridiated fuel. At the present time, no federal repository exists. Other means of permanent storage are also being considered.

2. Solidification

Although solidification does include some of the techniques for converting the radioactive materials to an insoluble form, as described in Section IV.C, Methods of Treatment, the concept here has further impli-

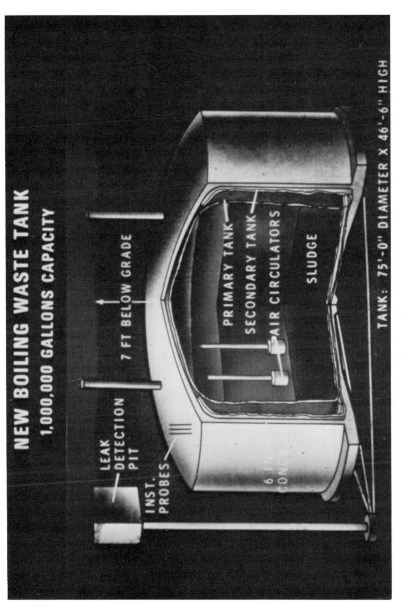

Fig. 6. High-level waste storage tank.

cations. Solidification as considered here involves the conversion of liquids, slurries, sludges, and relatively small particulate matter into an encapsulated form that will prevent further dispersal of the radioactive particles. One of simplest means of accomplishing this is to use the radioactive liquid as the water for forming concrete or plaster of paris. The water (along with the small particulates) is taken up as water of crystallization and forms a permanent body of material that can be handled with relative ease. In some instances, the concrete is poured into plastic or metal drums that further protect the concrete from abrasion or chemical attack.

Numerous methods have been studied and developed for the solidification of high-level radioactive liquid wastes [26]. Three methods of calcining that have been developed are pot solidification, spray solidification, and fluidized bed solidification. The calcining process involves converting the radioactive materials to a less soluble oxide form. Another process called phosphate glass solidification combines glass forming chemicals with the wastes and evaporates the liquid to form a melt that, when cooled, appears to be a glass-type material. Another process involves solidification in various clays that then are glazed and fired to form a seal around the outside of the container, thus preventing escape of the radioactive materials.

The goal of solidification is to convert the radioactive materials into a more concentrated form and then to convert the whole material into an easily handled solid that essentially isolates the materials from the rest of the environment. Provisions must still be made for disposal of the solid material.

3. Land Application of Liquids

When soluble materials are applied to the land they may be removed from solution by processes involving adsorption, ion exchange, and mineralization. In addition various crops may take up the materials as a source of nutrients. Radioactive materials may be removed on the same basis as stable materials. When using this system, precautions must be taken so as not to contaminate any groundwater supply. In addition, if the radioactivity is transferred to the crop, precautions must be taken in the ultimate disposal of the crop. If trees are grown, there is less chance of passing on the radioactivity through the food chain. Land application is most practical for low-level liquid wastes and should be practiced only where extensive monitoring can be provided in order to assure there is no contamination of the environment.

4. Land Burial

Once the radioactive materials have been converted to a solid state and are encapsulated, they must still be disposed of in a satisfactory way. In addition, there are wastes from building materials that have become contaminated and, in some instances, large pieces of equipment. Whereas it is a general policy to encapsulate these radioactive materials at a minimum in cartons, it may be impossible to encapsulate such things as building rubble, equipment, and so on. The procedures for land disposal are similar to those involved in a sanitary landfill. The materials may be deposited in a trench that is covered with a minimum of one foot of soil at the end of each working day. Larger sites involve a wider trench or bank, but are still covered at the end of each day's operation. Caution should be taken to include a minimum of organic material in such a landfill. Sealing the material disposed of will prevent rodents and other vermin from getting into the area. Covering not only protects materials, but also keeps out people, thereby preventing them from coming in close contact with the radioactivity. Where extremely high levels of radiation are encountered, greater cover depth may be required to attenuate these radiations. Precautions should be taken to prevent contamination of groundwater. In general, burial sites should be in impervious soil and precautions should be taken to minimize the flow of surface and groundwater around and into the disposal site itself.

Land burial sites are considered permanent installations. These will have to be monitored for long periods of time, again presenting the problem of permanent storage with changes in engineering technology and politics. Accurate records must be kept and maintained showing where the radioactive materials have been disposed of so that in future years digging in the area will not expose individuals to excessive levels of radiation. Long-term monitoring of these disposal sites is essential.

5. Sea Burial

Sea burial is confined to solidified wastes usually further encapsulated in barrels. The principle is to deposit the encapsulated radioactive materials in an area that will not be involved with the environment until such a time as the radioactive materials contained therein have decayed to sufficiently low levels that they would not be harmful to the environment. There are two schools of thought regarding this type of disposal. One is to deposit the material in soft-bottomed portions of the ocean where the containers will sink into the ocean floor and therefore will not be able to move into other parts of the environment. On the other hand, this type of disposal site cannot be readily monitored as it is difficult to locate the individual containers that have been deposited there. The other concept is to deposit the containers on a hard-bottomed area of the ocean. In this way, the containers may be observed periodically to determine their integrity and their stability. An instance has been recorded of a barrel floating up onto the beach in England after having been deposited in the ocean off that area.

Depositing this type of material in extreme depths of the ocean presents problems of integrity of the container at these great depths and also precludes the opportunity to monitor the container in the future. At the present time, very little ocean disposal is being conducted, although the method still has potential as a means of disposal of radioactive waste materials.

6. Mines

The use of mines for the disposal of radioactive materials has received much thought and consideration. Mines may either be made specifically for the disposal of radioactive wastes, or existing abandoned mines may be used for this purpose. An advantage of mines is that the location is normally remote and far from the expected paths of persons on the surface of the earth. Their location can also be established and they are not subject to movement except by earthquake activity. Furthermore, provisions can be made to re-enter the mines for inspection or possible future removal. Precautions must be taken to control underground water movement that could leach the radioactive materials and carry them into a water supply system. Also with such long-term storage it is difficult to anticipate what minerals may be found in the area that would be profitable for future mining.

One of the most seriously considered mine sites was a former salt mine near Lyons, Kansas. This had many features that were desirable for the permanent storage of high-level radioactive wastes. First of all the mine had already been completely mined-out and was available for storage of any type of material. Another advantage was the salt remaining in the mines. The high level radioactive wastes could be encapsulated in the salt with the decay heat melting the salt, forming a molten matrix. With further additional cooling this would solidify, holding the radioactive materials permanently in this salt cave. Individual rooms or vaults would be provided, each of which could be separated and monitored individually. Heavy equipment could be used to move the radioactive materials in and/or around in the mine. Provisions could be made for retrieving the radioactive materials in the event of a change of ideas in the future. Figure 7 shows the concepts involved in such a mine [20]. The project was temporarily abandonded because of potential hazards from a nearby water injection salt mine actively in operation, and for political reasons. The concept of storage in abandoned salt mines still has great possibilities.

7. Undersea Disposal

As part of Project Mohole, consideration was made of the disposal of radioactive wastes in the floor of the ocean. Project Mohole involved the drilling of holes in the bottom of the ocean from ships floating on the surface. Drilling could be conducted in ocean depths of up to 3 mi and a

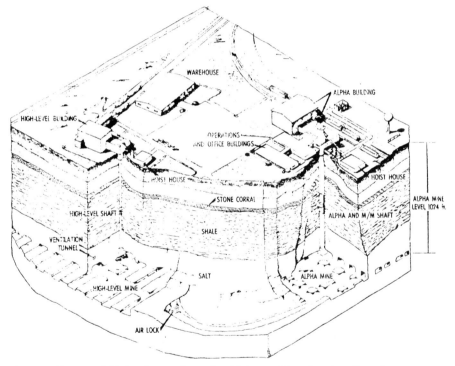

Fig. 7. Proposed Federal repository for high- and low-level radioactive waste near Lyons, Kansas. The facility, served by a railroad spur, would handle commercial wastes from nuclear power plants through the year 2000, storing them 1000 ft underground in bedded salt formation (Oak Ridge National Laboratory) [20].

1000 ft-deep hole drilled in the floor of the ocean could be filled with radioactive materials, sealed with a concrete plug, and identified for future location. The radioactive waste disposed of would be in the solid state, and would be encapsulated in such a manner that even if the plug should inadvertently become removed from the hole, the radioactive materials would not be leached from the system and therefore would not gain access to the environment. This would provide a disposal site that would be remote and unaccessible, but yet could be reached in the event of change of conditions. It was felt that one drilling rig could drill three holes 1000 ft-deep each day, which would provide enough capacity to take care of all of the radioactive wastes from one reactor for a year.

Although this procedure is still considered a viable means for ultimate disposal of high-level radioactive wastes, it has not received much attention recently because funding for Project Mohole has been discontinued. Since the project has not been carried to completion, there is no way of knowing how successful this system would be nor whether it could be used for disposal of radioactive wastes.

8. Underground Fixation

The counterpart to undersea disposal is the concept of underground fixation. Here large holes would be drilled several miles deep into the earth into a strata that would be impermeable and not affected by any groundwater. The shafts would be deep enough so that there would be no potential future use of the minerals or any water located at that depth. Solid radioactive wastes would be placed in the hole and the entire hole sealed with concrete. The plug at the top of the radioactive materials would prevent contact with the surface of the ground and any overlying aquifers. This plug would still be quite deep in the earth and not necessarily near the surface. The well could be located with a surface marker, making it readily identifiable and available in the event of need for future monitoring, inspection, or removal. Two disadvantages of this system are its high cost and the undesirability of locating these wells "in my back yard."

In a modified method of depositing radioactive materials in ground formations, hydrofracturing of shale is accomplished by forcing liquids between the layers of shale at a high pressure. A cement slurry containing the radioactive materials is then forced between the now open layers of the shale. This again results in a permanent means of disposing of liquid radioactive wastes.

9. Deep-Well Injection

High-level liquid radioactive wastes may be discharged under pressure in deep wells several miles under the ground. The concept here is to force the liquid into an aquifer that is of such high salt content and so remote that it would "never" be used for any conceivable human purposes. A typical installation of a deep-well disposal system is shown in Fig. 8.

Normally, the injection well is comprised of a double casing in order to provide additional safeguards for the injection procedure. The entire double pipe extends through all potentially potable aquifers into the injection aquifer. The entire casing is encased in concrete to prevent any liquid from being forced up around the well casing into one of the potable aqui-

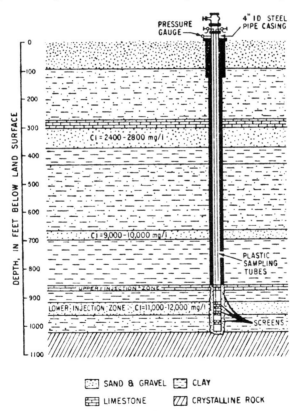

Fig. 8. Typical disposal well.

fers. A screen is located on the bottom of the casing to allow the injected liquid to pass into the disposal zone. The waste to be disposed of is pumped by pressure through the inner pipe. The space between the two casings is normally filled with a fluid that is monitored for pressure. In the event of any leakage in the inner casing or back pressure in the entire system, this increase in pressure would be observed immediately in the outer casing fluid. In addition to the monitoring equipment, high pressure pumps are normally part of this system. Under certain circumstances pretreatment of the wastes may be required before injection so as to make them compatible with the aquifer and the liquid therein. Precautions must be taken to prevent any chemical reactions, biological reactions, gas formation, or physical clogging problems within the disposal aquifer.

A feature of this type of disposal system is that it is continuous in that liquids may continue to be discharged to such a system for a period of years. The feasibility of employing underground formations for the retention or permanent storage of radioactive wastes is dependent upon the chemical and physical compatibility of the wastes with the receiving formation and the degree of certainty of predicting velocity and direction of flow of the more hazardous components of the waste in the ground [28]. This method is presently in use at several sites throughout the United States.

10. Dust Bin Rockets

A proposal has been set forth to place radioactive materials in rockets that would be fired to the sun. Here they would be removed completely from the earth, never to create a problem. There are two major problems with this method of removal of radioactive materials. The first is the limited payload of the rockets with respect to the amount of radioactive materials that must be disposed of within a certain period of time. Many rockets would have to be provided and this would result in an extremely high cost. The second problem is that of the reliability of the rocket. At the present time we have not developed large-scale rocket production and the number of rockets needed times the fraction of failure would indicate the potential for the failure of a rocket loaded with radioactive materials, allowing it to fall back onto the earth at some undesirable location. Thus this method of disposal of radioactive wastes has not been attempted up to this time.

11. Heat

It must be remembered that the ultimate dissipation of energy from radiaoctivity is the production of heat. It was already indicated that high-level radioactive liquid wastes may self-boil for many years. Also, solidification in salt would involve the melting of the salt because of the heat produced by the high-level radioactive materials. Provisions must be made to dissipate this heat in an acceptable manner. In some cases, cooling coils may be utilized in order to remove a portion of the heat. Concentrations of radioactive materials placed in the ground should be such that the heat produced may be absorbed by the surrounding earth. In the particular case of deep-well injection, precautions must be made to prevent the boiling of the liquid deep under the ground because this would produce steam, thereby potentially destroying the disposal system. Solidified wastes would have to be such a level that the total encapsulated material does not rupture because of the expansion from the heat produced. Calculations can be made of the anticipated amounts of heat that would be produced by a certain concentration of radioactive materials. Based on this information, the amounts of radioactive materials that may be disposed of in any one type of system can be controlled.

12. Federal Repository

Appendix F of AEC Regulation 10 CFR50 requires that the inventory of high-level liquid wastes at fuel reprocessing plants be limited to that produced in the prior 5 yr and that it be converted to solid form and transferred to a federal repository within 10 yr of its separation from the irradiated fuel [15]. The location of a federal repository for long-term storage of high level wastes is still under investigation. However, as of January 1976 no positive action had been taken to create such a repository. Starting in 1982, until a long-term repository can be established, the federal government is providing for interim storage of high-level wastes. This facility is to be designed with capability for storage of wastes for a 100-yr period. The federal government will have the responsibility for the care and maintenance of any site at which high-level wastes are stored. Control must be maintained as long as any wastes are stored in a facility.

Although numerous sites have been discussed as a federal repository, none have been chosen as of the present moment. The site that received most attention was the former salt mine near Lyons, Kansas. However, at the present time further discussion of this site as a federal repository has been discontinued. In general, the problem of selection of a site is more a matter of politics than of engineering. It is the old story of "I'm in favor of it, but don't put it in my back yard." People must be shown that such a storage facility can be made safe and that there will be no harm to the local citizens nor to the environment as a whole. Steps must be taken to establish such a site and to be sure that it will be safe to all parties involved for long periods of time.

V. SURVEILLANCE

The basic principle of environment surveillance involves sampling wherever and whenever radioactive materials may be present. In addition, controls must be set up by measuring at locations and times where radioactivity is expected never to be present from the site under surveillance. These principles are involved whether the monitoring is to be performed inplant or in the environment; however, this discussion will be directed towards environmental surveillance since control within the plant is the responsibility of the plant management.

Environmental radiation surveillance programs conducted around nuclear power facilities should as a minimum provide data that may be used: (1) for population dose calculations that can be compared with federal and state standards; (2) for the evaluation of environmental radioactivity buildup, and (3) for public information purposes [22].

The off-site environmental surveillance program for nuclear power reactors should be established on the basis of the evaluation of the radionuclides discharged in the liquid and gaseous wastes and the environmental parameters that could affect their dispersion and dilution in the environment. An environmental surveillance program should include both preoperational and operational data. The preoperational data provide a baseline for evaluating any increases in radioactivity levels during the operation of the nuclear facility. The operational surveillance program will provide data needed for estimation of the population dose. In both types of monitoring program, a remote site must be chosen that will not be affected by the facility under study. This will provide a comparison so that if radioactivity levels are increased or changed because of any other factors, the effects of the nuclear facility can be evaluated in terms of values above the control levels. Otherwise, increases in the environment from other sources would be blamed upon the nuclear facility.

An acceptable duration of time for the preoperational survey is 1 yr prior to the shipment of any radioactive materials to the nuclear site. The program should consist of (1) identification of the probable exposure pathways (See Section III, Transport Mechanisms); (2) the population distribution in the area, with particular attention to population centers; (3) selection of the sampling equipment, media, and sample site locations; (4) the collection and analyses of the samples, and (5) interpretation of the data.

Table 17 indicates a guide for off-site surveillance of operating light-water cooled nuclear power facilities [22].

A minimum preoperational surveillance program to be undertaken 1 yr prior to the startup of the facility operations is outlined in the Environmental Radioactivity Surveillance Guide as follows [22]:

1. Make gamma radiation dose rate measurements (i.e., TLD, or pressurized ion chamber) at locations identified for direct radiation measurement in Table 17. The locations may be chosen on the basis of meteorological data supplied with the Preliminary Safety Analysis Report for the facility.
2. Make *in situ* quantitative gamma spectrometric measurements at the stations in item 1. Analyze the spectra to apportion the total gamma dose rate among the various contributing radionuclides. Guidance and procedures for performing these measurements have been provided. Laboratory analysis of soil and other terrestrial materials contributing to ambient gamma dose levels may be substituted for the *in situ* measurements where practical.
3. Collect low volume air samples at one station for 6 mo before startup and determine the gross beta activity. Perform gamma isotopic analyses of a monthly composite of these samples.
4. Identify the critical population in the plant environs. Collect relevant demographic data for the area within 50 mi of the facility.
5. Collect samples of water, food, and biota along the critical dose pathways. Perform gamma isotopic analyses. The samples should be collected and analyzed quarterly where appropriate to identify seasonal variations.
6. Long-lived alpha-emitting radionuclides such as radium-226, thorium-232, and plutonium-238/239, though not normally attributed to light-water-cooled power reactor operations, have been detected in environmental samples. Additional gross or specific alpha analyses during the preoperational phase may be required to fully document the population radiation exposure situation in the vicinity of the nuclear facility.

Gross alpha and/or gross beta screening of environmental samples may be substituted for gamma spectroscopy during the preoperational phase.

TABLE 17

Offsite Surveillance of Operating Light-Water-Cooled Nuclear Power Facilities [22]

Operation or sample type	Approximate number of samples and their locations	Collection frequency	Analysis type[a] and frequency
Air particulates	1 sample from the 3 locations of the highest offsite ground level concentrations 1 sample from 1-3 communities within a 10-mile radius of the facility 1 sample from a location greater than a 20-mile radius in the least prevalent annual wind direction[d]	Continuous collection—filter change as required	Gross long-lived β at filter change[b] Composite for gamma isotopic analysis and radiostrontium analysis[c] quarterly
Air iodine	Same sites as for air particulates	Continuous collection—canister changes as required	Analyze weekly unless absence of radioiodine can be demonstrated
Direct radiation	2 or more dosimeters placed at each of the locations of the air particulate samples which are located at the 3 highest offsite ground level concentrations 2 or more dosimeters placed at each of 3 other locations for which the highest annual offsite dose at ground level is predicted[e] 2 or more dosimeters placed at each of 1-3 communities within a 10-mile radius of the facility[f] 2 or more dosimeters placed at a location greater than a 20-mile radius in the least prevalent annual wind direction[d]	Quarterly	Gamma dose quarterly
Surface water[g]	1 upstream 1 downstream after dilution (e.g., 1 mile)	Monthly (Record status of discharge operations at time of sampling)	Gross β, gamma isotopic analysis[h] monthly. Composite for tritium and radiostrontium analysis[c] quarterly
Ground water	1 or 2 from sources most likely to be affected	Quarterly	Gross β, gamma isotopic analysis[h] and tritium quarterly
Drinking water	Any supplies obtained within 10 miles of the facility which could be affected by its discharges or the first supply within 100 miles if none exists within 10 miles	Continuous proportional samples[i]	Gross β, gamma isotopic analysis[h] monthly. Composite for tritium and radiostrontium analysis[c] quarterly
Sediment, benthic organisms and aquatic plants	1 directly downstream of outfall[j] 1 upstream of outfall[j] 1 at dam site downstream or in impoundments[j]	Semiannually	Gamma isotopic analysis semiannually

MANAGEMENT OF RADIOACTIVE WASTES

Milk	1 sample at nearest offsite dairy farm in the prevailing downwind direction 1 sample of milk from local dairy representative of milkshed for the area	Monthly	Gamma isotopic analysis and radiostrontium analysis monthly[c]
Fish and shellfish	1 of each of principal edible types from vicinity of outfall 1 of each of the sample types from area not influenced by the discharges	Semiannually	Gamma isotopic analysis semiannually on edible portions
Fruits and vegetables	1 each of principal food products grown near the point of maximum predicted annual ground concentration from stack releases and from any area which is irrigated by water in which liquid plant wastes have been discharged 1 each of the same foods grown at greater than 20 miles distance in the least prevalent wind direction	Annually (At harvest)	Gamma isotopic analysis annually on edible portions
Meat and poultry	Meat, poultry, and eggs from animals fed on crops grown within 10 miles of the facility at the prevailing downwind direction or where drinking water is supplied from a downstream source	Annually during or immediately following grazing season	Gamma isotopic analysis annually on edible portions
Quality control[k]	Samples as required for accurate sampling and analysis		Minimum frequency—annually

[a] Gamma isotopic analysis means identification of gamma emitters plus quantitative results for radionuclides that may be attributable to the facility.
[b] Particulate sample filters should be analyzed for gross beta after at least 24 hours to allow for radon and thoron daughter decay.
[c] Radiostrontium analysis is to be done only if gamma isotopic analysis indicates presence of cesium-137 associated with nuclear power facility discharges.
[d] The purpose of this sample is to obtain background information. If it is not practical to locate a site in accordance with the criterion, another site which provides valid background data should be used.
[e] These sites based on estimated dose levels, as opposed to ground level concentrations where the dose may be affected by sky shine, high plumes, or direct radiation from the facility being monitored.
[f] These locations will normally coincide with the air particulate samplers used in the monitored communities.
[g] For facilities not located on a stream, the upstream sample should be a sample taken at a distance beyond significant influence of the discharges. The downstream sample should be taken in an area beyond the outfall which would allow for mixing and dilution. Upstream samples taken in a tidal area must be taken far enough upstream to be beyond the plant influence when the effluent is actually flowing upstream during incoming tides.
[h] If gross beta exceed 30 pCi/liter.
[i] Drinking water samples should be taken continuously at the surface water intake to municipal water supplies. Alternatively, if a reservoir is used, drinking water samples should be taken from the reservoir monthly. If the holding time for the reservoir is less than 1 month, then the sampling frequency should equal this holdup time. Increases in concentration of activation and/or fission products at these sources necessitate the analysis of tap water for the purpose of dose calculations. Additional analyses of tap water may be necessary to satisfy public demand.
[j] See figure 6 for locations on a stream. For facilities located on large bodies of water, sampling sites should be located at the discharge point and in both directions along the shore line.
[k] The Analytical Quality Control Service of the Surveillance and Inspection Division (SID) provides low-level radiochemical standards and interlaboratory services to State and local health departments. Federal and international agencies, and nuclear power facilities and their contractors. The Service operates several types of cross-check programs for the analysis of radionuclide in environmental media, such as milk, food, water, air, and soil. The samples are submitted on a routine schedule designed to fit the needs of each laboratory. Technical experiments are undertaken to permit detailed analyses of the accuracy and precision obtained by participating laboratories. In addition, low-level radioactivity standards are provided to the agencies participating in the various programs. Primary and secondary standardization is also performed as needed on those radionuclides not used on a routine basis.

The operational surveillance program should commence when radioactive materials are delivered to the plant. The same basic principles for monitoring during the preoperational period should be followed during the operational monitoring. AEC regulation [29] require that each nuclear power facility operator reports semiannually to the Commission on the quality of each of the principal radionuclides released to the environment in liquid and gaseous effluents. The recommended minimum level environmental surveillance program is shown in Table 17. All critical pathways should be identified and monitored. In addition, air particulates, direct radiation, and radioactivity in surface waters should be monitored at all times even though they may not be critical or predominant pathways of exposure. As experience is gained with the monitoring operation, changes in the frequency of sampling, the location of samples, and the type of analyses performed may be made. However, these should be such that comparison can be made of any changes in radioactivity both at the area influenced by the reactor and at the remote site. Under certain circumstances, certain other materials may have to be evaluated, and sampling techniques and schedules may have to be established for these.

VI. PROTECTION

A. Principles

The basic principles for protection from radioactivity are time, distance, and shielding. The concept of time is that radioactive materials decay with time and after sufficient time the radioactivity will become depleted. However, some radioactive materials have half-lives of thousands of years and obviously it would not be practical to wait this length of time prior to approaching a source of radioactivity. Thus, time is a practical factor in protection with only relatively short-lived radioisotopes. As for distance, the level of radioactivity follows what is known as the inverse square law, or the amount of radioactivity at any location is equal to one over the distance squared $1 \div D^2$. Thus, as distance is increased, the amount of radioactivity decreases. The protection principle involved here is the ability to escape from, or not be near, any sources of radioactivity. This is the principle of locating a nuclear reactor at some relatively remove location. Shielding involves the ability of materials to absorb radiation. Alpha particles are absorbed by very thin layers of materials. Beta particles are stopped by a slightly thicker shielding material. Gamma rays and X-rays, on the other hand, may pass through even very heavy lead shields although the amount of radiation passing through a shield may be negligible. The shield becomes a barrier for the radiation and is known as an absorber. Thus a shield may be provided to prevent radiation from one source falling upon a target or a human at some other location.

B. Population Protection

Not all of these methods are applicable to population protection. In some instances they are utilized in laboratory studies involving high levels of radioactivity. It is not always possible to wait a sufficient time for a radioactive material to decay before it may be released to the environment. Distance becomes a significant factor, but in the case of a nuclear power reactor, extreme distances require longer electrical transmission lines, which become more expensive and less efficient. The use of shielding for individual environmental protection is not practical. On the other hand, shielding of the reactor itself becomes a very important factor in preventing radioactivity from reaching the environment.

The best protection for the environment is to prevent any radiation from reaching the environment. This involves practical control of the radioactive materials wherever they may be in the nuclear fuel cycle, in the laboratory, in industrial and development systems, and in medical uses. Any discharges containing radioactive materials, such as radioactive wastes, should be treated to remove as much as of the radioactive material as possible before release. Precautions must be taken to prevent accidental release of radioactive material to the environment. Thus, extra precautions are necessary to provide safety at a nuclear power reactor.

The time factor is taken into account by the holdup systems for the disposal of radioactive gases and liquids. These allow the radioisotopes time to decay before they are discharged to the environment. They are incorporated as part of the treatment of radioactive wastes. Time is also factored into the containment system for the nuclear power reactor. In the event of a major catastrophe, the containment is designed to leak at a slow rate to provide both time for other safety precautions to take effect and dilution of the radioactive wastes themselves. In extreme cases, this time factor could allow the evacuation of persons from the environment if necessary.

Distance factors have been established for the general population in the area of a nuclear reactor [30]. These are spelled out in Table 18. There are basically three zones involved. First is the exclusion distance

TABLE 18
Minimum Site Distances for Reactor Location [30]
(Exclusion Distances)

Reactor power level, thermal megawatts	Exclusion distance,[a] mi	Low population zone distance,[b] mi	Population center distance,[c] mi
1500	0.70	13.3	17.7
1200	0.60	11.5	15.3
1000	0.53	10.0	13.3
900	0.50	9.4	12.5
800	0.46	8.6	11.5
700	0.42	8.0	10.7
600	0.38	7.2	9.6
500	0.33	6.3	8.4
400	0.29	5.4	7.2
300	0.24	4.5	6.0
200	0.21	3.4	4.5
100	0.18	2.2	2.9
50	0.15	1.4	1.9
10	0.08	0.5	0.7

[a]Exclusion zone: An area of such size that an individual located at any point on its boundary for 2 hr immediately following the onset of the postulated fission product release would not receive a total radiation dose to the whole body in excess of 25 rem or a total radiation dose in excess of 300 rem to the thyroid from iodine exposure.

[b]Low population zone: An area of such size that an individual located on its outer boundary who is exposed to the radioactive cloud resulting from the postulated fission product release (during the entire period of its passage) would not receive a total radiation dose to the whole body in excess of 25 rem or a total radiation dose in excess of 300 rem to the thyroid from iodine exposure.

[c]Population center distance: A distance of at least 1.33 times the distance from the reactor to the outer boundary of the low population zone.

that is the area immediately surrounding the power plant and should be in complete control by the systems at the power plant itself. Second is the low population zone, which is the area immediately surrounding the reactor site and which should have a minimum population so that in the event that there are any releases of radioactivity from the site, there may be a minimum of hazard to the general public. The distance to a major center of population is the third distance criterion and is basically one and one-third times the low population zone distance. This is a relatively arbitrary value based on knowledge of dispersion and the inverse square law. The concept is to prevent the radiation exposure to large populations.

Shielding is accomplished primarily at the reactor site. A reactor core is sealed in the containment vessel that is surrounded by thick concrete shielding. This containment vessel is designed for the minimum release of radioactivity to the environment. The reactor building itself becomes a secondary shield to minimize any discharge of radioactivity to the environment. All wastes, gases, liquids, and solids, are treated and disposed of in a manner so as not to contaminate the environment.

In general, population protection is built into the entire nuclear system. Shielding and safety precautions are designed into a nuclear power plant and waste treatment is provided for all of the wastes produced at the plant. Large buffer zones are provided around the plant to allow for dispersion and some decay due to time. With all these safety precautions built into the nuclear power plant, the only time protection becomes an acute problem is during and immediately after a reactor accident. Extreme precautions are taken to minimize these also. Thus, population protection is built into the nuclear power system and additional precautions are not normally needed to protect individuals from this radioactively.

It is the intended goal of the nuclear industry to limit the radiation from power plants to the general population to less than 1% of the normal background radiation (~ 130 mrem/yr). This is less than the radiation dose received in a 1 wk trip in the mountains, the difference between the radiation received by persons living in a wooden house compared to a masonry house, or the radiation received in a transcontinental airplane flight. This limit is considered not to have any measurable effects on any individual.

VII. COSTS

A. Various Treatment and Disposal Methods

The costs of treatment and handling of various forms of radioactive wastes can be an entire study in itself. Each individual method has its own cost, and combinations of methods have various combined costs. In order not to overwhelm the reader with detailed cost data, but at the same time to present some feeling for the cost of radioactive waste treatment, some specific costs will be cited here and reference will be made to further information that will aid one in determining the costs of treatment methods.

In general the costs of treating and handling radioactive wastes are similar to the same methods for treating and handling nonradioactive wastes. For low-level radioactive wastes, no additional special handling is required; therefore, the costs are essentially the same as for treating a similar nonradioactive waste. Where the radioactive levels are high, additional precautions must be taken from the standpoint of shielding and protection of the treatment plant operators. In many instances, additional treatment is required over and above the normal treatment procedures for nonradioactive wastes. This is because a greater percentage of removal or purification of the radioactive material is frequently required. This is reflected in additional costs for the greater degree of treatment.

One of the biggest costs of radioactive waste handling is the accumulated high-level radioactive wastes. Rowe and Holcomb [31] estimated that by the year 2000 there would be approximately 757,000 m^3 (200 million gallons) of high level wastes accumulated, 11.2 million m^3 (400 million ft^3) of low-level wastes, and 2.4 million m^3 (85 million ft^3) of alpha wastes. Committed costs for handling these wastes in the year 2000 were estimated to be $7 billion. Because of the high cost involved, it is recommended that interim storage facilities being proposed should become permanent as soon as possible.

Daiev [32] studied the cost of absorbing radioactive wastes on natural absorbants and then immobilizing the materials in bitumens at a cost of $285/$m^3$ for wastes ranging in activity from 10^{-4} to 10^{-9} Ci/L.

The possbility of using transmutation to destroy the radioactivity of long-lived radionuclides such as ^{90}Sr and ^{137}Cs was reported in a review by Harries [33]. It was concluded that the use of protons, electrons, or gamma rays rquired more electrical energy than the reactor produced in creating fission products.

Some older data [34] estimated that for a reactor burnup of 5000 MW-d heat/ton, the processing volume is 1200 gal/ton, and the allocated cost for waste disposal is 2%, or 0.16 mils/kWh. This results in an allowable cost for waste treatment of $4.00/gal. With a basic life of 3 yr, it was estimated that the cost of ground disposal would amount to 0.1¢/original gallon. The cost of operating burial sites ranges from $1.52 to $9.40/$yd^3$ of material, not including sample containers, processing, and handling. The cost of disposal in the Irish Sea at Seacale, England, including packaging and handling, varied from a low of $.30/lb along the east coast to $0.80–$1.00/lb along the west coast. Experience was reported in the demolition of an alpha-contaminated building of 48,000 ft^2. The cost of the demolition and burial was approximately five times the cost of demolishing a normal building.

At Hanford, Washington, nonboiling wastes in 500,000 to 1 million gallon tanks can be stored in reinforced concrete tanks with mild steel liners at a cost of $0.20–$0.25/gal [35]. Self-concentrating wastes may be stored at a cost of 40–50¢/gal of tank space. This amounts to approximately 0.01–0.05 mills/kW of electrical power produced [34].

An extensive study was made [36] comparing the costs of reprocessing and recycling fuel as opposed to single use and disposal. Some of the assumptions used in this study reveal some of the anticipated costs of waste treatment and handling. The cost of uranium will increase from approximately $35/lb of U_3O_8 to $85/lb as the cumulative short tons of U_3O_8 mined increases from 0 to 4 million short tons. Enrichment service charges are based on $100/SWU. Based on a 1500 MTU/yr fabrication plant costing $130 million, the base case UO_2 fuel fabrication price was estimated at $90/kg. The construction cost of a 1500 MTU reprocessing complex was assumed to be $1 billion with an annual operation cost estimated to be in the order of $50 million. Reprocessing services were estimated to cost $280/kgU. A fabrication cost for mixed oxide fuel was estimated to be $5/g or $260/kg total at 3.2% fissile Pu. The cost of final ultimate waste disposal including isolation repository disposition of high-level wastes, hulls, and other TRU-contaminated solids is estimated at $25/kg of fuel reprocessed. For the once through (without reprocessing) cycle, a cost of $20/kg was assumed for storage in basins to reduce the heat load to the desired level. Based on a 10-yr storage period and $70/kg encapsulation and isolation charge, the overall storage cost estimate was $90/kg of fuel discharged. Based on these assumptions, the total fuel cycle expenditures without recycle through the year 2000 amounted to $241.4 billion versus $224.8 billion with recycle. Assuming the generated power cost to be about 75% reactor cost and 25% fuel cost, this was expected to affect the generation cost of power by 2%, or approximately 0.5 mills/kWh. Thus, the assumption was made that reprocessing would be more economical than once-through use of nuclear fuel. This does not take into account the additional saving in the uranium resources and the potential increase in price of uranium with increased use. Certainly the conservation of uranium is desirable without putting a price value on this.

B. Relation to Benefits

The relation of costs to benefits in using nuclear fuel as a source of energy to generate electricity is difficult to evaluate. There are no set costs for the production of electricity by either nuclear or conventional fuel. Many of the earlier predictions for the cost of getting a nuclear power plant on line have had to be increased greatly because of the cost of environmental impact statements, the cost of answering and fighting environmental lawsuits, and the additional cost of greater safety precautions. On the other hand, the cost of fossil fuels is also increasing greatly for the same reasons. Coal, although abundant, is difficult to mine and as the readily mineable coal is removed, the more expensive and deeper coal veins must be tapped. The use of coal is also subject to high transportation costs and there are constantly increased costs in air pollution control. The increasing costs of air pollution control may be related to the increase in costs of nuclear reactor safety precautions. An estimate has been made [37] that between 1975 and 1990 the United States will have to put into operation at least 210 coal mines each with a production of 5 million tons/yr. This will require 2700 unit trains of 100 cars each. There will have to be seven coal gasification plants each capable of producing 250 million SCFD of synthetic natural gas as well as three coal liquifaction plants producing 50,000 barrels of synthetic crude oil. To supplement all of this it will be necessary to open 30 uranium mines each producing 2 million lb/yr, 450,000 oil and gas wells, 700 offshore platforms, and 31 refineries each producing 150,000 bbl/d. It is estimated that this will involve about $28 billion/yr of construction and technology costs based on 1975 dollars.

As for comparison with oil, the price of oil is extremely unpredictable at the present time. With the large imports of oil to the United States, not only is the increase in cost a problem, but the total output of United States dollars to other countries in a negative balance of trade is bad for the economy of the United States. Furthermore, both oil and gas are nonrenewable resources whose supplies are limited. An interesting comment was recently heard at a seminar at which the speaker stated that natural gas was considered an undesirable byproduct in oil wells and over the period through approximately 1950 more gas was wasted and flared off than remains available for use today. Thus, it may be seen that needs of the economy existing at the time a decision is made very frequently carry more weight than the availability and/or usefulness of the resource in the future.

In conclusion it may be stated that, whereas cost is an important factor in determining the use and expansion of the use of nuclear fuel as a power source for electrical generation, the overall problem must be considered in terms of the environmental impact of the power generating facility itself, the sources of fuel (i.e., mining techniques), and the dependence of one nation upon another for its major source of energy. For the immediate future, it appears that the best solution is the use of nuclear energy. In the long range, additional studies should be made to determine other sources of energy that are compatible both in terms of cost and environmental impact.

VIII. ANTICIPATED FUTURE PROBLEMS

A. Long-Lived Isotopes

A particular concern in radioactive waste management is the long-lived isotopes that result from the nuclear industry. Table 19 shows the long-lived isotopes of particular interest and their half-lives. It may be seen that some of these isotopes decay very slowly and will be around for a long time. Therefore, considerations must be made to handle the materials for extremely long periods of time, extending through several generations and in many cases through different political leaderships.

Of particular concern is plutonium, which in addition to being radioactive, is highly toxic. Since it has a very long half-life (24,390 yr), this toxic material will be present in the environment for extremely long periods of time and can therefore present a potential hazard to coming generations.

One of the particular concerns of long-lived radioactive isotopes is that their relative abundance tends to increase. Although a particular

TABLE 19
Long-Lived Isotopes of Concern
and Their Half-Lives

Isotope	Half-life, yr
^{3}H	12.3
^{39}A	260
^{85}Kr	10.7
^{90}Sr	28.9
^{129}I	1.7×10^7
^{137}Cs	30.2
^{235}U	7.1×10^8
^{238}U	4.51×10^9
^{239}U	2.44×10^4

long-lived radioisotope may be present as a very low percentage of the total radioactive waste discharged from a reactor, after a period of time, the short-lived radioisotopes disappear, leaving the long-lived radioisotopes in a relatively higher abundance. It must be pointed out that during this time, the long-lived radioisotopes are decreasing very gradually in total amount, but since the release of radioisotopes is continuous there is a buildup in these long-lived radioisotopes over a period of time. Thus, even though the relative amount of a long-lived radioisotope is small immediately upon discharge from a reactor, the relative amount will increase as continuous operation and more and more reactors are put into operation. The relative amounts of the significant isotopes discharged from a reactor after various times are shown in Table 20 [20,38].

It is these long-lived radioisotopes for which special long-term geological storage requirements are necessary. Provisions must be made to isolate them from the ecological system until their radioactivity has decayed to safe low levels. Being stored in a geological formation does not remove them from the environment, but does isolate them from an ecological system. At the present time, means exist for handling and storing these radioisotopes, but are not in effect. Long-term storage facilities or disposal methods must be developed now so that a critical situation does not arise in the near future.

B. High-Temperature Gas-Cooled Reactors

High-temperature gas-cooled reactors (HTGCR) have been in operation in Great Britain since the mid 1950s [20]. The first reactor built in the United States has been operated by the Philadelphia Electric Co. at Peach Bottom, PA, since 1967. In the HTGCR, the reactor core is constructed from hexagonally shaped graphite fuel elements within which the fuel is contained as rods of highly enriched (approximately 93% ^{235}U) uranium dicarbide and thorium carbide. The reactor is cooled by helium maintained under pressure. One of the characteristics of the HTGCR is that it is operated at higher temperature and pressure and is, therefore, more efficient thermodynamically (~40%) than a BWR or PWR (30%) and also discharges less waste to the environment.

The radioactive waste problems from a HTGCR are similar to those of PWRs and BRWs, with the possible exception of the potential production of radioactive gases from impurities present in the helium coolant. Special provisions are made for holup of the off-gases with treatment at the reactor site before discharge to the environment.

TABLE 20
Significant Fission Product Radioisotopes Remaining After Various Times [20, 38]

Isotope	Half-life	50 d	100 d	200 d	1 yr	2 yr	10 yr	30 yr	100 yr
^{85}Kr	10.7 yr	<0.1	<0.1	<0.1	<0.1	<1	1.7	<1	<0.1
^{89}Sr	54 d	8	10	8	4.5	<1	<0.1	<0.1	<0.1
^{90}Sr, ^{90}Y	28 yr	<0.1	<0.1	<1	1.8	5	24	23	18
^{95}Nb	35 d	10	20	25	14	0.6	<0.1	<0.1	<0.1
^{95}Zr	63 d	12	15	14	7	0.6	<0.1	<0.1	<0.1
^{103}Ru	39 d	8	7	4	<1	<0.1	<0.1	<0.1	<0.1
^{106}Ru, ^{106}Rh	1 yr	<1	<1	4	3	3	<1	<0.1	<0.1
^{131}I	8 d	1	<1	<0.1	<0.1	<0.1	<0.1	<0.1	<0.1
^{137}Cs, ^{137}Ba	26.6 yr	<0.1	0.2	0.45	1.5	4	18	23	32
^{140}Ba	13 d	6	0.1	<0.1	<0.1	<0.1	<0.1	<0.1	<0.1
^{140}La	40 hr	8	1	<0.1	<0.1	<0.1	<0.1	<0.1	<0.1
^{144}Ce, ^{144}Pr	290 d	3.4	8	13	26	30	0.1	<0.1	<0.1
^{147}Pm	2.6 yr	0.4	1	2.5	6	15	18	<1	<0.1
^{151}Sm	87 yr	<0.1	<0.1	<0.1	<0.1	<1	2.6	2.5	1.2

C. Liquid-Metal Fast-Breeder Reactors

All BWRs and PWRs, which are currently predominant in the nuclear power generation field, are inherently inefficient, converting only 1–2% of the potentially available energy of the uranium into heat. In contrast, the liquid metal breeder (LMFBR) can economically use up to about 75% of the energy contained in uranium, thereby achieving efficiencies about 40 times greater than LWRs [20]., The advantage of the LMFBR is that it actually produces more fissionable materials than it uses. A conventional 1,000 MW(e) LWR during a 30-yr lifetime will require 3900 metric tons of natural uranium and will produce 3870 metric tons of depleted uranium and 5300 kg of plutonium. A similar sized LMFBR will use 100 metric tons of depleted uranium and 2300 kg of plutonium, but in addition to the power generated, it will produce 50 metric tons of depleted uranium and 7700 kg of plutonium.

Sodium is normally used as the heat exchange medium in the LMFBR. However, the sodium becomes highly radioactive in passing through the reactor core, and therefore, a secondary heat exchanger is provided, also with a sodium cycle. This secondarily heated sodium is then directed to a tertiary heat exchanger to generate steam to drive a turbine generator. Precautions must be taken to prevent the contact of the sodium with water, which results in an exothermic reaction. The intermediate heat exchanger prevents the contact of the radioactive sodium, which is passed through the core, with water. Thus, if there is any leakage of sodium into the water in the heat exchanger, it would be from nonradioactive sodium rather than the radioactive sodium.

Another concern with the LMFBR is the production and inventory of plutonium. Plutonium more importantly, but along with ^{233}U and ^{235}U, is an essential element in the construction of nuclear weapons. Thus, any nation having this type of reactor would more readily be able to produce sufficient amounts of plutonium to build its own atomic arsenal, and thereby potentially threaten the rest of the world. Another concern is that of theft. An individual or group of individuals could potentially arrange to steal plutonium from the reactor, the fuel reprocessing site, or during the element's transport between these two areas to build its own nuclear weapon, and thereby be enabled to extort from groups or a nation whatever it may desire. Whereas this is a concern, it is not a problem that will be taken up in this discussion.

Discharges of radioactive materials from a LMFBR have been reported by Erdman and Reynolds [39]. They have shown the average plutonium inventory in the LMFBR to be significantly greater than in a uranium-fueled LWR of the same size, but it is less than an order of magnitude greater than that for the LWR. The average plutonium inventory of an LWR fueled with recycled plutonium will be about half of the inventory of an LMFBR, thereby posing similar toxicity control problems.

The fission products inventory per metric ton of metal is much lower for the LWR than for the LMFBR [39]. A slightly greater quantity of radioactivity will be shipped annually in tons of fuel for reprocessing from the LMFBR than from the PWR, assuming equal cooling time after removing from the reactor. All reactors generating the same amount of electrical power generate fission products at roughly the same rate, with differences being introduced by differences in thermal efficiency and in energy released and fission products generated per fission. The higher overall fission product activity in the discharged LMFBR fuel results primarily from the shorter residence time of the fuel in the reactor (i.e., 540 d for the Al LMFBR vs 3 yr for the PWR).

Tritium produced in an LMFBR comes from two principle sources: ternary fission in the fuel and boron reactions in B_4C control rods. The estimated annual tritium production rate from a 1000 MW(e) LMFBR has been estimated at 20,000 Ci/yr compared with approximately 15,000 Ci/yr in a similar sized LWR. The HTGCR is estimated to produce approximately 7000 Ci/yr of tritium from reactions with the boron and 2500/Ci/yr from reactions with lithium. Comparable values for a PWR indicate 700 Ci from boron and 750 Ci from lithium. Tritium production rates in control rods of a BWR are higher, but the tritium does not escape from the rods.

The sodium coolant may be activated to form both ^{24}Na and ^{22}Na. The ^{24}Na has a short half-life of 15 hr and decays to a low value shortly after shutdown. The ^{22}Na poses a maintenance or sodium storage problem because of its 2.6-yr half-life.

The gaseous activation products may include ^{41}Ar, ^{39}Ar, and ^{23}Ne. The ^{41}Ar has a 1.3-hr half-life and the ^{23}Ne has a 37.6 s half-life. The latter is a very minor problem, whereas the former merely presents a problem in holdup time before release of the gas. On the other hand, ^{39}Ar, may be produced by activation of naturally occurring ^{39}K in the sodium coolant. Since this has a 260-yr half-life, it presents a potential problem. It was estimated [39] that 0.13 Ci/d of ^{39}Ar would be produced in an LMFBR if the potassium concentration in the coolant is 1000 ppm.

Most of the discussion of waste releases from LMFBRs is based upon calculated estimates of releases because insufficient data are presently available from operating LMFBRs to provide more precise measurements of actual waste releases.

D. Krypton-85

Krypton-85 is a particular problem in the reprocessing of spent fuel. It has been estimated [20] that approximately 10,000 Ci of ^{85}Kr are released with about 43,000 ft^3 of waste gas for each ton of uranium reprocessed. The particular problem is a combination of the 10.7 yr half-life and its almost total chemical inertness. Since it does not combine with other materials in the atmosphere, it tends to accumulate there and may present a future problem.

By late 1970, the ^{85}Kr concentration had increased to 16×10^{-6} pCi/cm^3 [20]. If no provisions are made for noble gas removal during fuel reprocessing, based on an estimated world-wide nuclear generating capacity of 50 billion kW by the middle of the 21st century, it is estimated that the ^{85}Kr concentrations in the atmosphere could increase by several orders of magnitude, possibly reaching a concentration of 0.3 pCi/cm^3. This could present an external radiation dose of 300 Mrem/yr or about three times the existing normal background radiation. Thus, this could be a significant problem in the future if provisions are not made to remove this product. This is now little considered as a need in radioactive waste management.

Since krypton does not enter into many reactions, its chemical removal is quite difficult. At present the best means of removal is in a cryogenic trap maintained at an extremely low temperature at which the krypton would be liquidified and could be separated from other gases that would also be liquidified at this temperature. Such a procedure would require considerable amounts of energy and the krypton removed would ultimately have to be stored in gas cylinders for long periods of time. Thus, consideration must be given at the present time to how best to control the release of ^{85}Kr to the atmosphere, particularly from fuel reprocessing plants.

E. Fusion

The next major step in the production of power is the use of nuclear fusion [40]. Whereas the commercial use of fusion for production of electricity is still probably 20 or more years in the future, the first continuous fusion reaction was maintained at the Tokamak Research Facility at Princeton in 1976. The major problem in utilizing this almost unlimited source of energy is controlling the heat of the reaction. At the present time, there are no construction materials available that could stand the intense heat. Thus, other means of supporting the reaction must be provided. None of these have been subjected to any significant heat intensity as of the present time. Another problem is the overall size of a reactor. It is estimated at the present time that an area on the order of the size of a football field would be necessary to provide the reaction space.

The two major waste problems from fusion are the production of tritium and the activation of the containment vessel. The lifetime of the containment vessel is estimated to be in the order of only a few years. The decontamination and disposal of such a large facility at such a great frequency would result in the necessity for dismantling an disposal of tremendous amounts of contaminated solid wastes. The tritium on the other hand would most likely occur as a waste in the forms of gas and water. Both forms are difficult to dispose of. It is difficult to separate the tritium from the normal hydrogen in water. Thus, large volumes of water would have to be disposed of or placed in storage. Because of the long (12.3 yr) half-life of tritium, this represents a significant problem in waste management. Since fusion reactors are still in the planning stage, it is possible to consider planning for management of tritium wastes along with planning for the construction of the facility.

IX. SUMMARY

The atomic age is upon us. In our desire for greater and greater amounts of energy we have committed ourselves to the use of atomic energy to provide our needs. A significant amount of our electricity is already being provided by nuclear energy. In order to provide for the increases in electrical demand for the next 50 yr, nuclear energy with coal and conservation are the only sources readily expandable in the amounts needed. Thus, additional nuclear power facilities will have to be built in the near future.

The public is rightly concerned over the problems of accumulation and disposal of radioactive wastes. This is also of concern to the nuclear regulating agencies. Plans have been made and facilities constructed to handle the known radioactive wastes. There is still room for improvement in methods for waste management and for identifying wastes that presently and in the future will need proper management. As our knowledge increases, we learn not only of new problems, but of new ways to solve these problems. These ways must be implemented in order to prevent the proliferation of radioactive wastes to the environment. The methods for waste handling are not simple, nor are they inexpensive. On the other hand, means for controlling environmental pollution from other energy sources are also complicated and expensive. The public must become aware that energy in the future will continue to increase in cost and will be a continued cause of concern. Studies should be initiated to establish other more reliable, less expensive, and less hazardous forms of energy. However, at the present time, we do not have these, and reliance will have to be placed on nuclear energy. The problem is with us; it will not go away. The best we can hope for is an adequate and reliable means of handling and managing the wastes that are produced by the nuclear industry.

REFERENCES

1. Safety of Nuclear Power Reactors (Light Water Cooled) and Related Facilities, WASH-1250, July 1973.
2. G. M. Chauncey, *Amer. J. Physica* **14,** 226 (1946).
3. C. J. Craven, "Our Atomic World", Understanding the Atom Series, US AEC, Washington, DC, 1969.
4. E. Gloyna and J. O. Ledbetter, *Principles of Radiological Health,* Dekker, NY, 1969.
5. S. Glasstone and A. Sesonnke, *Nuclear Reactor Engineering,* Van Nostrand, Princeton, 1953.
6. International Commission on Radiation Protection, "Radiosensitivity and Spatial Distribution of Dose," Publication 14, Pergamon Press, Oxford, 1969.
7. National Energy Outlook, Federal Energy Administration, 1976.
8. M. K. Hubbert, *Sci. Amer.* **224** (3), (September 1971).
9. D. B. Aulenbach, "Lesser Known Energy Sources", Transactions of Environmental Aspects of Non-Conventional Energy Resources—Topical Meeting, *TANSO* **23,** Supplement No. 1, 22 (1976).
10. "Certain Background Information for Consideration when Evaluating the "National" Energy Dilemma", Staff of the Joint Committee of Atomic Energy, Congress of the US, 1973.
11. J. Barnard, L. F. Fidrych, A. S. Gibson, K. M. Horst, P. M. Murphy, & B. Wolfe, *Nuclear News* **15** (12), (December 1972).
12. "Experts Mull Over Radioactive Waste Disposal," Report on the International Symposium on the Management of Wastes from the LWR Fuel Cycle, USERDA, Denver, Colorado, July 1976, *Chem. Eng. News* **54** (32), 21 (August 2, 1976).
13. J. O. Blomeke, "Projections of Future Requirements in Waste Management," paper presented at the Health Physics Meeting, NY, NY, July 14, 1971.

14. W. G. Belter, "Nuclear Wastes Management: An Overview of Current Practice," presented at the 16th Annual Health Physics Meeting, NY, July 14, 1971.
15. Environmental Survey of the Nuclear Fuel Cycle, AEC, WASH 1248.
16. "The Effects on Population of Exposure to Low Levels of Ionizing Radiation" (BEIR Report), National Academy of Sciences, National Research Council, November 1972.
17. Administration of the Federal Metal and Non-Metallic Mine Safety Act, P.L. 89–577, Annual Report of the Secretary of the Interior, 1975.
18. "Reactor Safety Study, An Assessment of Accidental Risks in US Commercial Nuclear Power Plants", USAEC, WASH-1400 (NUREG75/014), October 1975.
19. N. D. Dudey, "Review of Low-Mass Atom Production in Fast Reactors", ANL-7434, Clearing House for Federal Scientific and Technical Information, Springfield, VA, 22151, 1968.
20. M. Eisenbud, "Environmental Radioactivity", Academic Press, NY, 1973.
21. W. P. Bishop and F. J. Miraglia, Jr., "Environmental Survey of the Re-Processing of Waste Management Portions of the LWR Fuel Cycle", NUREG-0116, (Supplement to WASH-1248), October 1976.
22. "Environmental Radioactivity Surveillance Guide," USEPA, Office of Radiation Programs, ORP/SID 72-2, June 1972.
23. S. L. Williams, D. B. Aulenbach, and N. L. Clesceri, "Sources and Distributions of Trace Metals in Aquatic Environments," Chapter 2 in *Aqueous Environmental Chemistry of Metals*, A. J. Rubin, ed., Ann Arbor Science Publishers, 1974, p. 107.
24. J. F. Fletcher and W. L. Dotson, "HERMES—A Digital Computer Code for Estimating Regional Radiological Effects from the Nuclear Power Industry", USAEC, HEDL-TME-71-168, US-80, Reactor Technology, December, 1971.
25. NRC Regulatory Guide 1.109, "Calculation of Annual Doses to Man from Routine Releases of Reactor Effluents for the Purpose of Evaluating Compliances with 10CFR50, Appendix I".
26. G. G. Eichholz, *Environmental Aspects of Nuclear Power*, Ann Arbor Science, 1977.
27. "Typical Disposal Wells", R-11, Reference Number, *Water Works and Sewage*, April 30, 1975.
28. D. B. Aulenbach, in *State of Delaware Intrastate Water Resources Survey*, Dover, Delaware, 1959.
29. Code of Federal Regulations, "Technical Specifications on Effluents from Nuclear Power Reactors", 10CFR50, Section 50.36A, Superintendent of Documents, US Government Printing Office, Washington, DC 20402.
30. Code of Federal Regulations, "Reactor Site Criteria", 10CFR100, Superintendent of Documents, US Government Printing Office, Washington, DC 20402.
31. W. D. Rowe and W. F. Holcomb, *Nuclear Technol.* **24** 286 (1974).
32. K. Daiev, "Study of Bulgarian Bitumens and Natural Sorbents for Use in the Processing and Disposal of Radioactive Wastes", Report 1-AEA-R-428-F, International Atomic Energy Agency, Vienna, 1972.
33. J. R. Harries, "Transmutation of Radioactive Reactor Waste", Report AAEC/E-326, Australian Atomic Energy Commission Research Establishment, Lucus Heights, Australia, 1974.
34. H. Heukelekian, *Sewage Ind. Wastes* **29**, 613 (1957).
35. D. P. Grandquist and R. W. Tomlinson, "Estimated Magnitude of Waste Disposal

Problems," US Atomic Energy Commission Document, #TID, 7517, Part IB, 400, 1956.
36. "Benefit Analysis of Reprocessing and Recycling Light Water Reactor Fuels," ERDA-76/121, December 1976.
37. "Outlook for Coal: Bright But With Problems," *Chem. Eng. News* **55** (No. 7), 24 (Feb. 14, 1977).
38. J. J. Sabo, J. E. Martin, and R. F. Grossman, "Fission Product Nuclides, Beta-Gamma Energies According to Age," personal communications, September 1959.
39. C. A. Eridman and A. B. Reynolds, *Nuclear Safety* **16** (No. 1), 43, (Jan-Feb. 1975).
40. D. Steiner and A. P. Fraas, *Nuclear Safety* **13,** 353 (1972).
40. T. G. Spiro and W. M. Stigliani, *Environmental Science in Perspective,* State University of New York Press, Albany, NY, 1980, pp. 51–66.

8
Drying and Evaporation Processes

George P. Sakellaropoulos
Department of Chemical Engineering, University of Thessaloniki, Thessaloniki, Greece

I. INTRODUCTION

Water removal from municipal and industrial effluent streams constitutes an important step in wastewater and sludge treatment. The purpose of such treatment is to concentrate, separate, dispose, or utilize wastes and pollutants and to regenerate and return clean water to the environment. In this context, the discussion here will be limited only to industrial and municipal sludge dewatering, evaporation, and drying.

Although sand bed dewatering of sludges has been popular in small communities, heat drying or evaporation have proved feasible only in few instances. The reason is primarily economical. Water evaporation and heat drying are currently expensive and require fuel consumption to remove the water. They become feasible when the dried sludge can be sold as a fertilizer or used as a vitamin and protein enriched animal feedstock. Such possibilities depend not only on the regular market, but also on the attitude of the public to demand recycling of wastes and accept the associated cost. Sludge return to the environment in a dry form and utilization of its nutrient content may be ecologically more attractive than the current trend towards incineration. Although the latter uses the heat content of the sludge to accomplish combustion, the problem of ash disposal remains.

Recent incineration practice usually involves prior removal of excess water and sludge thickening through dewatering and drying steps. Preliminary drying of the sludge may take place either in a separate unit or in the first section of the incinerator. Furthermore, incineration equipment is often designed so as to permit flexibility of operation during either drying or combustion. The significance of the water removal steps in various sludge and wastewater treatment designs, the associated energy demands and costs, and the possible improvement of the market for dried sludge warrant analysis and further evaluation of drying and evaporation processes.

A rigid, sharp distinction among evaporation, dewatering, and drying does not exist. All involve water removal to some extent, and also reduction of the weight and volume of the fluid sludge to that of a concentrated sludge or a moist solid. Usually the water content of a dewatered sludge is higher than that of a dried sludge. However, sand beds used for physical dewatering are termed "drying beds," despite the considerable water content of the remaining cake, in order to distinguish them from other mechanical dewatering systems. In general, water is removed in drying beds and lagoons by natural drainage and evaporation. Drying to low water content requires high-temperature water removal in mechanical dryers. Depending on the drying temperature simultaneous sludge sterilization can be achieved.

The origin of the effluent stream, the final desirable moisture content, and the end use of the sludge determine whether dewatering is sufficient or heat drying is necessary. Since thermal drying is usually more expensive than physical or mechanical water removal, dewatering prior to heat drying is desirable.

Before attempting any theoretical analysis of drying and evaporation processes, the design and operating characteristics of some basic units will be presented. This will familiarize the reader with the operation of physical and mechanical dryers and evaporators and will permit a rational theoretical treatment of these systems. Drying and evaporation consist of a combination of mass and heat transfer processes common to all dryer designs for a given sludge. Therefore, the theory of evaporation and heat drying will be treated uniformly for all systems. Initial dewatering in drying beds is essentially a filtration process and will be examined in a separate section.

II. NATURAL DEWATERING

In natural dewatering units, water is removed by drainage and/or evaporation. Two systems of particular importance for wastewater treatment will be discussed, i.e., drying beds and drying lagoons. Since the design and operation of such systems is presently more of an art than a science, a number of empirical design criteria and variables will be discussed here rather than deferring them to the theory discussion.

A. Drying Sand Beds

1. Sand Bed Design

The popularity of sand bed dewatering of sludge appears to be undisputed. Burd reports that over 6000 sewage plants use covered or uncovered sand beds in the United States [1]. About 38% of the cities with population over 100,000 and at least 60% of smaller cities use this technique [2]. The same method is also widely used for dewatering municipal sludges in Europe and for treatment of many industrial sludges.

A schematic diagram of a drying bed is shown in Fig. 1. It consists of successive layers of gravel and sand on which the sludge is deposited. The sand layer has a thickness of 4–9 in. Typical effective size of the sand is 0.3–1.2 mm with a uniformity coefficient of less than 5.0 [1, 3, 4, 4a]. The gravel layer consists of graded gravel or stone, ⅛ to 1.0 in., with a bed thickness of 8–18 in. The water drained from the sludge is collected

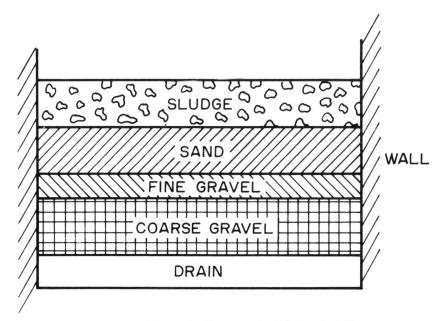

Fig. 1. Schematic diagram of a "drying bed."

in underdrains often made of vitrified clay laid with open joints. Underdrain piping, of about 1% minimum slope, is usually 4 in. in diameter and spaced 8–20 ft apart.

The Ten State Standards recommend the following design considerations [3].

- Use of ⅛ to ¼ in. gravel for the top 3 in. of gravel
- Gravel should extend at least 6 in. over the top of the underdrains
- The gravel layer should be 12 in. deep
- The sand layer should be at least 6–9 in. deep
- Underdrains should be no more than 20 ft apart
- Bed walls should be watertight and extend 15–18 in. above and at least 6 in. below the surface
- Outer walls should be curved to prevent soil from washing into the bed
- The influent pipes should terminate at least 12 in. above the surface, with concrete splash plates provided at discharge points
- Pairs of concrete truck tracks at 20-ft centers should be provided for all beds
- At least two beds should be provided

Sludge is usually charged from 6 to 12 in., depending on solid content, its physical characteristics and operating variables. It is important to distribute the sludge evenly over the bed for uniform dewatering. The dewatered sludge is hand removed or lifted mechanically. Mechanical lifting is feasible with sludges of 20–30% solids while hand removal requires 30–40% solids [4].

Sand beds can be open or covered depending on climatic conditions. Glass-covered sand beds are protected from rain and cold, improve the appearance of waste treatment plants, and control odors and insects. To control humidity and optimize water evaporation, the enclosure should be well ventilated. Covered beds require only ⅔ to ¾ of the area required for an open bed. However, mechanical sludge removal is more difficult in the former.

Dewatering on sand beds proceeds via two different mechanisms; filtration and evaporation. A theoretical analysis of these mechanisms will be presented with the general theory of drying in a later section. Water drainage is most important during the first 1–3 d leaving solid concentrations as high as 15–25% [5, 6]. Further water removal occurs by evaporation. Evaporation is facilitated by horizontal shrinking of the sludge and exposure of additional sludge areas [3]. Vogler and Rundolfs estimated that 60% of the water is drainable [7]. Up to 85% of the water of secondary sludges can be lost by drainage [8, 9]. In general, the higher the initial water content, the larger the fraction of drainable water [10].

2. Operating Variables

The described mechanisms for water removal impose a number of operating variables that affect the design of drying beds, such as:
  i. Water conditions
  ii. Sludge characteristics
  iii. Soil permeability
  iv. Land availability and cost

Air drying of sludge is sensitive to weather conditions. Rain lengthens the drying time, but its effect is less important once shrinkage and cracking has started [1]. However, it reduces the sludge fertilizing value by dissolving valuable nutrients[1].

Air temperature, relative humidity, percentage of sunshine, and wind velocity also affect the rate of water evaporation. In the summer or at high temperature and humidity, the rate of drying is 2–3 times faster than in the winter or at low temperatures [11–13]. It is noteworthy that in many wastewater treatment plants sludge is stored in digesters in the winter and dried only in the summer [1].

Raw sludge does not dry as easily as digested sludge. Its odor and attraction of insects are also serious drawbacks. Therefore, sand bed drying is usually restricted to well-digested sludges. Entrained gases in digested sludges float solids, leaving a layer of clear water that drains through the sand easily [3]. Haseltine reported, however, that overdigested sludges dry slower [14].

Primary sludges dry faster than secondary ones. Aged sludges dry slower than new ones. Oil and grease tend to clog the bed pores and retard drying. On the other hand grit facilitates drying [1].

The drying time is affected by the initial concentration of solids in the sludge. Vankleeck reported a doubling of drying time for an increase of solids in a sludge from 5 to 8% [15]. Laboratory results show that the amount of drainable water decreases approximately linearly with increasing solid concentration, as shown in Fig. 2 [7]. Similarly, the drainage time and the dried cake moisture increase with increasing solid content, as shown in Figs. 3 and 4 [7].

For optimum drying the solids loading of the sand bed should not

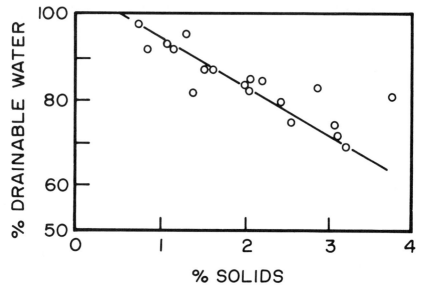

Fig. 2. Effect of solids concentration on drainable water [7].

exceed 15 lb of dry solids/ft² for uncovered beds and 25 lb/ft² for covered beds [16].

The rate of drying depends on the depth of the sludge charge over the sand bed [17]. Drainage time and cake moisture at 24 h increase with increasing depth of the charge [7, 13], as shown in Figs. 5 and 6.

The drying rate can be increased by sludge conditioning with organic or inorganic coagulants–flocculants. Chemical conditioners permit higher sludge porosity, decrease solids compression, and result in reduced sand bed maintenance [1]. Inorganic chemicals such as alum, ferric chloride, sulfuric acid, anthracite, and activated carbon have been used [18]. Several organic polyelectrolytes have been tested [18, 19] sometimes with conflicting results.

Sperry found alum the most effective and economic conditioning additive [20]. Alum, aluminum chlorohydrate, and ferrous sulfate are in common use in Britain [21]. Recommended alum dosage is 1 lb/100 lb of sewage sludge [15, 22]. For thin sludges (<3% solids), one-half to one-third of this dosage suffices [20].

Trivalent iron salts are also effective flocculating agents for sewage sludge [20, 23], although somewhat higher dosages may be necessary. Ferric chloride is the most commonly used inorganic coagulant in the United States. A dosage of 90 lb/ton (dry) results in a liftable dried sludge after 10–20 h [23].

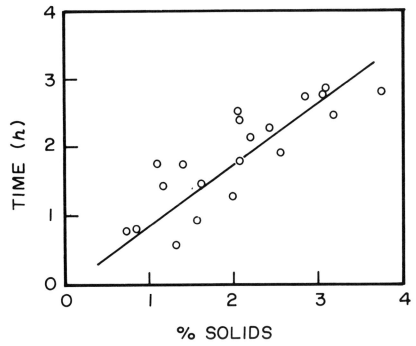

Fig. 3. Dependence of drainage time on solids concentration [7].

The amount of inorganic ionic coagulants required for effective sludge drying depends on the sludge pH. Since metal ion conditioning additives lower the pH, highly buffered sludges require large amounts of coagulants before any noticeable pH change occurs. Weber discusses the reactions, the effect, and the mode of operation of ionic coagulants [18].

In contrast to the high dosages of inorganic additives, organic polyelectrolytes are effective at dosages less than 1% of the solids weight. However, they are appreciably more expensive than the inorganic salts.

The efficiency of organic polymers in promoting drying decreases in the order [24]:

raw primary sludge > digested sludge > digested activated sludge

Inorganic sludges are less susceptible to conditioning with organic polyelectrolytes. Gates and McDermott found some improvement with certain polymers [25]. Neubauer observed no effect on inorganic sludge drying [26].

Organic flocculants do not only increase the rate of drying by as

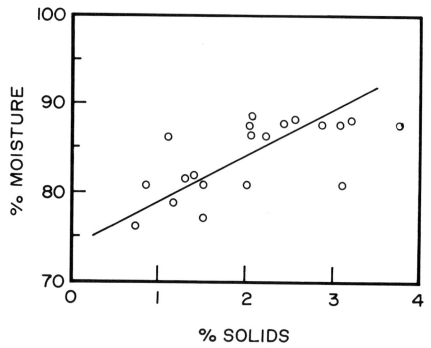

Fig. 4. Cake moisture dependence on solids concentration [7].

much as 30 times, but they also permit heavy bed loadings up to 3 ft [2]. Figures 7 and 8 show the effect of polymer dosage and bed depth on the solids concentration with time.

Most of the discussion above referred to conventional beds, built directly on soil. Attempts to use mechanical cleaning of sand beds or to increase drying rates, led to alternative bed designs such as paved beds, wedge-wire beds, and heated beds. Evaporation becomes important with some of these bed designs. Asphalt and concrete-paved beds proved that bed performance was not impaired by the pavement [28–32]. Shorter drying times were reported, since mechanical lifting permitted removal of the sludge with higher water content compared to hand cleaning [29].

Wedge-wire beds have operated successfully in England [1, 33, 34] and in the US [4]. A diagram of a wedge-wire bed is shown in Fig. 9. Initially, support water is introduced to prevent blinding of the filtering medium. After the sludge is applied to the bed, this free water is drained and the sludge is allowed to dry by drainage and evaporation. The following difficult sludges have been dewatered successfully over wedge-wire

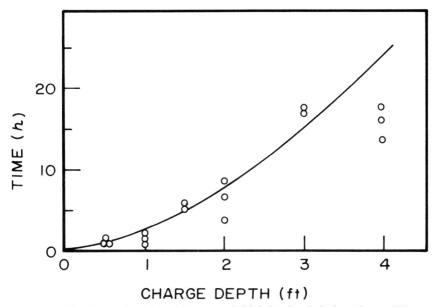

Fig. 5. Drainage time dependence on initial depth of sludge charge [7].

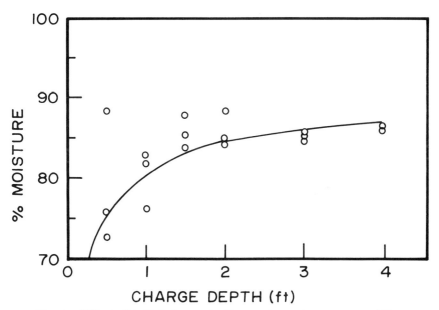

Fig. 6. Effect of initial charge depth on cake moisture content [7].

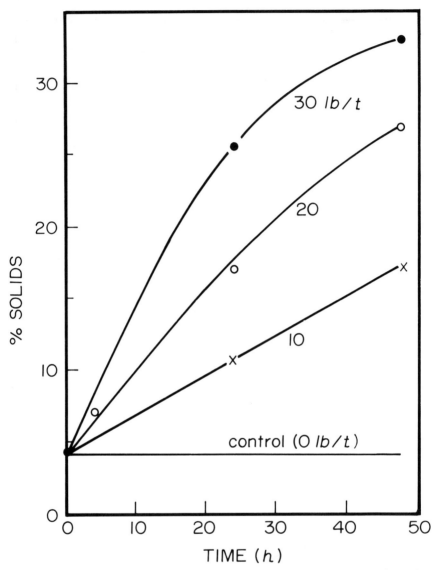

Fig. 7. Dewatering of digested sludge treated with varying amounts of a polymeric flocculant [2]. Reprinted with permission from the Dow Chemical Company.

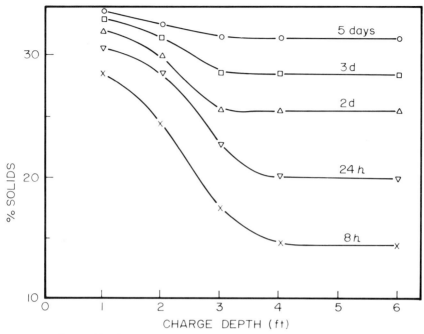

Fig. 8. Effect of initial sludge depth on solids concentration in the cake for an 8% sludge treated with a polymeric flocculant dose of 60 lb/ton [2]. Reprinted with permission from the Dow Chemical Company.

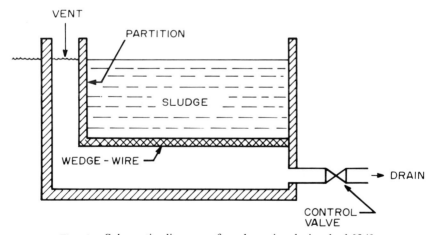

Fig. 9. Schematic diagram of wedge-wire drying bed [34].

beds: tannery sludges, slag fines, bacterial slimes, vegetable wastes, and hydroxide sludges [35]. Some typical performance data with wedge-wire beds are shown in Table 1.

Wedge-wire beds appear to be resistant to clogging, less susceptible to adverse weather, easy to clean and maintain. Bed capacity is increased because of shorter drying and cleaning times.

Heating sludge with hot water, circulated into covered beds through heating coils, improves the bed drying capacity appreciably [1, 36]. Heating probably accelerates biological decomposition to form gases that float the solid particles and allow water drainage [8]. It also breaks down the colloidal structure of the sludge and promotes particle coalescence [37]. Finally, heating decreases the sludge viscosity, thereby improving drainage rates.

3. Land Requirements

The rate of sludge drying, dependent on the previously discussed parameters, determines the minimal land area required for the sand bed installation. Open beds require a larger area than covered beds; see Table 2. In warm climates the area requirements may be lower than those of Table 2. Furman recommends only 0.3 ft^2/capita area for drying of digested primary sludge and of trickling filter humus in Florida [38].

Eckenfelder and O'Connor recommend the solid loadings per unit bed area for sewage sludge drying given in Table 3 [5]. The effect of sludge properties on bed area requirements is shown in Table 4.

The area requirements decrease for chemically conditioned sludges. This results from the greater solids loadings permitted for conditioned sludges, the application of deeper sludge loads, and the faster drying rates of such sludges.

TABLE 1
Performance Data of Wedge-Wire Beds [34]

Type of sludge	Feed solids, %	Cake solids, %	Time	Solids capture, %
Primary	8.5	25.0	14 d	99
Trickling filter humus	2.9	8.8	20 h	85
Digested primary, activated	3.0	10.0	12 d	86
Fresh, excess activated	0.7	6.2	12 h	94
Thickened, excess activated	2.5	8.1	41 h	100

TABLE 2
Recommended "Ten States Standards" Sand Bed Areas
for 40° and 45°N Latitude [3]

Type of sludge	Bed area, ft^2/capita	
	Open bed	Covered bed
Primary	1.00	0.75
High-rate trickling filter	1.50	1.25
Activated sludge	1.75	1.35

Land unavailability limits application of sludge drying on sand beds at the vicinity of large cities. However, sand beds will continue to be important sludge drying facilities in medium and small municipalities and plants. Pipe line transportation of fluid sludges to rural areas may prove economical and advantageous for sludge disposal.

B. Drying Lagoons

Operation of drying lagoons is similar to that of sand beds, but the drying process is generally slower, proceeding mainly through evaporation. Because of the longer retention of water in lagoons, sludge stabilization is necessary to minimize noxious odors.

Factors discussed in relation to sand bed design, such as climatic conditions, sludge properties, subsoil permeability, sludge load, and so on, also determine the design of lagoons.

At least two [3] or even three [40] lagoons should be provided, of a maximum depth of 2 ft. Depending on the climate and the sludge characteristics, 1–4 ft^2/capita are required for sludge drying. A solid loading rate of about 2.4 lb/ft^3 yr is recommended [3].

TABLE 3
Solids Loading Per Unit Bed Area

Type of sludge	Sludge loading rate, dry solids lb/ft^2 yr
Primary	27.5
Primary, trickling filter	22.0
Primary, activated	15.0
Chemically precipitated	22.0

TABLE 4
Recommended Sand Bed Areas
for England [39]

Type of sludge	Open bed area, ft^2/capita
Primary	1.30
Trickling filter	1.50
Digested, mixed	1.00
Undigested, mixed	2.25
Greasy sludge	3.00

Lagoon dewatering of sludges does not usually result in fork-liftable sludge. Dewatering from 5% solids to 40–45% solids lasts 2–3 yr [1, 4, 16], and a 3-yr cycle is usually recommended for lagoon dewatering [1, 4, 40]. Sludge is first dewatered in a lagoon for one year. The lagoon is then allowed to dry for 12–18 mo, followed by a rest period of 6–12 mo.

Comparison of drying times, land requirements, and solids loading rates between sand beds and lagoons obviously favors the first dewatering technique. However, lagoons are quite commonly used for sludge drying where inexpensive land is available, because of their simple, low cost operation. Burd [1] and Howells and Dubois [41] have discussed in detail the operating cost of drying lagoons.

Several large cities have used lagoon drying successfully for several years [1]. Many inorganic, industrial sludges are often dried in lagoons since offensive odors are minimal. Lagooning has proved economical for oil and metal finishing sludges for which vacuum filtration is difficult.

III. THERMAL DRYING

Heat drying of sludges in mechanical dryers aims primarily at reclaiming their nutrient content by using them as agricultural fertilizers or enriched feedstock. Dried sludge fertilizers and soil conditioners can increase significantly crop yields [42] and the content of organic matter moisture and nitrogen of the soil [43]. However, for best balance, combination with inorganic fertilizers is advisable.

Despite the value of organic sludges as soil conditioners, little public promotion of dried sludge has been done. This and certain health precautions necessary for sludge fertilizer handling [44] have probably resulted in a relatively small market for heat dried sludge. Yet its use as a fertilizer represents a natural disposal of solids in the environment and a recycling of natural resources.

Three large cities, Chicago, Houston, and Milwaukee, have successfully promoted disposal of heat-dried activated sludge as a fertilizer. Over 200,000 tons of dried sludge, of 5–6% nitrogen content, have been marketed by these cities at an average price of $11–20/ton. The price depended on the type of raw materials, the method of preparation, the location, and the public attitude.

Organic sludges from municipalities and food processing plants contain appreciable amounts of proteins, grease, and vitamins. Concentrated or carefully dried sludges of this nature have been used as animal feed. Pilot plant tests at Milwaukee indicate that about 300 lb of pure vitamin B_{12} can be produced yearly from this city's 70,000 tons/yr sludge supply [45]. Other investigators [28, 46, 47], have confirmed the value of heat-dried activated sludge as vitamin and protein supplement for animal feeds. Food waste sludges have been successfully and economically used as supplement feedstock after filtering or multistage evaporation and heat drying [48, 49]. Burd [1] discusses nonfertilizer uses of sludges from various food industries in detail.

Inorganic industrial sludges can be dewatered or calcined to recover and recycle useful materials. Paper pulp sludges [50], waste tars, spent catalyst [51], and alumina processing muds [52] have been treated thermally at high temperatures. The first stage of such recovery treatments involves dewatering, evaporation, or drying.

Heat drying is also employed as a preliminary dewatering stage in sludge incineration to improve the thermal efficiency of the incineration processes. Flash, spray, and rotary dryers and multiple effect evaporators have been used as precursors for drying of municipal sludges prior to incineration [52a].

The driving forces for drying in mechanical, thermal dryers are the temperature difference and the water content difference between the drying air and the solids. Five major designs of dryers are discussed here. These designs have been used to some extent for sludge drying and some are potentially useful for specialized applications to recover heat-sensitive materials from sludges. Further description of mechanical dryers commonly used in chemical processing is available in standard references [53–56].

The same dryer designs can be used for removal of liquids other than water. Recently, a laboratory scheme has been proposed (BEST—Basic Extractive Sludge Treatment) to extract water from the sludge with triethylamine (TEA), followed by TEA removal in a continuous dryer and by recycling. Because of the low latent heat of vaporization of TEA (309 kJ/kg or 133 BTU/lb, compared to 2320 kJ/kg or 1000 BTU/lb for water), this proposed method would require less energy for drying than with conventional drying. Economic evaluation of the BEST process for Seattle's Metropolitan Engineers and by the LA/OMA project showed that it is one of the most expensive sludge disposal systems [56a]

A. Tray Dryers

A schematic diagram of a batch tray dryer is shown in Fig. 10. Sludge is loaded on trays and placed in racks (H) mounted on truck wheels (I). Fan (C) circulates air, with the aid of motor (D), passing over heating elements (E). Baffles (G) distribute the heated air over the trays. Some moist air is continuously vented (B), replenished by fresh air entering through

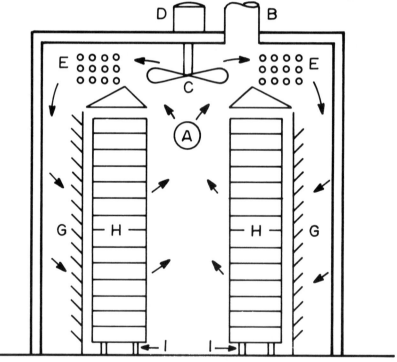

Fig. 10. A batch tray dryer: A, fresh air inlet; B, vent; C, circulation fan; D, motor; E, heating elements; G, distribution baffles. Reprinted with permission from ref. [53]. Copyright 1955 McGraw Hill.

inlet (A). At the end of the drying period the trucks are removed from the drying chamber and a new batch is introduced.

The intermittent operation of a tray dryer can be modified for continuous operation by designing the drying chamber as a tunnel and introducing a train of trucks through it.

For proper air distribution in the dryer, high air velocities, 5–15 ft/s, are necessary. This results in low fresh air requirements, about 10–20% of the total circulated air. Drying across the stationary layers of solid is slow and the drying cycles are long.

Tray dryers are suitable only for low rates of sludge processing. The operating costs of these dryers are directly proportional to the loading and unloading labor costs.

B. Rotary Dryers

Figure 11 shows a diagram of a direct-heated rotary drier. It consists of a cylindrical shell placed at a slight angle, rotating on rollers (H) at about 4–8 rpm. Sludge is fed through (N) and is driven towards the opposite end of the dryer. Flights (Q) transport, lift the material, and shower it inside the dryer in order to facilitate contact with hot air or gases drawn in the dryer by fan (P). The necesssary heat may be derived from fuel combustion or from condensing steam at the air inlet. The use of a blower at the air inlet can help maintain a pressure inside the dryer close to atmospheric, thereby minimizing cold air leakages into the dryer. Rotary dryers can have diameters of 1–10 ft and can dry materials at the rate of up to 5000 $lb/ft^2 \cdot h$.

Rotary dryers are suitable for relatively dry materials that do not agglomerate or adhere to the walls. Sludge should be relatively dry prior to introduction into a rotary dryer. If the sludge has a water content of over 50%, it has the tendency to form balls as it passes through the dryer.

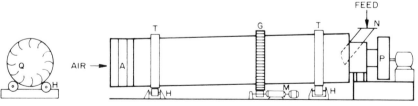

Fig. 11. A rotary dryer: A, air heaters; G, drive gear; H, rollers; M, motor and speed reducer; N, feed; P, fan; Q, flights; T, tires.

Recirculating part of the dried sludge and mixing with incoming material can lower the water content at the inlet.

The first rotary dryer was installed in Houston in 1919 to produce a fertilizer mix from activated sludge [57]. Today, rotary dryers are used for municipal sludge drying in Basel (Switzerland), Largo (Florida), Milwaukee (Wisconsin), and Stamford (Connecticut). In the latter, drying constitutes a dewatering step for sludge incineration. Rotary dryers are also used in brewery and distilling industries to reduce the sludge moisture down to 8% [58]. High-temperature rotary kilns are used for incineration of sludges in paper and alumina processing [50, 52] and in the disposal of liquid tars [59].

Some typical data for the Largo facility would demonstrate the operating conditions of rotary dryers. The wastewater treatment plant has a capacity of 1400m^3/h (9 MGD) with an actual dryer throughput of 270 kg/h of dry solids and ~2500 kg/h of water. A thickened aerobic sludge, containing 1.1% solids, undergoes filtration of 10–12% solids and, then, it is mixed with dried product before entering the dryer. Drying is accomplished with hot flue gases, entering at 650°C (1200°F) and exiting at 150°C (300°F). The sludge stays in the dryer for 20–60 min and exits the unit at 80–90% solids with about 5% humidity. Installation costs were $850,000 in 1977 and the operating costs are $134/ton [52a].

The Jones Island Plant at Milwaukee has operated ten counterflow rotary dryers, each 8 ft (2.4 m) in diameter by 57 ft (17.4 m) long, for over 20 yr. Air temperatures are similar to those of the Largo plant, and the dryers can evaporate about 10,000 lb/h (4540 kg/h) of water each. With the aid of a system of gas turbines and waste heat recovery boilers, about 70% of the required drying energy is recovered. The plant produces 67,000 tons of granular dried sludge as a fertilizer (Milorganite), processing about 31,500 m^3/h (200 MGD) of activated sludge [52a].

C. Flash Dryers

A diagram of a flash drying unit is shown in Fig. 12 [60]. The unit consists of a furnace (A) producing hot gases, a cage mill (B) for size reduction and drying, and a cyclone (C) for separation of solid particles. Partially dewatered sludge is blended in a mixer (D) with dried sludge to reduce the water content and improve pneumatic conveyance. This mixture is contacted with hot gases at 1300°F (705°C) and is introduced in the cage mill (B) where drying is complete within a few seconds. The resulting dry sludge with a moisture content of 8–10% is separated from the

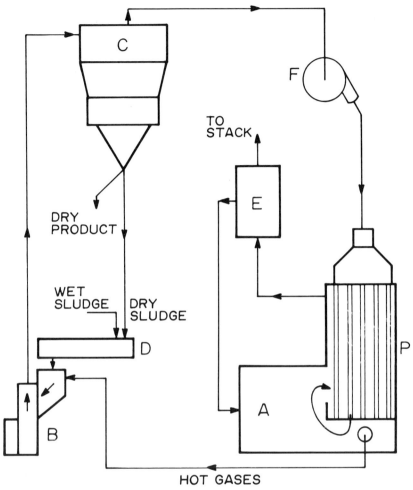

Fig. 12. Flash drying unit [60]: A, furnace with deodorizing preheater (P); B, cage mill; C, cyclone; D, mixer; E, combustion air preheater; F, vapor fan.

gases in a cyclone (C) at about 220°F (104°C). Part of the dry product is mixed with the wet sludge as described above. The finished sludge can be sent to the furnace for incineration or it can be used as a fertilizer.

To generate hot gases, the combustion of a fuel is necesssary. This can be coal, gas, oil, or dried sludge. Primary combustion air is preheated and introduced into the furnace where fuel is combusted. The furnace also serves as a heat source for deodorizing the drying gas after its exit from the cyclone. The deodorized gases are used to preheat the combustion air.

Flash drying was first used for sludge drying at the Chicago Sanitary District in 1932 [4]. Presently the city of Houston, Texas, is using the C-E Raymond process [60] to dry 150 ton/d of sludge for use as a fertilizer. Typical operating values are shown in Table 5. The dry product contains ferric chloride from the chemical conditioning of the sludge.

The Blue Plains plant in Washington, DC uses a jet mill instead of a cage mill for flash drying of sludge. The jet mill has the advantage of having no moving parts and of accomplishing drying and particle classification in one step [4].

The City of Schenectady, NY uses flash drying of sewage sludge to produce fertilizer [61]. Digested primary sludge is first vacuum filtered to 26% solids before entering the flash dryer. Drying costs have been estimated at $32/ton [61].

The Krefeld wastewater plant by Düsseldorf, W. Germany, uses a vertical shaft flash dryer to prepare raw sludge for injection into an incinerator. The sludge, containing 5% solids, is first centrifuged to 25% solids and then flash dried at 1500°F (816°C). The facility has been operated for 5 yr, treating 40–45 tons of solids daily [61a, b].

The cost of flash drying is relatively high. The sludge should be pretreated to attain reasonable nutrient contents and balance. Fuel needs for sludge drying are about 8000 BTU/lb. However, flash drying gives the flexibility of either drying or incinerating the sludge in the same unit. The Laboon process uses flash drying–incineration of concentrated primary and biologically floated sludge [62]. The dryer is fed with a mixture of thickened sludge, 18% in solids, and previously dried material.

When the sole use of dried sludge is for incineration, flash drying

TABLE 5
Material Contents of Flash-Dried Sludge [60]

Contents	Percentage
Fluid sludge	
Solids	2.8–4.0
Dried sludge	
Moisture	5.0
Nitrogen	5.3
Phosphoric acid	3.9
Ash	26–44
Ferric chloride	3.8

appears to be expensive. The Metropolitan Denver Sewage Disposal District No. 1 abandoned a flash-drying unit treating a sludge of about 100 mgd. Air pollution by fine particles and explosions in the unit were claimed [4]. The cost of flash drying at Chicago was $45/ton of dry sludge and the process was also abandoned [4]. However, flash drying appears attractive when the dried sludge is used as a fertilizer. The city of Houston reported a revenue of $21/ton in 1972 [60].

D. Spray Dryers

A spray drying unit is shown schematically in Fig. 13. The dryer (A) consists of a cylindrical drying chamber tapering to a conical bottom for collection of the dried material. Slurry is fed into the dryer at (B) with the aid of a spray disk or nozzle. For slurries with significant amounts of solids a spray disk is recommended, rotating at 5000–10,000 rpm. The incoming slurry is atomized into tiny drops that move radially into a stream of hot gases, entering at the top of the dryer (C) and following a spiral motion downwards. The dried product is removed from the bottom of the conical chamber. Gases are drawn by a fan into a cyclone (D) to remove entrained solid particles.

The sprayed slurry from the nozzle or the spray disk and the wet

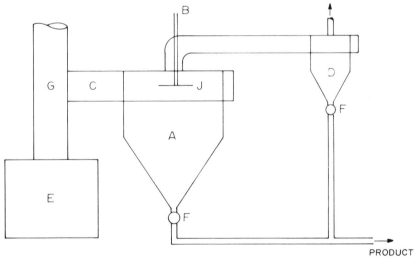

Fig. 13. A schematic diagram of a spray dryer unit: A, dryer; B, slurry feed; C, hot gas entrance; D, cyclone; E, furnace; F, rotary valve; G, stack; J, spray disc or nozzle.

solid particles should not be allowed to strike the walls of the dryer before drying is complete. Therefore, large diameters, 8–30 ft, are common for the drying chamber.

The residence time of the droplets in the drying chamber is only a few (2–30)s, and the dried solids are not heated much above the wet-bulb temperature. Because of the low drying temperature and the short drying time, hot gases can be used in a spray dryer to dry heat-sensitive materials, such as food processing slurries and phamaceuticals. Spray drying gives quite uniform, hollow particles, because surface drying proceeds faster than moisture diffusion outwards from the interior of the particles [53, 63].

The performance of a spray dryer depends on the residence time of the droplets in the drying chamber. The residence time is a complex function of the size and velocity of the droplets, the velocity and flow pattern of the drying gases, and the size and geometry of the drying chamber [63]. Large drops may be underdried, while small ones may be overdried.

Spray drying has been applied to sludge treatment for drying, evaporation, pyrolysis, and incineration [16, 64, 65]. Sludge is thickened to about 8% solids, ground to reduce the particle size, and atomized in the spray dryer. If the particles remain long enough in the dryer, above 600°F, ignition and incineration takes place.

The Ansonia, Connecticut wastewater treatment plant processes dewatered sludge through a spray dryer with hot flue-gases (1300°F or 705°C) as the drying medium [52a]. The dried sludge with more than 90% solids is given away as a soil conditioner instead of being incinerated at the plant incinerator.

The city of Milwaukee investigated municipal sludge spray drying to obtain vitamin B_{12}. After washing the sludge, the extracted liquor was concentrated by vacuum evaporation before it was spray dried to obtain the vitamin [45]. Atomized spray drying is also suitable for the treatment of food industry sludges to recover vitamins and proteins [48]. Spray drying is a standard technique for recovering spent catalysts in the petrochemical industry.

Atomized spray drying provides flexibility of operation to either dry or incinerate sludge. However, equipment is expensive and it is justified only if large amounts of material are dried or when heat sensitive materials are processed. Sludge abrasiveness can cause material problems (wear and corrosion) with spray-drying equipment.

E. Toroidal Dryers

This new dryer design by UOP consists of a toroidal jet mill (A) with no moving parts, in which transport, lump break-up, and drying are accomplished by high-velocity air, Fig. 13a. Dewatered sludge (B) with 35–40% solids is mixed with dried sludge, to decrease the moisture contents, and it is propelled into the dryer by an air-jet stream. In the drying zone, air and sludge have a velocity of 100 ft/s (30 m/s), which results in size reduction of solid particles by collision with other particles and the walls. Solids stay in the dryer until disintegrated and dried. Dried particles are separated from the air stream in a cyclone (C) and part of them is mixed with wet sludge.

A 240 ton H_2O/d toroidal unit at the Blue Plains plant, Washington DC, operated for about 3 yr with raw, digested primary, and activated sludges. The system is currently non-operative [52a].

F. Multiple Hearth Furnaces

Multiple hearth furnaces are common sludge incineration equipment with potential use as dryers. In fact, the upper section of multiple hearth furnaces serves as a dryer of incoming dewatered sludge prior to its reaching the incineration section.

A typical multiple hearth furnace design is shown in Fig. 14. The furnace consists of a cylindrical refractory shell divided into several hori-

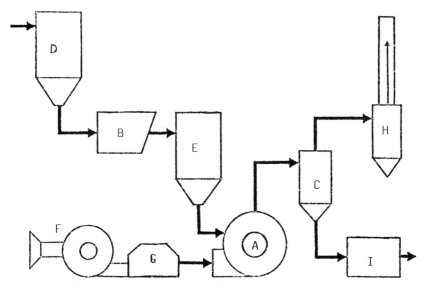

Fig. 13a. Sludge drying process using a toroidal dryer (A), mechanical dewatering (B), and cyclone (C) separation of solids; (D) wet sludge storage; (E) secondary storage; (F) blower; (G) air heater; (H) air emission control; and (I) product finishing. [52a].

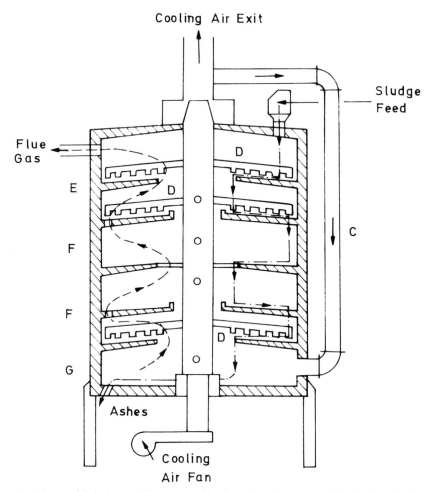

Fig. 14. Multiple hearth furnace and incinerator: (A) gas exit; (B) ash exit; (C) air return; (D) ramble arms; (E) drying zone; (F) combustion zone; (G) cooling zone.

zontal, refractory hearths. A central shaft rotates a set of toothed arms in each hearth that agitate and move the sludge. The sludge enters at the periphery of the top section of the furnace and is moved toward the center, where it passes through openings into the next hearth. The openings of successive hearths are located alternately at the periphery and the center so that the material is transported across each hearth for best drying and incineration.

Sludge is dried in the upper section of the hearth furnace before it enters the middle combustion section which operates at 1400–1700°F. The lower section serves as a cooling zone of the combusted sludge.

Hearth furnace drying and incineration has gained popularity in the US during the last decade [1, 67].

IV. EVAPORATION

Detailed discussion of evaporation equipment is beyond the scope of this article. However, thermal evaporation precedes sludge drying in several processes. Therefore, a brief description of some basic evaporators will be given here. Standard, chemical engineering treatises examine evaporation in further detail [53–55].

Steam-heated evaporators can be distinguished into single-effect and multiple-effect evaporators. Single-effect evaporators are further subdivided into short-tube, long-tube, and agitated-film evaporators. Here the common, vertical, short-tube evaporator will be discussed, which also constitutes the basic repeated unit in multiple-effect evaporators.

A vertical, short-tube evaporator is shown in Fig. 15. A bundle of short tubes (A), 4–8 ft long and 2–4 in. in diameter is placed in a vertical shell (B) in which the evaporating liquor is introduced. Steam condenses outside the tubes causing boiling of the liquor. The liquor spouts upwards inside the tubes and returns through the downtake. Concentrated liquor is removed from the bottom of the evaporator (C) and liquid vapor is removed at (D). The cross-sectional area of the downtake is 25–40% of the total cross-sectional area of the tubes.

Short-tube evaporators are replaced today by long-tube evaporators to achieve a higher heat transfer coefficient [53–55].

Single-effect evaporators can be combined in series for multiple-effect operation. Connection is arranged so that the vapor from one evaporator serves as the heating medium for the next one, as in Fig. 16. A vacuum is established in the last stage to remove noncondensed vapor from the system. Steam is supplied to the first stage. This arrangement results in spreading of the pressure difference between inlet steam and final outlet condensate over all stages. The first stage operates at the highest pressure and the last one at the lowest.

Each stage operates as a single-effect evaporator with its own temperature driving force and heat transfer coefficient, corresponding to the pressure drop in that stage. At steady-state operation, the temperature,

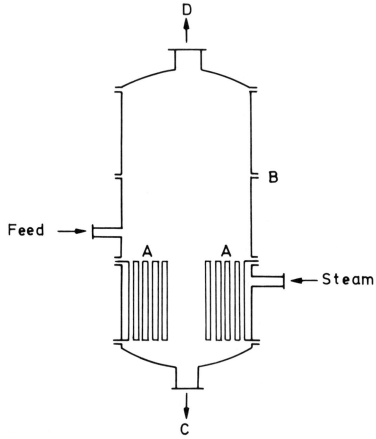

Fig. 15. A vertical short-tube evaporator: (A) bundle of tubes; (B) shell; (C) exit of concentrated liquor; (D) vapor exit.

concentration, and flow rate of the feed are fixed. The inlet steam pressure and the outlet condensate pressure are also fixed. Operating conditions within each stage are uniquely established. The composition of the final concentrated liquor can be changed only by adjusting the flow rate of the feed. By reducing the feed flow rate the thick liquor concentration is increased and a new steady-state operation is reached eventually.

Evaporators often operate under vacuum to decrease the boiling point of water or solvent. This results in a larger temperature gradient between evaporating liquid and heating medium and in a smaller heat exchange area than if atmospheric pressure were maintained. Vacuum evap-

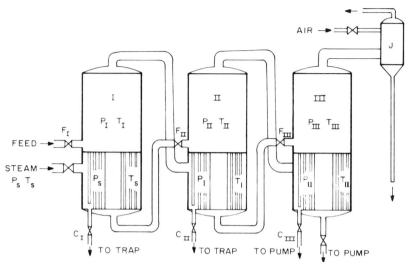

Fig. 16. Triple-effect evaporator. C_I, C_{II}, C_{III}, condensate valves; F_I, F_{II}, F_{III}, feed valves; J, air injector and condenser; P and T, pressure and temperature for each effect.

oration is important for heat-sensitive materials not only to achieve better heat transfer, but also to avoid decomposition or alteration of these materials at elevated boiling temperatures. Food and pharmaceutical slurries are, therefore, evaporated under vacuum.

Vacuum evaporation has been used in vitamin B_{12} production from sewage sludge in Milwaukee [45]. The Carver-Greenfield dehydration system uses a triple-effect evaporation step in recovering grease from sewage and industrial wastes [1]. The Bell-Fons process uses vacuum evaporation to precipitate ferrous sulfate monohdyrate and to recover sulfuric acid from the pickling liquor of steel mills [68]. Similarly, calcium chloride is recovered from the industrial wastes of Columbia-Southern Chemical Company and is marketed for ice-making and highway dust control [68]. A solar evaporation process has been developed by Celanese Chemical Corporation at Bishop, Texas for treatment of its wastes from the production of organic chemicals.

Some typical operating data for the Carver-Greenfield process are given here. Over 65 plants exist worldwide, including a 180,000 gal/d (682 m^3/d) plant for treating a 4%-solids activated sludge from the Coors Brewery and a 264,000 gal/d (1000 m^3/d) plant for a 2%-solids sewage effluent at Hiroshima [1, 52a, 68a]. Mixing of the sludges with oils (e.g., No. 2 fuel) helps maintain fluidity of the sludge through all stages and minimizes corrosion and scale formation in the equipment. Steam requirements have been estimated at ~0.45 lb/lb H_2O at about 50 psig for a four-effect unit. Energy requirements, including steam production, are about 675 BTU/lb water, compared to 1200–2000 BTU/lb water for other dryers [52a].

V. THEORY OF DRYING

Removal of water in thermal dryers and in drying sand beds proceeds via two different mechanisms: Heat and mass transfer (evaporation) occur in thermal dryers while filtration predominates in sand beds during the initial dewatering period. Here, the theory of thermal drying and some gravity dewatering principles will be reviewed.

A. Water Evaporation and Thermal Drying

1. Moisture Content Equilibrium

The ability of a moist solid to lose water into drying air depends on the moisture contents of the solid and the air, at the operating temperature, pressure, and air flow rate. Long exposure of the moist solid to air will result in an "equilibrium moisture content" for the specific operating conditions. Further exposure to the same air will leave the solid moisture content unchanged.

The moisture content can be expressed on a *wet basis* as the weight of water per unit weight of solids plus water, or on a *dry basis* as the weight of water per unit weight of dry solids. The latter definition is more convenient for drying calculations and will be adopted here unless specified otherwise.

Under equilibrium conditions, the water partial pressure of the solid equals the water vapor pressure in the air. The latter is usually expressed as relative humidity, $\mathcal{H}$, defined as the percent ratio of the water partial pressure, P_w, at given temperature and total pressure, to the saturation vapor pressure of water, P_s, under similar conditions.

$$\mathcal{H} = \frac{P_w}{P_s}(100) \qquad (1)$$

An equilibrium curve, then, relates the equilibrium moisture content of a solid on a dry solid basis to the relative humidity of air, as shown in Fig. 17. At zero air humidity, all solids have a zero equilibrium moisture con-

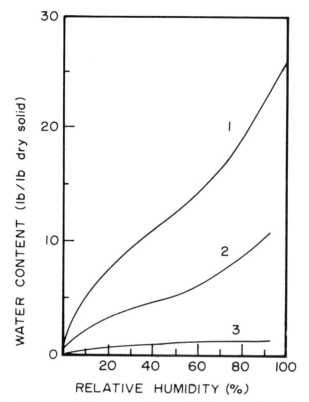

Fig. 17. Equilibrium moisture curves for wool (1), paper (2), and kaolin (3) at 25°C. Reprinted with permission from ref. [54]. Copyright 1976 McGraw-Hill.

tent. The equilibrium moisture content decreases with increasing air temperature. Here, the desorption equilibrium curves are of interest.

If a solid of moisture content X_o is brought into contact with air of such relative humidity that the equilibrium moisture content of the solid is X_e, the solid will lose water

$$X = X_o - X_e \qquad (2)$$

The amount X is called the "free moisture" content and it is this amount that is of interest in drying calculations. Free and equilibrium moisture depend on the drying conditions.

By extending an equilibrium curve to 100% relative humidity, the amount of "bound water" is obtained; this exerts a vapor pressure lower than that of liquid water at the same temperature [53, 54]. This is a result of water retention in the fine capillaries of porous solids, dissolution of solids in the water, or physical and chemical binding of the water with organic structures. Substances containing bound water are called hygroscopic.

Any water retained in excess of the bound water content is called unbound water. Unbound water is held in the voids of the solid and exerts its full vapor pressure. The quantities of bound and unbound water depend on the nature of the solid material.

2. Drying Rate

The design, operation, and capacity of thermal dryers depends on the rate of mass and heat transfer. The free moisture content, i.e., the difference between the total moisture and the equilibrium moisture content, is the driving force for mass transfer. Depending on the solid and the drying conditions, moisture flows as a liquid or a vapor through the solid and is removed as a vapor from the surface of the solid into the air or gas stream.

Water vaporization and removal requires the supply of the heat of vaporization to the evaporation zone. This is accomplished by contacting the moist solid with hot air or gases and/or hot surfaces. If heat is supplied only by hot gases, the process is called adiabatic drying. Rotary, flash, and spray dryers operate under adiabatic conditions, with heat transferred by convection and conduction. In high-temperature dryers, heated by flue gases or flames, radiative heat transfer may also be important.

In the presence of unbound water, the temperature at the vaporization zone is the equilibrium temperature between air and liquid water. In the drying of hygroscopic materials containing only bound water, the water vapor pressure is lower than the vapor pressure of liquid water at the temperature of the vaporization zone and the water vapor is superheated. At the end of water removal from hygroscopic materials, the temperature of the solid approaches that of the air.

3. Estimation of Drying Rate

The rate of drying can be correlated to the rate of mass and heat transfer between a wet solid and air. Here, steady-state, adiabatic drying is assumed at constant rate, for a solid containing unbound water. Heat transfer by radiation or conduction is neglected.

The rates of mass and heat transfer at the solid surface, in contact with air, are:

$$N = k_w A(P_i - P_a) \tag{3}$$
$$Q = hA(T_a - T_i) \tag{4}$$

where N = is the molar rate of evaporation
Q = the rate of heat transfer
k_w = the molar mass transfer coefficient
h = the heat transfer coefficient
P_i = the partial pressure of water vapor at the interface
P_a = the partial pressure of water vapor in the bulk air
T_i = the temperature of the wetted surface
T_a = the bulk temperature of air
A = the exposed solid area

The temperature of the wetted surface in this process is the wet-bulb temperature corresponding to the air conditions. Assuming that no mass transfer resistance exists at the water–air interface, P_i will be the vapor pressure of liquid water at the interface temperature.

At steady-state, the rate of heat transfer will be equal to the rate of water removal times the latent heat of vaporization per unit mass, ΔH_v

$$Q = AR\Delta H_v \tag{5}$$

where R is the rate of water removal per unit exposed area

$$R = -\frac{1}{A}\frac{dm}{dt} = \frac{N}{A}M \tag{6}$$

where m is the total mass of moisture in the solid and M is the molecular weight of water. From Eqs. (3)–(6) the drying rate then becomes

$$R = k_w M(P_i - P_a) = \frac{h}{\Delta H_v}(T_a - T_i) \tag{7}$$

Use of the heat transfer equation to evaluate R is more reliable, since an error in determining the interface temperature T_i affects the temperature driving force, $T_a - T_i$, less than it affects the pressure driving force $P_i - P_a$.

For constant air velocity, the air temperature and humidity affect the drying rate through the driving forces for heat and mass transfer, while h and k_w remain constant. Velocity changes will cause a change in the heat and mass transfer coefficients. The heat transfer coefficient can be correlated to the air flowing conditions through the Colburn j_H-factor [69]

$$j_H = f(\text{Re}) = \left(\frac{h}{C_p G}\right)\left(\frac{C_p \mu}{k}\right)^{2/3} = \frac{h}{C_p G} Pr^{2/3} \tag{8}$$

where G = the mass velocity of air
 Pr = the Prandtl number = $C_p \mu/k$
 C_p = the air heat capacity at constant pressure
 μ = the air viscosity
 k = the air thermal conductivity

Several functions $f(Re)$ have been proposed [53, 70]. For flow parallel to flat plates the appropriate $f(Re)$ combined with Eq. (8) yields for the heat transfer coefficient

$$h = 0.0128 G^{0.8} \qquad (9)$$

With air flow perpendicular to a flat surface

$$h = 0.37 G^{0.37} \qquad (10)$$

A review of other equations proposed for various conditions is given by Peck and Wasan [70].

4. Drying Curves

When a moist solid is introduced into a dryer, there is a short preliminary time period during which the temperature of the material adjusts itself to the drying conditions. Beyond this period, the temperature of the solid and the drying rate remain constant for an appreciable length of time, as shown in Fig. 18. It is this constant rate period that will be examined here in detail.

During the constant rate period, the solid particle surface in contact with the air remains completely wetted by a continuous film of water. Evaporation, then, proceeds at the same rate as from a free liquid surface of similar area under the same conditions, as if the solid were not present. However, the solid surface roughness increases the liquid surface area and can enhance the evaporation rate [71]. With nonporous solids, only superficial surface moisture is removed during the constant rate period. In a porous material, water must be transported from the interior porous structure to the surface before it is removed.

As water is evaporated, the liquid water on the surface may become insufficient to form a continuous film over the entire surface. At this point, called the first critical point (point B, Fig. 19), the constant-rate period ends, the drying rate decreases, and the surface temperature starts increasing.

The critical point for nonporous solids corresponds to the evaporation of the superficial moisture. In porous solids, the falling rate period

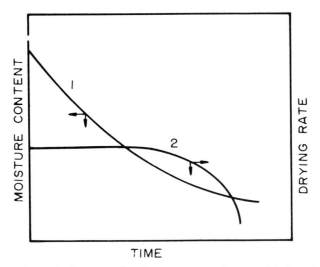

Fig. 18. A schematic diagram of moisture content change with time during drying (1) and of a drying curve (2).

(line BC, Fig. 19) is reached when the rate of water transport in the solid becomes slower than the rate of evaporation on the surface. If the initial moisture is below the critical moisture, the constant rate period is not observed.

In the first falling rate period, the water in the pores forms a continu-

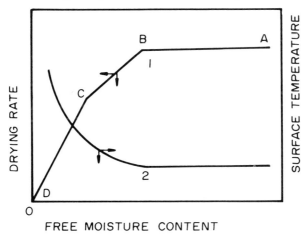

Fig. 19. A schematic diagram of a drying rate curve (1) and of the surface temperature of the solid (2).

ous film. The vaporization zone is close or at the surface, and the evaporation mechanism is similar to that for the constant-rate period. Therefore, factors controlling the drying rate in the constant-rate period continue to control the rate of drying during the first falling-rate period.

The critical moisture is normally determined experimentally. It is not only a property of the material, but it also depends on the thickness of the solid and on the drying conditions. Figures 20 and 21 show the change of the critical point and of the drying rate for different bed thicknesses and air velocities for sand and asbestos pulp.

During the falling-rate period, a discontinuity often appears, as in point C of Fig. 19. This is called the second critical point and the drying period between C and D is called the second falling-rate period. The beginning of the second falling-rate period corresponds to a completely dry solid surface. Water is evaporated inside the pores and the water–air interface recedes into the pores. When the remaining water is insufficient to form a continuous film across the pores, surface tension in the pore capillaries causes the formation of isolated pools of water. Air becomes, then, the continuous phase in the pores. The required heat of vaporization is transmitted through the solid by conduction and the solid surface temperature approaches the dry-bulb temperature of the air.

In the second falling-rate period, water vapor is removed from the

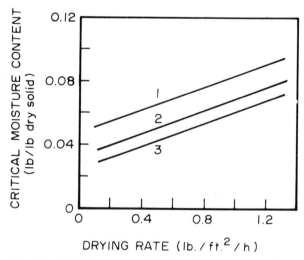

Fig. 20. Effect of bed thickness on critical moisture content for sand drying. Curve (1): 3 in. bed thickness; (2): 2 in. bed thickness; (3): 1 in. bed thickness. Reprinted with permission from ref. [71]. Copyright 1951 American Chemical Society.

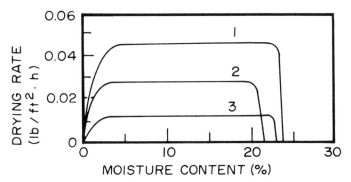

Fig. 21. Effect of air velocity on drying rate for asbestos pulp. Curve (1): air velocity 1370 ft/min; (2): 640 ft/min; (3): 300 ft/min. Reprinted with permission from ref. [72]. Copyright 1938 American Chemical Society.

pores by diffusion. The drying rate follows the diffusion law for fine pores and the drying rate curve is concave upward. With solids of high porosity, the diffusion equations do not apply and the second falling rate period is represented by a straight line.

Hygroscopic porous solids contain both unbound and bound water. When the unbound water has been removed, vaporization of the bound water below the solid surface may progress via diffusion of water vapor through the solid [73]. Under these conditions only the first critical point is observed. The drying rate depends on the cake thickness, but no simple relation holds, as indicated by the curves of Fig. 22.

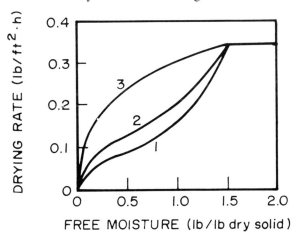

Fig. 22. Effect of cake thickness on drying rate for drying paper pulp slabs [73]. Curve (1): thickness, 2.37 cm; (2): 0.648 cm; and (3): 0.108 cm.

Coackley and Allos [74] obtained drying rate curves for raw primary and secondary digested sludge, as shown in Fig. 23. These curves show a constant rate period to about 70% water (wet basis), corresponding to a dry basis moisture content of about 250%. All tested sludges showed a second critical point at a moisture content of about 100%.

Nebiker [75] studied the drying behavior of several sludges at different sludge depths, as shown in Fig. 24. Results were presented in terms of an evaporation ratio defined as the ratio in percent of sludge weight loss by drying to the simultaneous evaporation of the water for a given container [75]. Considerable scattering of data do not allow exact determination of critical moisture contents. The data were interpreted in terms of a linear falling-rate period and the second falling-rate period was ignored. Because of the small equilibrium moisture content (8%), lines were drawn through the origin.

Example 1

A sludge cake is to be dried in air of dry-bulb temperature of 350°F and of wet-bulb temperature of 100°F at a velocity of 25 ft/s. A drying

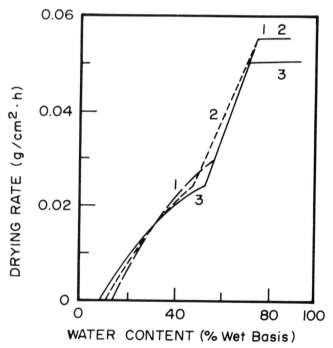

Fig. 23. Drying curves for raw sludge (1), primary digested sludge (2), and secondary digested sludge (3) [74].

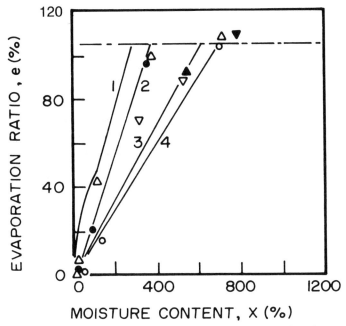

Fig. 24. Evaporation ratio of activated sludge as a function of moisture content (dry basis) [75]. Curve (1): after Coackley and Allos [74]. Curve (2): 5 and 10% solids; sludge depth, 10 cm. Curve (3): 2.5 and 5% solids; sludge depth, 20 cm. Curve (4): 2.5% solids; sludge depth, 10 cm.

test at the same air velocity and wet-bulb temperature, but at 260°F dry-bulb temperature, gave a constant drying rate of 0.8 lb/ft² · h. Estimate the operating drying rate in a large scale dryer, neglecting heat transfer by radiation or conduction.

Solution: The wetted surface temperature will be the wet-bulb temperature of air, $T_i = 100°F$. Since the air velocity is the same for the experimental and the large scale drying, the heat transfer coefficient is the same. Equation (25), then gives

$$\text{Experimental: } R_e = \frac{h(T_{a,e} - 100)}{\Delta H_v} \qquad \text{(i)}$$

$$\text{Large-scale: } R_s = \frac{h(T_{a,s} - 100)}{\Delta H_v} \qquad \text{(ii)}$$

where subscripts "e" and "s" indicate experimental or large-scale conditions. Assuming negligible change of ΔH_v with temperature, Eqs. (i) and (ii) give

$$R_s = R_e \left[\frac{T_{a,s} - 100}{T_{a,e} - 100} \right] = 0.8 \times \frac{350 - 100}{260 - 100} = 1.25 \text{ lb/ft}^2 \cdot \text{h}$$

Example 2

A municipal sludge has been dewatered on a drying sand bed from 4% solids initially, to 10% solids. From this point on, drying proceeds only by evaporation from the free surface. The sand bed has an area of 100,000 ft² and the sludge loading was initially 8 in. high. Calculate the drying time to a final solid content of 32% assuming constant drying rate. The average air velocity is 2 ft/s. The dry-bulb temperature is 90°F and the wet-bulb temperature is 75°F. Sludge density 62.4 lb/ft³ initially.

Solution Initial loading of sludge

$$L = 100{,}000 \times \frac{8''}{12''} \, 62.4 = 4.16 \times 10^6 \text{ lb}$$

Solids loading: $S_o = 0.04L = 166{,}400$ lb (dry)
Initial water: $W_o = 0.96L = 3{,}993{,}600$ lb

$$\text{Sludge after drainage: } S_d = \frac{S_o}{0.1} = \frac{166{,}400}{0.1} = 1{,}664{,}000 \text{ lb}$$

$$\text{Cake at end of evaporation: } S_e = \frac{S_o}{0.32} = \frac{166{,}400}{0.32} = 520{,}000 \text{ lb}$$

Water to be removed: $W = S_d - S_e = 1{,}144{,}000 \text{ lb} = -m$

The latent heat of vaporization of water is 1051 BTU/lb [54]. Evaluate the mass velocity of air, assuming air as ideal gas

$$G = \rho v = \frac{PM_{air}}{\mathcal{R} T_a} v = \frac{14.7 \times 144 \times 29}{1545 \times (460 + 90)} \times 2 \times 3600 = 520.1 \text{ lb/ft}^2 \cdot \text{h}$$

The heat transfer coefficient [Eq. (9)] is

$$h = 0.0128 G^{0.8} = 0.0128 \, (520.1)^{0.8} = 1.906 \text{ BTU/ft}^2\text{h°F}$$

For evaporation of W lbs. of water from one side of the cake, Eqs. (6) and (7) give

$$t = \frac{\Delta H_v W}{Ah(T_a - T_i)} = \frac{1051 \times 1{,}144{,}000}{100{,}000 \times 1.906 \times (550 - 535)}$$

or

$$t = 420.5 \text{ h}$$

5. Shrinkage

Rigid solids do not shrink appreciably whether they are porous or nonporous. Shrinkage becomes important with colloidal and fibrous solids. Shrinkage results from loss of bound moisture from the outer layers of the solid faster than it is transported outwards from the interior of the solid. The outer layers, then, shrink against an unyielding constant volume core. This shrinkage further impedes moisture transport to the surface and ultimately a hard impervious-to-moisture skin is formed. Shrinkage can be controlled by adjusting the air humidity so that a steep moisture gradient is avoided between air and solid surface.

6. Moisture Transport Mechanisms

Moisture transport in porous solids depends on the solid structure, the nature of the liquid and the operating conditions such as temperature, pressure and liquid concentration. Five mechanisms have been advanced to explain moisture transport.
  i. Liquid transport by capillary forces
  ii. Liquid diffusion due to a water concentration gradient
  iii. Liquid flow by gravity forces
  iv. Liquid or vapor transport by a pressure gradient from capillarity, shrinkage, or overheating of the interior of the solid.
  v. Vapor diffusion because of a partial pressure gradient

These mechanisms have been consolidated into three transport theories [70]:
  a. The diffusion theory
  b. The evaporation–condensation theory
  c. The capillary theory

Only the diffusion theory has been used in sludge drying [75]. However, Hougen et al. [76] have discussed the limitations of the diffusion theory and suggested capillarity as the driving force for moisture transport in porous solids. Evaluation of other transport theories for sludge drying would, therefore, be desirable.

a. The Diffusion Theory. Several investigators have considered diffusion as the main process for moisture transport from the solid interior to the surface [73, 77, 78]. Sherwood [79] realized the limitations of the diffusion theory and the significance of capillary movement. The success of the diffusion equations with wood and clay drying was attributed to the method of calculation of drying time by integration, which compensated for errors in the moisture profile equation.

Krischer [80] gave a comprehensive analysis of the diffusion theory of drying, based on the time dependent diffusion equation

$$\frac{\partial x}{\partial t} = \frac{\partial}{\partial z}\left[D\frac{\partial x}{\partial z}\right] \qquad (11)$$

where D is the diffusivity of water.

For constant rate drying from the surface of a slab of thickness, l, Eq. (11) can be solved for $t > 0$ and constant diffusivity, with the following boundry conditions

$$\text{At } z = 0 \text{ (bottom)}, \quad \frac{\partial X}{\partial z} = 0 \qquad (12)$$

$$\text{At } z = l \text{ (surface)}, \quad \frac{\partial X}{\partial z} = \frac{R_c}{\gamma_o} \qquad (13)$$

where R_c is the constant drying rate and γ_o is the weight of skeletal solids per unit initial volume [75]. Assuming a uniform moisture distributation initially (at $t = 0$), Eqs. (11)–(13) give for large times [75]

$$X = X_i + \frac{lR_c}{D\gamma_o}\left\{\frac{1}{6} - \frac{1}{2}\left(\frac{z}{l}\right)^2 - \frac{Dt}{l^2}\right\}$$

where X_i is the initial moisture content.

The average moisture content at a given time is then

$$\langle X \rangle = \frac{1}{l}\int_o^l X dz = X_i - \left(\frac{R_c}{\gamma_o l}\right)t \qquad (15)$$

If the first critical moisture point occurs when only hygroscopic moisture exists at the surface, Eq. (14) gives the time that the hydroscopic moisture content, X_h, is reached

$$X_h = X_i - \frac{lR_c}{D\gamma_o}\left(\frac{1}{3} - \frac{Dt}{l^2}\right) \qquad (16)$$

At that time, the average moisture content, given by Eq. (15) is equal to the first critical moisture content, X_c. By eliminating time between Eqs. (15) and (16) and setting $\langle X \rangle = X_c$, here,

$$X_c = X_h + \frac{lR_c}{3D\gamma_o} \qquad (17)$$

This equation, derived for thin, isothermal slabs permits a simple evaluation of the critical moisture content if all previous assumptions hold.

Hougen et al. [76] tested a number of simple materials such as sand, soap, clays, and paper pulp and proved that the diffusion theory does not apply to them. Even when diffusion was a factor, integration of the diffusion equation was difficult in cases where water diffusivities were not constant or where the initial water content varied because of shrinkage.

b. The Evaporation–Condensation Theory According to this theory, moisture is transferred in the pores in the gas phase [81, 82]. This assumption seems to be correct when the system is subjected to a temperature gradient, even at relatively high pore saturation [83, 84]. Harmathy [85] described mathematically the simultaneous mass and heat transfer processes during drying based on this model. The theory has been applied only to a limited number of systems.

c. The Capillary Theory The moisture distribution in several porous materials and their drying characteristics led Ceaglske and Hougen to propose that moisture is transported by capillarity and gravity forces [86]. Moisture transfer by capillarity applies to water not held in solution and to all water above the equilibrium moisture content [76]. However, if an adequate temperature gradient exists, water vapor diffusion can also be important.

The capillary theory suggests that water evaporates from small menisci at the solid surface [87]. The small curvature of these menisci results in capillary suction of water from other passages of larger curvature. Water drawn to the surface is replaced by air. Corben and Newitt demonstrated the applicability of the capillary theory for the drying of granular porous solids [88].

The existence of capillary forces in moisture transport can explain the actual water transfer from a region of low concentration to a region of high concentration, if the high concentration region consists of fine pores. Such a situation arises when fine sand is placed over coarse sand and drying occurs at the surface of the fine sand [86]. This behavior cannot be explained by the diffusion theory.

Peck and Wasan applied recently some concepts of capillary water transport to drying in the falling rate period [70]. The derived three-parameter model described well the drying characteristics of thin slabs.

7. *Design Equations—Drying Time*

Consider a slab of solid of thickness l and density ρ_s (mass per unit volume of dry material). The rate of drying per unit area is given by Eq. (6) above,

$$R = -\frac{1}{A}\frac{dm}{dt} \qquad (6)$$

The mass of moisture is related to the free moisture content X (in mass of free moisture per unit mass of dry solid) through Eq. (18).

$$m = Al\rho_s X \qquad (18)$$

Ignoring possible shrinkage of the solid during drying and subsequent density change, the rate of drying becomes, in terms of X:

$$R = -l\rho_s \cdot \frac{dX}{dt} \qquad (19)$$

Equation (19) can be integrated to obtain the drying time if a relationship between R and X is known:

$$t = -l\rho_s \int_{X_1}^{X_2} \frac{dX}{R} \qquad (20)$$

In two situations, analytical solution of Eq. (20) is possible. During the constant rate period, R is a constant and the drying time is proportional to the amount of moisture removed.

Constant Drying Time

$$t_c = l\rho_s \int_{X_1}^{X_2} \frac{dX}{R_c} = -\frac{l\rho_s}{R_c}(X_2 - X_1) \qquad (21)$$

During the falling-rate period, the drying rate is often a linear function of the moisture content

$$R_f = a + bX \qquad (22)$$

where a and b are constants.

From Eqs. (20) and (22), then

$$t_f = -l\rho_s \int_{X_1}^{X_2} \frac{dX}{a + bX} = -\frac{l\rho_s}{b}\ln\left(\frac{a + bX_2}{a + bX_1}\right) = -\frac{l\rho_s}{b}\ln\frac{R_2}{R_1} \qquad (23)$$

If R_c and X_c are the rate and free moisture content, respectively, at the first critical point and R_{cs}, X_{cs} are the rate and free moisture content at the second critical point, a and b can be evaluated from the falling rate portion of the drying curves (cf., Fig. 19). Then

$$R = \left(\frac{R_{cs}X_c - R_cX_{cs}}{X_c - X_{cs}}\right) + \left(\frac{R_c - R_{cs}}{X_c - X_{cs}}\right)X \qquad (24)$$

and, Eq. (23) gives

Falling Rate Drying Time

$$t_f = -\frac{l\rho_s(X_c - X_{cs})}{R_c - R_{cs}} \ln \frac{R_2}{R_1} \qquad (25)$$

If the initial moisture content is higher than that of the first critical point, and drying continues to a point between the first and the second critical point, the total drying time will be given by the sum of Eqs. (21) and (25)

$$t_T = t_c + t_f = l\rho_s\left(\frac{X_1 - X_c}{R_c} + \frac{X_c - X_{cs}}{R_c - R_{cs}}\ln\frac{R_c}{R_2}\right) \qquad (26)$$

For an arbitrary shape of the drying curve, Eq. (20) can be integrated graphically or numerically to estimate the drying time.

Example 3

A dewatered sludge with 27% solids is thermally dried to 7% final moisture using hot air of 200°F dry-bulb temperature and of 90°F wet-bulb temperature. The critical moisture content corresponds to about 32% solids in the sludge. Estimate the drying time of sludge cakes 2 in. thick, in an air stream of 10 ft/s over the slabs. What is the constant rate mass transfer coefficient? The dry cake density is 110 lb/ft^3. Assume negligible equilibrium moisture.

Solution. Drying will proceed initially at a constant rate until the critical moisture content is reached. Neglecting the equilibrium moisture content and any second critical point, the drying rate in the falling rate period will be proportional to the moisture content

$$R_f = bX$$

until the final moisture is reached.

$$X_1 = \text{Initial moisture content:} \quad \frac{1 - 0.27}{0.27} = 2.70$$

$$X_c = \text{Critical moisture content:} \quad \frac{1 - 0.32}{0.32} = 2.12$$

$$X_2 = \text{Final moisture content:} \quad 0.07$$

The mass velocity is

$$G = \rho v = \frac{PM_{air}}{\Re T} \cdot v = \frac{14.7 \times 144 \times 29}{1545 (460 + 200)} \times 10 \times 3600 = 2167 \text{ lb/ft}^2 \cdot \text{h}$$

Thus

$$h = 0.0128(2167)^{0.8} = 5.969 \text{ BTU/ft}^2 \cdot \text{h} \cdot °\text{F}$$

The constant drying rate becomes then

$$R_c = \frac{h(T_a - T_i)}{\Delta H_v} = \frac{5.969 (660 - 550)}{1043} = 0.629 \text{ lb/ft}^2 \cdot \text{h}$$

From Eq. (21), the time for constant rate drying can be estimated

$$t_c = -\frac{l\rho_s}{R_c}(X_c - X_1) = -\frac{(2/12) \cdot 110}{0.629}(2.12 - 2.70)$$

or $t_c = 16.9$ h.

For the falling rate period, the time, t_f, is

$$t_f = -l\rho_s \int_{X_c}^{X_2} \frac{dX}{bX} = -\frac{l\rho_s}{b} \ln\left(\frac{X_2}{X_c}\right) \quad \text{(i)}$$

where the slope of the drying curve is given by

$$b = \frac{R_c}{X_c} = \frac{0.629}{2.12} = 0.297 \text{ lb/ft}^2 \cdot \text{h} \quad \text{(ii)}$$

Thus

$$t_f = -\frac{l\rho_s X_c}{R_c} \ln\left(\frac{X_2}{X_c}\right) = -\frac{(2/12) \times 110 \times 2.12}{0.629} \ln\left(\frac{0.07}{2.12}\right)$$

or

$$t_f = 210.7 \text{ h}$$

Total Drying Time

$$t_T = t_c + t_f = 227.6 \text{ h}$$

The long drying time in the falling rate period results from the large amount of moisture that should be removed at a decreasing rate. Therefore, thermal drying should be applied after appreciable sludge dewatering.

The mass transfer coefficient during the constant rate period can be evaluated from Eq. (7).

$$k_w = \frac{h(T_a - T_i)}{M_w \Delta H_v (P_i - P_a)} \qquad (7)$$

where

$P_i = 0.6982 \text{ lb}_f/\text{in.}^2$ at $T_i = 90°\text{F}$ [54]
$P_a = 0.0088 \text{ lb}_f/\text{in.}^2$ obtained from humidity charts

Then

$$k_w = \frac{5.969 \times (200 - 90)}{18 \times 1043(0.6982 - 0.0088)} = 0.0507 \text{ lb/ft}^2 \cdot \text{h} \cdot \text{psi}$$

8. Theory of Rotary Dryers

The operation of rotary dryers can be analyzed mathematically in simple cases and a design equation can be derived. In the theoretical analysis here, the following assumptions apply:

i. No heat transfer takes place between the dryer and the surroundings
ii. All required heat is supplied by the hot air
iii. No evaporation and drying takes place during the preliminary and final heatup periods
iv. Only free moisture is present (unbound)
v. All drying takes place at the wet-bulb temperature

The temperature profile of air and solids in a countercurrent rotary dryer is shown in Fig. 25. Solids enter the dryer and they are first heated up to the wet-bulb temperature. They remain at this temperature nearly to the exit of the dryer (at z_f). Because of usual overdesign of the dryer length, to achieve the desired degree of drying, the moisture evaporates short before the dryer exit. Therefore, the temperature of the solids rises beyond z_f, tending to reach the inlet air temperature.

Air cools through its course in the dryer by supplying the necessary heat of vaporization and by heating the solids.

A rotary dryer is shown schematically in Fig. 26. A heat balance for the solids and the air across dz gives

$$dQ = -G_s A_c C_s dT = -G_a A_c f dT_a \tag{27}$$

where G_s and G_a are the mass flux of solids and air through the dryer

A_c = the dryer cross-sectional area
C_s = the specific heat of the solids
f = the humid heat of the air

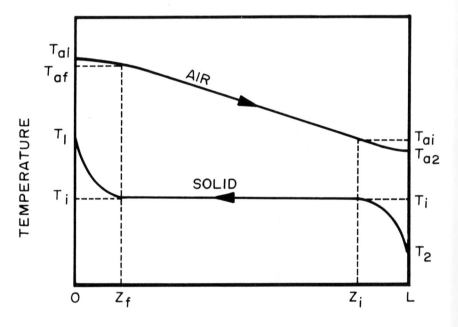

Fig. 25. Temperature profiles for air and solid in a countercurrent flow rotary dryer.

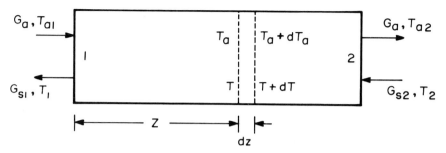

Fig. 26. A schematic diagram of a rotary dryer. Mass and heat balances.

T = the temperature of the solids
T_a = the temperature of the air

The humid heat is the amount of BTU required to raise the temperature of 1 lb of dry air and its water vapor content by 1°F.

If the surface area of the solid particles per unit volume of drying is A_s and U is an overall heat transfer coefficient, then

$$dQ = UA_sA_c(T_a - T)dz \qquad (28)$$

From Eqs. (27) and (28),

$$-G_aA_cfdT_a = UA_sA_c(T_a - T)dz \qquad (29)$$

Assuming constant f and UA_s, Eq. (29) can be integrated between T_{af} and T_{ai}, z_f and z_i

$$-\int_{T_{af}}^{T_{ai}} \frac{dT_a}{T_a - T} = \int_{z_f}^{z_i} \frac{UA_s}{G_af} dz = \frac{UA_s(z_i - z_f)}{G_af} \qquad (30)$$

In the evaporating zone, the temperature of the solids is the wet bulb temperature, T_i, which is constant. Therefore,

$$-\int_{T_{af}}^{T_{ai}} \frac{dT_a}{T_a - T_i} = \ln\left(\frac{T_{af} - T_i}{T_{ai} - T_i}\right) \qquad (31)$$

From Eqs. (30) and (31), the *constant temperature zone equation* becomes

$$\frac{UA_s(z_i - z_f)}{G_af} = \ln\left(\frac{T_{af} - T_i}{T_{ai} - T_i}\right) \qquad (32)$$

For the preliminary heatup zone, Eq. (27) still holds. If f and C_s are constant with temperature

$$dT = \frac{G_a f}{G_s C_s} dT_a \tag{33}$$

i. e., T is a linear function of T_a between T_2 and T_i (Fig. 25). Under these conditions, integration with respect to T_a gives

$$-\int_{T_{ai}}^{T_{a2}} \frac{dT_a}{T_a - T} = \frac{T_{ai} - T_{a2}}{(T_a - T)_m} \tag{34}$$

where $(T_a - T)_m$ is the logarithmic average temperature

$$(T_a - T)_m = \frac{(T_{ai} - T_i) - (T_{a2} - T_2)}{\ln[(T_{ai} - T_i)/(T_{a2} - T_2)]} \tag{35}$$

Equations (30), (34), and (35) yield for the *heatup zone*

$$\frac{UA_s(L - z_i)}{G_a f} = \frac{T_{ai} - T_{a2}}{(T_a - T)_m} \tag{36}$$

A similar equation holds for the final heating section of the dryer. In this section, the air temperature will vary between the inlet temperature T_{a1} and the temperature T_{af} that corresponds to the end of the constant temperature section (Fig. 25):

$$\frac{UA_s z_f}{G_a f} = \frac{T_{a1} - T_{af}}{(T_a - T)_m} \tag{37}$$

where now $(T_a - T)_m$ refers to a new set of temperatures for the two streams (Fig. 25).

The total dryer length will be the sum of the lengths for all sections above.

Sludges contain bound water, in contrast to assumptions (iv) and (v) above. Therefore, the theory holds only for the initial heatup period and for constant temperature drying, until the unbound water has evaporated.

The holdup time of material in the dryer depends on the diameter, length, flight size, and air velocity [89–91].

Recently Peck and Wasan [70] reviewed the retention time equations. A steady-state momentum balance on the particles in the dryer yields

$$\rho_s V_s \frac{dv}{d\tau} + \rho_s g V_s \sin\theta - F_d = 0 \qquad (38)$$

where ρ_s and V_s are the density and volume of the particle, v is its velocity along the dryer, τ the time of fall of the particle, F_d the drag force, and θ the angle of the dryer with the horizontal.

The drag force can be approximated by

$$F_d \simeq C_o V_s u_a^n \qquad (39)$$

where C_o and n are constants and u_a is the air velocity. Solution of these equations with the boundary condition

$$\text{at } \tau = 0, \ v = 0 \qquad (40)$$

gives the particle velocity

$$v = \frac{\tau}{\rho_s}(C_o u_a^n - \rho_s g \sin\theta) \qquad (41)$$

A similar analysis for the falling particle in the dryer yields finally the average length traveled by a particle per revolution [70]

$$\langle z \rangle = 1.85 D_d \left(\frac{C_o u_a^n}{\rho_s g \cos\theta} - \tan\theta\right) \qquad (42)$$

where D_d is the dryer diameter.

For a dryer of length L, rotating at a rate of ω (rpm), the retention time is

$$t = \frac{L}{1.85 D_d \omega \left(\dfrac{C_o u_a^n}{\rho_s g \cos\theta} - \tan\theta\right)} \qquad (43)$$

This expression is similar to the equation of Saeman and Mitchell [92]

$$t = \frac{L}{C' D_d \omega (\theta - u_a)} \qquad (44)$$

where C' is a constant

Example 4

A rotary dryer is considered for drying of 10,000 lb/h of a partly dewatered municipal sludge. The sludge enters the dryer at 70°F with 25% solids and exits when the unbound moisture is evaporated, which corresponds to 33% solids. Air of wet-bulb temperature of 75°F is heated to 260°F before entering the dryer countercurrently to the sludge. Evaluate the required dryer length. The specific heat of the solids is $C_s = 0.8$ BTU/lb; °F.

> *Note:* Sludge drying to the extent described by this problem is not of practical importance. However, the previous theory applies only within this region and the example is used to demonstrate application of the concepts discussed.

Solution. It is assumed that the solids are raised fast to the wet-bulb temperatures and remain at this temperature during drying. No heating of the solids takes place above the wet-bulb temperature.

The final temperature can be selected on economic considerations based on dryer costs and fuel costs. Empirical recommended values are given [53, 54]

$$\ln\left(\frac{T_{a1} - T_i}{T_{a2} - T_i}\right) = 1.5 \tag{i}$$

where $T_{a1} = 260°F$ and $T_i = 75°F$.

Solving for T_{a2}, a value of $T_{a2} = 116.3°F$ is obtained

Energy and Mass Balance

Inlet material: Solids: m_s: $0.25 \times 10{,}000 = 2500$ lb/h
                     Water: $0.75 \times 10{,}000 = 7500$ lb/h

Outlet material: Total: $\dfrac{2500}{0.33} = 7576$ lb/h
                     Water: $m_w = 0.67 \times 7576 = 5076$ lb/h

Water to be removed: $W = 7500 - 5076 = 2424$ lb/h
Heat required to evaporate water ($\Delta H_v = 1051$ BTU/lb)

$$Q_v = W\Delta H_v = 2424 \times 1051 = 2.547 \times 10^6 \text{ BTU/h}$$

From Eq. (27), the amount of air required can be calculated

$$Q_v = -G_a A_c f(T_{a2} - T_{a1})$$

Humidity charts give $f = 0.265$ BTU/lb°F at 75°F. Therefore,

$$G_a A_c = -\frac{2.547 \times 10^6}{0.265(116.3 - 260)} = 6.69 \times 10^4 \text{ lb dry air/h}$$

Dryer Diameter To account for heat losses, a 10% excess air will be considered, or $G_a A_c = 7.36 \times 10^4$ lb dry air/h. The amount of water in this air can be neglected. Assuming a superficial mass velocity of 1000 lb/h · ft² [89], the dryer cross-sectional area becomes

$$A_c = \frac{7.36 \times 10^4}{1000} = 73.6 \text{ ft}^2$$

corresponding to a diameter $D_d = 9.68$ ft or $D_a \simeq 10$ ft and $A_c = 78.5$ ft². Therefore,

$$G_a = \frac{7.36 \times 10^4}{78.5} = 937.6 \text{ lb/ft}^2\text{h}$$

Heat Transfer Coefficient Friedman and Marshall [93] suggest

$$UA_s = 15\frac{G_a^{0.16}}{D_d} = 15\frac{(937.6)^{0.16}}{12} = 3.74 \text{ BTU/h°F}$$

Dryer Length From Eq. (32), the length for the constant temperature zone, L_c, can be calculated

$$L_c = \frac{G_a f}{UA_s} \ln\left(\frac{T_{af} - T_i}{T_{a2} - T_i}\right) = \frac{937.6 \times 0.265}{3.74} \ln\left(\frac{260 - 75}{116.3 - 75}\right) = 99.6 \text{ ft}$$

where $T_{a2} = T_{ai}$ and $T_{af} = T_{a1}$ here.

Note that in this length the preheating zone has been neglected. The ratio

$$\frac{L}{D_d} = \frac{99.6}{10} = 9.96$$

is within the L/D_d ratio, 4–10, suggested by Perry [55].

B. Theory of Gravity Dewatering

Dewatering of sludge on sand beds and lagoons starts with gravity drainage of water, a process similar to cake filtration and flow through porous media under constant pressure. Under these conditions the flow is laminar. However, the Hagen-Poisseuille equation [94] cannot describe the rate of drainage fully, since the resistance to flow depends on the shape and size of the particles, the fraction of voids, the length of the tortuous path of the liquid through the cake and the resistance of the supporting bed.

The rate of drainage can be expressed as

$$\text{Rate} = \frac{\text{driving force}}{\text{resistance}} \quad (45)$$

The resistance to flow consists of two terms: the sludge cake resistance and the supporting bed resistance. The driving force is the overall pressure drop across the sludge and the bed.

Kozeny [cf.95] proposed the following equation for flow through packed beds (excluding support resistance).

$$u = \frac{1}{A}\frac{dV}{dt} = \left[\frac{\epsilon^3}{C(1-\epsilon)^2(s_s/v_s)^2}\right]\frac{\Delta P_s}{\mu L} \quad (46)$$

where
- u = the linear velocity of the fluid
- A = the cross-sectional area of the bed, normal to flow
- V = the volume of fluid
- C = a constant (dimensionless)
- ϵ = sludge porosity (volume of voids per volume of sludge cake)
- s_s = surface of a particle
- v_s = volume of a particle
- ΔP_s = pressure drop across the sludge cake
- L = sludge cake thickness
- μ = fluid viscosity

Equation (46) can be derived from the Hagen-Poisseuille equation, with $C = 2$. However, for incompressible beds, C is about 5 [95, 96].

A specific cake resistance, α, can be defined for the sludge as

$$\alpha = \frac{C(1-\epsilon)(s_s/v_s)^2}{\rho_s \epsilon^3} \quad (47)$$

where ρ_s is the density of the solids. The specific resistance has units of [L/M]. The weight of solids in an incompressible cake of volume $AL(1 - \epsilon)$ is

$$W_s = m_s g = \rho_s g(1 - \epsilon)AL = \rho g V X_s \qquad (48)$$

where X_s is the weight fraction of the initial solids. Combining Eqs. (46)–(48), the Kozeny equation (46) takes the form

$$\Delta P_s = \frac{\alpha \mu m_s u}{A} = \frac{\alpha \mu \rho X_s V u}{A} \qquad (49)$$

Similarly, a pressure drop across the supporting bed, ΔP_b, can be defined as [54]

$$\Delta P_b = \mu u R_b \qquad (50)$$

where R_b is the supporting bed resistance in $[L^{-1}]$ units. Old, used supporting media have higher resistance than new ones. The medium resistance is usually important during the initial filtration stage and it is usually taken as a constant. Values of R_b range from 0.05 to 0.1 of the specific resistance of the cake, α. A value of

$$R_b \simeq 0.1\, \alpha \qquad (51)$$

has been suggested for most calculations, except for thin cakes [97]. Note that Eq. (51) is only an empirical equation, and that the units of R_b and α are not the same.

Equations (49) and (50) give the total pressure drop across the cake and the supporting medium

$$\Delta P = \Delta P_s + \Delta P_b = \mu u \left[\frac{\alpha \rho X_s V}{A} + R_b \right] \qquad (52)$$

Substituting $dV/dt = uA$ (Eq. 46), the inverse of the rate of fluid flow through the bed becomes

$$\frac{dt}{dV} = \frac{\mu}{A \cdot \Delta P} \left[\left(\frac{\alpha \rho X_s}{A} \right) V + R_b \right] \qquad (53)$$

A similar expression has been proposed by Crackley and Jones [98] for vacuum filtration.

Integration of Eq. (53) for $t = 0$, $V = 0$ and t, V gives

$$t = \frac{\mu}{\Delta P} \left[\frac{\alpha \rho X_s}{2} \left(\frac{V}{A} \right)^2 + R_b \left(\frac{V}{A} \right) \right] \qquad (54)$$

Therefore, the effluent volume-time relationship is represented by a parabola.

Equation (54) permits evaluation of the resistance terms α and R_b by experimentally determining V and t at a specified pressure gradient. A plot of (t/V) vs V should give a straight line

$$\frac{t}{V} = \left(\frac{\rho\alpha\mu X_s}{2A^2 \cdot \Delta P}\right)V + \frac{\mu R_b}{A \cdot \Delta P} \tag{55}$$

From the slope, the specific resistance α can be obtained. The intercept gives a value for the supporting medium resistance R_b.

Derivation of Eq. (53) above was based on the assumption of cake incompressibility, which implies that the specific resistance is constant and independent of the pressure drop. The compressibility of sludge cakes can be taken into account through an empirical equation (56).

$$\alpha = \alpha_o(h_p/h_o)^\sigma \tag{56}$$

where h_p is the pressure head, α_o is the specific resistance at a reference pressure head, h_o, and σ is the compressibility coefficient of the cake [54]. Values of σ range from zero for incompressible cakes to 0.2 to 0.8 for compressible ones [54].

The compressibility coefficient, σ, is evaluated from Eq. (56), if α is determined for several pressure drops. Nebiker et al. [99] applied Eqs. (55) and (56) to determine the compressibility coefficients of sludges, as in Fig. 27. In that investigation the medium resistance was considered negligible. A media factor was introduced to account for obtaining h_o from a Buchner funnel test.

Equation (53) can be modified to express the rate of drainage in terms of the height of the sludge on the bed, by noting that

$$\frac{1}{A}\frac{dV}{dt} = -\frac{dL}{dt} \tag{57}$$

and

$$\frac{V}{A} = L_o - L \tag{58}$$

Here, L is the height of sludge at any time $t > 0$ and L_o is the initial height of the sludge. The pressure drop can be expressed as a pressure head, h_p,

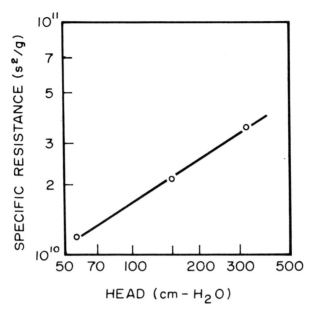

Fig. 27. Dependence of specific resistance on head, for a sludge containing 3.7% solids. Slope = compressibility coefficient = 0.63 [99].

$$h_p = \frac{\Delta P}{\rho g} = h_s + h_b \tag{59}$$

The pressure head due to sludge, h_s, is equal to the height, L ($h_s = L$). The pressure head from the medium is approximately constant. Equations (53) and (57–59) give then

$$-\frac{dL}{dt} = \frac{\rho g}{\mu} \frac{h_b + L}{\alpha \rho X_s(L_o - L) + R_b} \tag{60}$$

By setting $R_b = K\alpha$ [cf. Eq. (51)] and by expressing α in terms of the pressure head, Eq. (56), the height decrease with time becomes

$$-\frac{dL}{dt} = \frac{\rho g h_o^\sigma}{\mu \alpha_o} \left(\frac{(h_b + L)^{1-\sigma}}{\rho X_s(L_o - L) + K} \right) \tag{61}$$

Integration for

$$L = L_o \text{ at } t = 0$$
$$L = L \text{ at } t = t$$

gives

$$t = \frac{\mu\alpha_o(h_b + L_o)^\sigma}{\rho g h_o^\sigma} \left\{ \left(\frac{\rho X_s L_o + K + \rho X_s h_b}{\sigma} \right) \left[1 - \left(\frac{h_b + L}{h_b + L_o} \right)^\sigma \right] \right.$$
$$\left. - \frac{\rho X_s (h_b + L_o)}{\sigma + 1} \left[1 - \left(\frac{h_b + L}{h_b + L_o} \right)^{\sigma-1} \right] \right\} \quad (62)$$

If the pressure head in the supporting medium is relatively small compared to the total pressure head, i.e., $h_b \ll L$ (and L_o), Eq. (63) simplifies to

$$t = \frac{\mu\alpha_b X_s L_o^{\sigma+1}}{g\sigma(\sigma + 1)h^\sigma l_o +} \left\{ (\sigma + 1) \left[1 + \frac{K}{\sigma \rho X_s L_o} \right] \left[1 - \left(\frac{L}{L_o} \right)^\sigma \right] \right.$$
$$\left. - \sigma \left[1 - \left(\frac{L}{L_o} \right)^{\sigma+1} \right] \right\} \quad (64)$$

The supporting medium resistance is usually important during the initial stages of drainage. For long times or negligible medium resistance the term containing K can be neglected and Eq. (64) gives

$$t = \frac{\mu\alpha_o X_s L_o^{\sigma+1}}{g\sigma(\sigma + 1)h_o^\sigma} \left\{ (\sigma + 1) \left[1 - \left(\frac{L}{L_o} \right)^\sigma \right] - \sigma \left[1 - \left(\frac{L}{L_o} \right)^{\sigma+1} \right] \right\} \quad (65)$$

This expression is similar to the one obtained by Nebiker et al. [99] for negligible medium resistance.

Treatment of sludges with ionic coagulants may change the specific resistance of the sludges. Novak [100] reported that addition of magnesium ions results in precipitation of $Mg(OH)_2$ in the form of a gelatinous, highly hydrated floc that increases the specific resistance of the sludge.

In presenting the theory of gravity dewatering, the historical development of some empirical correlations was omitted. Nebiker [101] and Adrian [102] have reviewed a number of empirical equations for the design of drying beds. Blunk [103] suggested the following equation for bed area requirements to dewater 1 m³ of sludge

$$A = 0.8 - 0.02 S_o \quad (66)$$

where S_o is the initial percent solids content and A the bed area (in m²). A dewatering rate equation was proposed by Vater [104]

$$Y = 0.33 S_o^{1.6} \quad (67)$$

where Y is the rate of solids dewatered per unit area per day (kg/m² d). A different equation for Y was proposed by Haseltine [14].

$$Y = 0.157S_o - 0.286 \tag{68}$$

All these equations base the design of drying beds only on the initial solids concentration and fail to take into account other sludge parameters.

Volger and Rudolfs [7] realized that the initial depth, h_o, of the sludge application is important and suggested the following equation for dewatering time

$$t = S_o h_o^{1.6} \tag{69}$$

However, the specific resistance of the sludge was ignored. Sanwick [105] considered the effect of the sludge specific resistance on the dewatering rate and gave an equation

$$Y' = 10^7/\sqrt{R} \tag{70}$$

where Y' is expressed here in kg/m²yr and R is the specific resistance in s²/g. This equation does not consider the effect of the initial solids content and of the initial depth of application. These parameters were taken into account by Nebiker et al. [99], who proposed an equation similar to Eq. (65) above.

All equations ignore the effect of climatic conditions and particularly of rainfall. Nebiker and Adrian [106] considered this effect in conjunction with Eqs. (21), (25), and (65) in a computer simulation of drying and drainage processes. However, all weather effects were lumped into the drying rate term. A more flexible model can be built using variable weather conditions [102, 107].

Example 5

In an experimental investigation with a sludge, initially containing 3.7% solids, a column of 4 in. diameter was filled to an initial height of 21 inches. Table 6 shows some typical results of effluent volume with respect to time at a pressure drop of 0.53 atm. Estimate the specific resistance of the cake and of the supporting medium (sand). Given: Sludge density: 62.2 lb$_m$/ft³; viscosity: 6.18 × 10^{-4} lb/ft s.

Solution. Figure 28 shows a plot of t/V vs V[cf. Eq. (55)] constructed from the above data. The slope of this graph is

TABLE 6
Sludge Drainage on Sand
with Time

Time, d	Effluent volume, in.^{3}
0	0
3	14.5
5	18.8
8	24.5
10	28.4
15	34.5
20	39.5
25	45.8

$$\frac{\alpha\mu\rho X_s}{2A^2\Delta P} = 0.0112 \text{ (d/in.}^6) = 8.03 \times 10^5 \text{ (h/ft}^6)$$

where
$\mu = 2.22 \text{ lb}_m/\text{ft} \cdot \text{h}$ $A = 0.0872 \text{ ft}^2$
$\rho = 62.2 \text{ lb}_m/\text{ft}^3$ $\Delta P = 1122 \text{ lb}_f/\text{ft}^2$
$X_s = 0.037$ $g_c = 4.15 \times 10^8 \text{ ft} \cdot \text{lb}_m/\text{lb}_f\text{h}^2$

The specific resistance is, therefore

$$\alpha = 1.113 \times 10^{15} \text{ ft/lb}_m$$

The intercept of Fig. 28 is

$$\frac{\mu R_b}{A\Delta P} = 0.044 \text{ (d/in.}^3) = 1.82 \times 10^3 \text{ (h/ft}^3)$$

and

$$R_b = 3.33 \times 10^{13} \text{ (ft}^{-1})$$

The supporting medium resistance is about 3% of the slurry specific resistance and could be ignored here.

Example 6

The specific resistance of a sludge was determined experimentally at four different pressure heads, as in Table 7. Estimate the compressibility

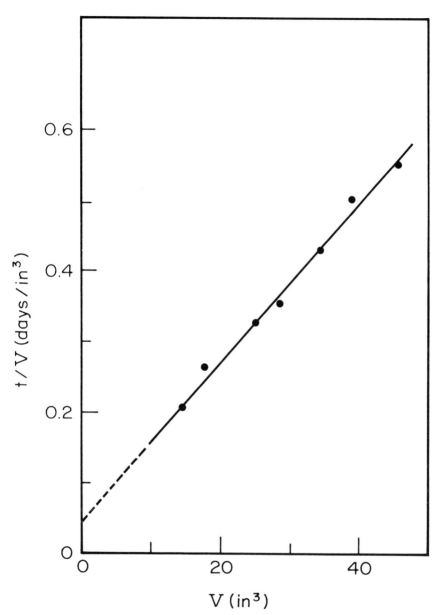

Fig. 28. An experimental plot of t/V vs V [Eq. (55)] to evaluate the specific resistance and the supporting bed resistance (see text).

TABLE 7
Variation of Sludge-Specific Resistance with Pressure

α, ft/lb	Pressure head, ft H_2O
2.25×10^{14}	2.29
3.50×10^{14}	4.92
4.2×10^{14}	6.56
5.0×10^{14}	9.84

coefficient of this sludge and the reference resistance α_o at 1 ft of water pressure head.

Solution. Equation (56) gives

$$\log \alpha = \log (\alpha_o/h_o^\sigma) + \sigma \log h_p$$

The slope of a log-log plot of the data of Table 7 or Fig. 29, gives σ

$$\sigma = 0.535$$

Then,

$$\frac{\alpha_o}{h_o^\sigma} = 1.48 \times 10^{14}$$

and

$$\alpha_o = 1.48 \times 10^{14} \text{ ft/lb}$$

The compressibility coefficient is, therefore, given by

$$\alpha = 1.48 \times 10^{14} (h_p)^{0.535} \text{ ft/lb}$$

where α_o is referenced at $h_o = 1$ ft.

VI. THEORY OF EVAPORATION

A. Heat Transfer

The design of evaporators depends on their required capacity and the required steam consumption. Heat is transferred from the steam to the

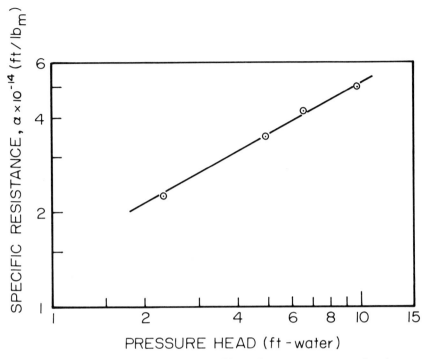

Fig. 29. Dependence of the specific resistance on pressure head.

evaporating liquid through a heating surface. If U is the overall heat transfer coefficient, the amount of heat transferred is

$$Q = UA\Delta T \tag{71}$$

where ΔT is the overall temperature drop between steam and evaporating liquid and A the heating surface area.

The transferred heat raises the temperature of the liquid to its boiling point, corresponding to the absolute pressure in the evaporator, and supplies the latent heat of vaporization of water. If the feed is at a temperature above the boiling point, flash evaporation occurs.

Dissolved substances in water tend to lower the vapor pressure of water at a given temperature. Conversely, the boiling point of solutions at a given pressure is higher than that of pure water. Boiling point elevation is particularly significant for strong solutions for which Dühring's rule applies [53, 54].

In an evaporator loaded with an appreciable depth of liquid, the boiling point increases with the depth because of the existing liquid head.

Therefore, the actual boiling point is higher than that corresponding to the pressure in the evaporator, resulting in decreased capacity [53, 54].

The heat transfer coefficient expresses the facility to heat flow for a particular design and operation. The overall resistance ($1/U$) is the sum of the resistances to heat transfer on the steam side, on the liquid side and across the tube wall.

$$\frac{1}{U} = \frac{D_s}{D_l h_l} + \frac{\delta}{k_T} \cdot \frac{D_s}{D_{\ln}} + \frac{1}{h_{os}} \qquad (72)$$

where D_s, D_l, and $D_{\ln}$ are the outside (steam), inside (liquid), and mean logarithmic tube diameters, δ is the tube wall thickness, k_T the wall thermal conductivity, h_l the heat transfer coefficient on the liquid side and h_{os} the heat transfer coefficient on the steam side. If scalings are formed inside and outside the tube walls, additional resistance terms should be added to Eq. (72). Some typical values of overall heat transfer coefficients are given in Table 8 for various evaporator designs.

B. Heat and Material Balance

A schematic diagram of a single-effect evaporator with all streams and their properties is shown in Fig. 30. A material balance in the evaporator gives

TABLE 8
Overall Heat Transfer Coefficients for Evaporators
[54, 108]*

Evaporator type	U, BTU/ft²h °F
1. Short tube	
a. Horizontal tubes	200–400
b. Calandria type	150–500
2. Long tube, vertical	
a. Natural circulation	200–600
b. Forced circulation	400–2000
3. Coil evaporators	200–400
4. Agitated-film, $\mu = 1$ cp	400

*Reprinted with permission. Copyright 1953, 1976 McGraw-Hill.

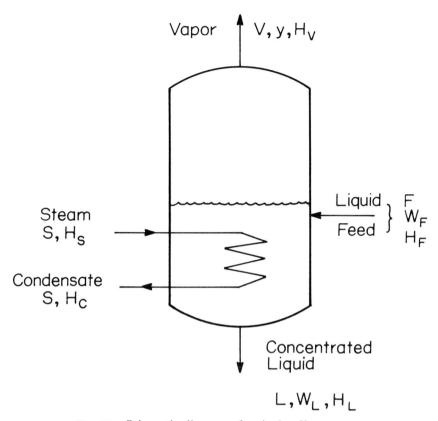

Fig. 30. Schematic diagram of a single-effect evaporator.

$$F' = L' + V' \tag{73}$$

where F' is the weight of feed, L' the weight of the resulting thick liquor, and V' the weight of the vapor phase.

If w_F is the weight fraction of water in the feed, w_L its weight fraction in the concentrated liquid, and y the weight fraction of evaporated water, a material balance for water yields

$$w_F F' = w_L L' + y V' \tag{74}$$

The heat supplied by the steam is

$$Q_s = S'(H_s - H_c) \tag{75}$$

where S' is the the weight of steam supplied to the evaporator and H_s and H_c are the enthalpies of steam and condensate, respectively. Complete condensation of saturated steam is assumed with no condensate cooling in the evaporator.

A heat balance for the condensing liquid gives

$$Q_L = \text{(Heat out, in vapor and thick liquid)} - \text{(heat in, in feed)} \quad (76)$$

$$Q_L = (L'H_L + V'H_v) - F'H_F \quad (77)$$

At steady state, $Q_L = Q_s$, and using Eq. (73),

$$S'(H_s - H_c) = L'(H_L - H_F) + V'(H_v - H_F) \quad (78)$$

Normally low pressure steam is used in evaporators. Although high pressure steam could provide a larger temperature gradient across the heating surface for given conditions, such steam is usually valuable for energy generation.

C. Multiple-Effect Evaporators

Equation (71) holds for heat transfer in each stage of a multiple-effect evaporator (cf. Fig. 16). For the first effect (I) then

$$Q_I = U_I A_I \Delta T_I \quad (79)$$

If the feed is at or close to the boiling point corresponding to the conditions in the first stage, essentially all Q_I goes into vaporizing water in this stage. At steady-state, this water vapor will condense around the tubes of the second stage to vaporize almost equal amount of water in stage II. The condensate in this stage is at about the same temperature as the vapor of the boiling liquid in the first stage.

The heat exchanged in Stage II is

$$Q_{II} = U_{II} A_{II} \Delta T_{II} \quad (80)$$

From the operation of this stage it follows that the amounts of heat exchanged in effects I and II are almost equal.

$$U_I A_I \Delta T_I \sim U_{II} A_{II} \Delta T_{II} \quad (81)$$

By similar reasoning, the same amount of heat is exchanged in the third effect, thus

$$U_i A_i \Delta T_i = Q_i = Q \quad (82)$$

where Q is a constant and the subscript denotes the effect.

Usually, the heating surface areas of all effects are equal for construction economy, therefore,

DRYING AND EVAPORATION PROCESSES 437

$$U_i \Delta T_i = Q/A = \text{constant} \tag{83}$$

Equation (83) suggests that the temperature drops are inversely proportional to the overall heat transfer coefficients in each effect, e.g.

$$\frac{\Delta T_I}{\Delta T_{II}} = \frac{U_{II}}{U_I} \tag{84}$$

It should be emphasized that Eq. (82) to (84) are only approximate.
  The total heat exchanged in an N-effect evaporator is the sum of all Q_i

$$Q_T = \sum_{i=1}^{N} Q_i = \sum_{i=1}^{N} U_i A_i \Delta T_i \tag{85}$$

If all A_i and U_i are equal to a constant value A and U, respectively, then

$$Q_T = UA \sum_{i=1}^{N} \Delta T_i = U_i A_i \Delta T_T \tag{86}$$

where ΔT_T is the total temperature drop across the system. Equation (86) suggests that the total heat exchanged in the multiple effect evaporator would be the same as the heat exchanged in a single effect evaporator with the same U and A as each effect, operating under a temperature gradient ΔT_T. Therefore, the capacity of a multiple-effect evaporator is no better than that of an equivalent single-effect unit. Here, capacity is defined as the total rate of water vaporization (lb/h). However, significant steam economy is achieved. Each pound of steam supplied to an N-effect evaporator vaporizes approximately N pounds of water. In a single-effect evaporator, each pound of steam vaporizes only about 1 lb of water. To arrive at these approximate relationships, liquid heating and any heat losses have been neglected.

Example 7

A sludge 8% in solids, is concentrated in a single-effect evaporator to 35% solids. The evaporator operates at 0.95 psia and uses steam at 10 psig. If the feed rate is 10,000 lb/h at 70°F and the heat transfer coefficient is 400 BTU/ft²h °F, calculate: (a) The amount of steam required; (b) the amount of water evaporated per pound of steam (economy); and (c) the heating surface area. Neglect the boiling point elevation and heat of dilution for the sludge. The specific heat capacity of the feed sludge is $C_{P,F} = 0.88$ BTU/lb°F

Solution

Material Balance

Water in feed: $\dfrac{92}{8} = 11.50$ lb water/lb solids

Water in concentrate: $\dfrac{65}{35} = 1.85$ lb water/lb solids

Water evaporated: $= 9.65$ lb water/lb solids

For a feed $F' = 10{,}000$ lb sludge/h, then,

$$V' = 10{,}000 \times 0.08 \times 9.65 = 7720 \text{ lb water/h}$$
$$L' = 10{,}000 - 7720 = 2380 \text{ lb liquor/h}$$

Steam Requirements The steam requirements are estimated from Eqs. (73) to (77) rearranged to give

$$S'\Delta H_v = (F' - L')H_v + L'H_L - F'H_F \qquad (i)$$

where ΔH_v is the latent heat of vaporization of steam at 10 psig. If the temperature of the concentrated liquor, T_L, is considered as a reference temperature ($T_L = T_R$), then

$$H_L = C_{P,L}(T_L - T_R) = 0 \qquad (ii)$$

and

$$H_F = C_{P,F}(T_F - T_R) = C_{P,F}(T_F - T_L) \qquad (iii)$$

The enthalpy of the vapor H_v with respect to the thick liquor represents now the latent heat of vaporization at the operating pressure in the evaporator. Equation (i) yields then

$$S'\Delta H_v = (F' - L')H_v + F'C_{P,F}(T_L - T_F) \qquad (iv)$$

From steam tables, 10 psig:

$$T_S = 239.4°F; \quad \Delta H_v = H_s - H_c = 952.6 \text{ BTU/lb}$$

At 0.95 psia,

$$T_L = 100°F \quad H_v = 1037 \text{ BTU/lb}$$

With the feed entering at $T_F = 70°F$, then the steam rate is

DRYING AND EVAPORATION PROCESSES

$$S' = \frac{(10{,}000 - 2380)\,1037 + 10{,}000 \times 0.88\,(100 - 70)}{952.6} = 8572 \text{ lb/h}$$

Steam Economy

$$\frac{V'}{S'} = \frac{7720}{8572} = 0.90 \text{ lb water evaporated/lb steam}$$

Heating Surface Area From Eq. (71) and (75)

$$A = \frac{Q}{U(T_S - T_L)} = \frac{S\Delta H_v}{U(T_S - T_L)} \qquad \text{(v)}$$

or

$$A = \frac{8572 \times 952.6}{400(239.4 - 100)} = 146.4 \text{ ft}^2$$

Example 8:

A sludge is concentrated in a triple-effect evaporator to recover vitamin B_{12}. The first effect operates at 8 psig, whereas the last effect operates at a temperature of 110°F. If the overall heat transfer coefficients are 400, 320, and 240 for the first, second, and third effect, respectively, estimate the liquor boiling temperature in each stage. Neglect any boiling point elevation.

Solution. Assuming that all effects have the same heating surface area, Eq. (83) and (84) give

$$\Delta T_I : \Delta T_{II} : \Delta T_{III} : (\Delta T_I + \Delta T_{II} + \Delta T_{III}) = \qquad \text{(i)}$$

$$\frac{1}{U_I} : \frac{1}{U_{II}} : \frac{1}{U_{III}} : \left(\frac{1}{U_I} + \frac{1}{U_{II}} + \frac{1}{U_{III}}\right)$$

Thus

$$\Delta T_i = \frac{\Delta T_T}{\sum_{i=1}^{3} \frac{1}{U_i}} \cdot \frac{1}{U_i} \qquad \text{(ii)}$$

where

$$\Delta T_T = \sum_{i=1}^{3} \Delta T_i = T_I - T_{III}$$

From steam tables, at 8 psig,

$$T_I = 235°F$$

and

$$\Delta T_T = 235 - 110 = 125°F$$

Thus

$$\Delta T_i = \frac{125}{\left(\dfrac{1}{400} + \dfrac{1}{320} + \dfrac{1}{240}\right)} \cdot 400 = 32°F$$

Similarly, $\Delta T_{II} = 39.8°F$ and $\Delta T_{III} = 53.2°F$ from which $T_{II} = T_I - \Delta T_I = 235 - 32 = 203°F$ and $T_{III} = T_{II} - \Delta T_{II} = 163.2°F$.

NOMENCLATURE

A	Exposed or cross-sectional area, ft²
A_c	Cross-sectional area of dryer, ft²
A_s	Surface area of solids per unit dryer volume, ft²/ft³
C	Constant (dimensionless), Eq. (46)
C_o	Constant, Eq. (39)
C_p	Heat capacity of air at constant pressure, BTU/lb °F
$C_{p,f}$	Heat capacity of feed fluid BTU/lb °F
C_s	Specific heat of solids, BTU/lb °F
D	Diffusivity, ft²/h
D_d	Dryer diameter, ft
D_I	Inside tube diameter, Eq. (72), ft
D_{ln}	Mean logarithmic diameter, Eq. (72), ft
D_s	Outside tube diameter Eq. (72), ft
F'	Mass of feed, lb
F_d	Drag force, lb_f
f	Humid heat of air
G	Mass velocity of air, lb/ft²h

DRYING AND EVAPORATION PROCESSES 441

G_a	Mass flux of air, lb/ft²h
G_s	Mass flux of solids, lb/ft²h
H	Enthalpy, BTU/lb
H_s	Entalpy of steam, BTU/lb
H_c	Enthalpy of condensate, BTU/lb
ΔH_v	Specific latent heat of vaporization, BTU/lb
$\mathscr{H}$	Relative humidity, %
h_l	Heat transfer coefficient, BTU/ft²h°F
h_b	Pressure head due to sludge, ft
h_o	Reference pressure head (ft)
h_{os}	Heat transfer coefficient on steam side, BTU/ft²h°F
h_p	Total pressure head, ft
h_s	Pressure head from the medium, ft
j_H	j-Factor for heat transfer, dimensionless
K	Constant, Eq. (61)
k	Thermal conductivity of air, BTUft/ft²h°F
k_T	Thermal conductivity of wall, BTU ft/ft² h °F
k_w	Molar mass transfer coefficient, lb-mol/ft² h atm
L	Dryer length, ft
L_o	Initial bed thickness
L'	Mass of concentrated liquid, lb
l, L	Length or thickness of slab, ft
M	Molecular weight of water
m	Total mass of moisture, lb
m_s	Mass of solids, lb
N	Molar rate of evaporation, lb-mol/h
P_a	Bulk partial pressure of water vapor, atm
P_i	Partial pressure of water vapor at the interface, atm
P_s	Saturation vapor pressure of water, atm
P_w	Partial pressure of water, atm
ΔP	Total pressure drop, atm
ΔP_b	Pressure drop across the supporting bed, atm
ΔP_s	Pressure drop across the sludge cake, atm
Pr	Prandtl number, dimensionless
Q	Rate of heat transfer, BTU/h
Q_L	Rate of heat transfer to liquid, BTU/h
Q_s	Rate of heat transfer by steam, BTU/lb
R	Rate of water removal per unit area, lb/ft²h
$\mathscr{R}$	Gas constant, BTU/lb-mol °R
R_b	Supporting bed resistance, ft⁻¹
R_c	Constant drying rate, lb/ft²h

R_{cs}		Drying rate at the second critical point, lb/ft²h
Re		Reynolds number (dimensionless)
S'		Mass of steam, lb
S_o		Initial solids content (o/o)
s_s		Surface of particle, ft²
T		Temperature of solids, °F
T_a		Bulk temperature of air, °F
T_{ai}, T_{af}		Air temperature in dryer (Fig. 25), °F
T_i		Temperature of wetted surface, °F
ΔT		Overall temperature drop, °F
t		Time, h
t_c		Drying time at constant rate, h
t_f		Drying time at falling rate, h
t_T		Total drying time, h
U		Overall heat transfer coefficient, BTU/ft²h °F
u		Linear velocity, ft/h
u_a		Air velocity, ft/h
V		Volume of fluid, ft³
V'		Mass of vapor phase, lb
V_s		Particle volume, ft³
v		Particle velocity, ft/h
v_s		Volume of particle, ft³
W_s		Weight of solids, lb$_f$
w		Weight fraction of liquid
X		Free moisture content
$\langle X \rangle$		Average moisture content
X_c		First critical moisture content
X_{cs}		Free moisture content at the second critical point
X_e		Equilibrium moisture content
X_h		Hygroscopic moisture content
X_i		Initial moisture content
X_S		Weight fraction of solids
Y'		Dewatering rate Eq. (70), kg/m²yr
y		Weight fraction of water vapor
z		Length coordinate, ft
z_f, z_i		Length in dryer (Fig. 25), ft
α		Specific resistance, ft/lb
α_o		Specific resistance at a reference pressure head, ft/lb
γ_o		Weight of skeletal solids per unit volume, lb/ft³
δ		Tube wall thickness, ft

ε	Sludge porosity
θ	Angle of dryer with horizontal
μ	Viscosity, $lb_m/ft\ h$
ρ	Density of fluid, lb/ft^3
$ρ_s$	Density of solids, lb/ft^3
σ	Compressibility coefficient
τ	Time of fall of particle, h
ω	Rate of dryer rotation, rpm

Subscripts

a	Air
i	Interface; stage of evaporators
F	Feed
L	Thick liquor
V	Vapor

REFERENCES

1. R. S. Burd, "A Study of Sludge Handling and Disposal", US Dept. of the Interior, Federal Water Pollution Control Administration, Publ. WP-20-4 (1968).
2. S. Kelman and C. P. Priesing, "Polyelectrolyte Flocculation-Sand Bed Dewatering," paper presented at the Michigan WPCA Conf., June, 1964.
3. "Recommended Standards for Sewage Works", a Report of the Committee of the Great Lakes-Upper Mississippi River Board of State Sanitary Engineers, Health Education Service, Albany, NY, 1968.
4. "Process Design Manual for Sewage Treatment and Disposal", US EPA Technology Transfer, EPA 625/1-74-006, October, 1974, Ch. 7.
4a. Water Pollution Control Federation, "MOP 8. Wastewater Treatment Plant Design," WPCA (1977).
5. W. W. Eckenfelder, Jr. and D. J. O'Connor, *Biological Waste Treatment*, Pergamon Press, New York, 1961.
6. W. W. Eckenfelder, Jr. and D. L. Ford, *Water Pollution Control*, Pemberton Press, Jenkins Publishing Co., New York, 1970, Ch. 20.
7. J. F. Vogler and W. Rundolfs, "Factors Involved in the Drainage of White-Water Sludge," Proc. of the 5th Purdue Industrial Waste Conf., 1949, p. 305.
8. J. D. Swanwick and R. C. Baskerville, "Sludge Dewatering on Drying Beds," paper presented at London Int. Engr. Exhib., April, 1965.
9. J. D. Swanwick, *J. Water Poll. Control Fed.* **34** (3), 239 (1962).
10. L. V. Carpenter, *Sewage Works J.* **10**(3), 503 (1938).
11. J. R. Fleming, *Public Works* **90**(8), 120, 202 (1959).
12. J. E. Quon and G. B. Ward, *Int. J. Air Water Poll.* **9,** 311 (1955).
13. "Sludge Drying" (editors), *Sewage and Ind. Wastes* **31** (2), 239 (1959).
14. T. R. Haseltine, "Measurement of Sludge Drying Bed Performance," *Sewage Ind. Wastes* **23,** (9), 1065 (1951).

15. L. W. VanKleeck, "Sewage Works Guide," reprint from *Wastes Engineering*, New York City (1961).
16. J. W. McLarren, *Can. Mun. Util.* **23**, 51(1961).
17. M. Bowers, *Sewage Ind. Wastes* **29** (7), 835 (1957).
18. W. J. Weber, Jr., *Physicochemical Processes for Water Quality Control*, Wiley-Interscience, New York, 1972, Ch. 2, Ch. 12.
19. R. L. Carr, Jr., *Water Sew. Works* **114**, R-64 (1967).
20. W. A. Sperry, *Sewage Works J.* **13** (5), 855 (1941).
21. R. S. Gale, *Filtr. Separ.* **5,** 2, 133 (1968).
22. "Operation of Wastewater Treatment Plants, Manual of Practice," No. 11, WPCF, Washington, DC, 1961.
23. J. K. Adams, *Sewage Works J* **15** (4), 704 (1943).
24. J. T. Burke and M. T. Dajani, "Organic Polymers in Treatment of Industrial Wastes", Proc. 21st Ind. Waste Conf., Purdue University, Ext. Ser. 121, 303 (1966).
25. C. D. Gates and R. McDermott, *J. Amer. Water Works Assoc.* **60**, 3, 331 (1968).
26. W. K. Neubauer, *J. Amer. Water Works Assoc.* **60**, 7, 819 (1968).
27. A. J. Kilbride, *Wastes Engr.* 568 (Oct. 1961).
28. "Advances in Sludge Disposal in the Period from Oct. 1, 1954 to Feb. 1, 1960", A progress report of ASCE-San. Engr. Div., **88** (SA2), 13 (1962).
29. W. T. South, *Water Sew. Works* **105,** 347 (1958).
30. C. W. Randall, *Water Sewage Works* **116,** 373 (1969).
31. C. W. Randall and C. T. Koch, *J. Water Pollut. Contr. Fed.* **4,** R215 (1969).
32. E. R. Lynd, *Sewage Ind. Wastes* **28** (5), 697 (1956).
33. R. B. Gauntlett and R. F. Pakcham, *Pub. Works* **102,** 90 (1971).
34. J. B. Crockford and V. R. Sparham, "Developments to Upgrade Settlement Tank Performance, Screening and Sludge Dewatering Associated with Industrial Wastewater Treatment", 27th Purdue Industrial Waste Conference, May, 1972.
35. V. H. Lewin, *The Surveyor* **121** (3680), 1521 (1962).
36. L. A. Lubow, *J. N. Carolina Sec. AWWA* **16**, 118 (1941).
37. J. Harrison and H. R. Bungay, *Water Sew. Works* **115,** 217 (1968).
38. T. Furman, *Sewage Ind. Wastes* **26** (6), 745 (1954).
39. "Treatment and Disposal of Sewage Sludge", Ministry of Housing and Local Government, London, 1954.
40. E. A. Jeffrey, "Laboratory Study of Dewatering Rates for Digested Sludge in Lagoons," Proc. 14th Ind. Water Conf. Purdue University, Ext. Ser. **104**, 359 (1959).
41. D. H. Howells and D. P. Dubois, *Sewage Ind. Wastes* **31** (7), 811 (1959).
42. M. S. Anderson, "Sewage Sludge for Soil Improvement", Circular No. 972, USDA, 1955.
43. H. A. Lunt, *Water Sewage Works* **100** (8), 295 (1953).
44. L. W. VanKleeck, *Water Sewage Works*, Ref. edition, p. R-203 (1953).
45. R. D. Leary, "Production of Vitamin B_{12} from Milorganite," Proc. of the 9th Purdue Industrial Waste Conf., 173, 1954.
46. W. Rudolfs and E. J. Cleary, *Sewage Works J.* **5**(3), 409 (1933).
47. E. Hurwitz, "The Use of Activated Sludge as an Adjuvant to Animal Feeds", Proc. of the 12th Purdue Industrial Waste Conference, 1957, p. 395.
48. M. H. Doughery and R. R. McNary, *Sewage Ind. Wastes* **30**(9), 1151 (1958).

49. S. A. Hart and P. H. McGauhey, *Food Tech.* 30 (1964).
50. W. Bottenfield and N. C. Burbank, *Ind. Water Wastes* **9** (1), 18 (1964).
51. J. T. Garrett, *Sewage Ind. Wastes* **31** (7), 841 (1959).
52. C. J. Dick, *Ind. Water Wastes* **6**(1), 1 (1961).
52a. "Process Design Manual for Sludge Treatment and Disposal", US EPA 625/1-79-011, Sept. (1979), Ch. 9–11.
53. W. L. Badger and J. T. Banchero, *Introduction to Chemical Engineering*, McGraw-Hill, New York, 1955, Ch. 10.
54. W. L. McCabe and J. C. Smith, *Unit Operations of Chemical Engineering*, McGraw-Hill, New York, 1976, Ch. 25.
55. J. H. Perry, ed., *Chemical Engineer's Handbook*, McGraw-Hill, New York, 1963.
56. G. Nonhebel and A. A. H. Moss, *Drying of Solids in the Chemical Industry*, CRC Press, Cleveland, Ohio, 1971.
56a. W. Davis and R. T. Hang, "Los Angeles faces Several Sludge Management Problems", *Water Wastes Eng.* April (1978).
57. G. L. Fugate "Mechanical Sludge Filters and Dryers", in "Manual for Sewage Plant Operation," L. C. Billings, ed., Texas State Department of Health, Austin, Texas, 1946, p. 234.
58. C. H. Lipsett, *Industrial Wastes—Their Conservation and Utilization*, Atlas Publishing Co., New York, 1951.
59. C. L. Sercu, "New Incineration Facilities at Dow, Midland," Proc. of the 14th Purdue Ind. Waste Conf., 612 (1959).
60. A. C. Bryan and M. T. Garrett Jr., "What Do You Do With Sludge? Houston An Answer", *Pub. Works*, *103*, 10 Dec. 1972.
61. C. E. Irving, *Water Works Wastes Engr.* **2** (9), 70 (1965).
61a. D. B. Sussman and H. W. Gershman, "Thermal Methods for the Codisposal of Sludges and Municipal Residues," Proc. 5th Conf. on Acceptable Sludge Disposal Techniques (1978).
61b. D. B. Sussman, "More Disposal Operations Mixing Sewage Sludge and Municipal Solid Wastes", *Solid Wastes Management*, August (1977).
62. J. F. Laboon, *Civil Engr.* **24** (1), 44, 100 (1954).
63. W. R. Marshall Jr., *Chem. Eng. Prog. Monogr. Ser.* **50** (2), (1954).
64. "Atomized Suspension Adopted for Sludge", eds., *Chem. Eng. News* (Sept. 14, 1964), 79.
65. K. L. Pinder and W. H. Gauvin, "Applications of the Atomized Suspension Technique to the Treatment of Waste Effluents," in Proc. of the 12th Purdue Ind. Waste Conf., 1957, 217.
66. T. Helfgott and P. Webber, *Water Works Wastes Engr.* **2** (9), 76 (1965).
67. P. J. Cardinal, *Waste Water Treat. J.* **12,** 62 (1968).
68. E. B. Besselievre, *The Treatment of Industrial Wastes*, McGraw-Hill, NY, 1969.
68a. J. H. Yamamota, J. F. Schnelle, Jr., and J. M. O'Donnell, "High Nitrogen Synthetic Fertilizer Produced from Organic Wastes", Public Works, 106 Jan. (1975).
69. A. P. Colburn, *Trans. AIChE* **29,** 174 (1933).
70. R. E. Peck and D. T. Wasan, *Adv. Chem. Engrg.* **9,** 247 (1974).
71. L. Wenzel and R. R. White, *Ind. Eng. Chem.* **43,** 1829 (1951).
72. C. B. Shepherd, C. Hadlock, and R. C. Brewer, *Ind. Eng. Chem.* **30,** 388 (1938).
73. D. W. McCready and W. L. McCabe, *Trans. AIChE* **29,** 131 (1933).
74. P. Coackley and R. Allos, *J. Proc. Inst. Sew. Purif.* **6,** 557 (1962).

75. J. H. Nebiker, *J. Water Poll. Control Fed.* **39** (4), 608 (1967).
76. O. A. Hougen, H. J. McCauley, and W. R. Marshall Jr., *Trans. AIChE* **36**, 183 (1940).
77. W. K. Lewis, *Ind. Eng. Chem.* **13**, 427 (1921).
78. T. K. Sherwood, *Ind. Eng. Chem.* **21**, 12, 976 (1929).
79. E. W. Comings and T. K. Sherwood, *Ind. Eng. Chem.* **26**, 1096 (1934).
80. O. Krischer, *The Scientific Principles of Drying Technology*, Springer-Verlag, W. Berlin, 1963.
81. P. S. H. Henry, *Proc. Roy. Soc. Ser. A* **171**, 215 (1939).
82. G. King and A. B. D. Cassie, *Trans. Faraday Soc.* **36**, 445 (1940).
83. C. G. Gurr, T. J. Marshall, and J. T. Hutton, *Soil Sci.* **74**, 335 (1952).
84. J. M. Kuzmak and P. J. Sereda, *Soil Sci.* **84**, 419 (1957).
85. T. I. Harmathy, *Ind. Eng. Chem. Fund.* **8**, 92 (1969).
86. N. H. Ceaglske and O. A. Hougen, *Trans. AIChE* **33**, 283 (1937).
87. T. K. Sherwood and E. W. Comings, *Ind. Eng. Chem.* **25**, 311 (1933).
88. R. W. Corben and D. M. Newitt, *Trans. Inst. Chem. Engrs. (London)* **33**, 52 (1955).
89. S. J. Friedman and W. R. Marshall, Jr., *Chem. Eng. Progr.* **45** (8), 482 (1949).
90. C. O. Miller, B. A. Smith, and W. H. Schuette, *Trans. AIChE* **38**, 841 (1942).
91. C. F. Prutton, C. O. Miller, and W. H. Schuette, *Trans. AIChE* **38**, 123 (1942).
92. W. C. Saeman and T. R. Mitchell, Jr., *Chem. Eng. Progr.* **50** (9), 467 (1954).
93. S. J. Friedman and W. R. Marshall, Jr., *Chem. Eng. Progr.* **45** (9), 573 (1949).
94. R. B. Bird, W. E. Stewart, and E. N. Lightfoot, *Transport Phenomena*, Wiley, New York, 1960.
95. P. C. Carman, *Trans. Inst. Chem. Engrs. (London)* **16**, 168 (1938).
96. H. P. Grace, *Chem. Eng. Progr.* **49** (6), 303 (1953).
97. H. P. Grace, *Chem. Engr. Progr.* **49** (7), 367 (1953).
98. P. Coackley and B. R. S. Jones, *Sew Ind. Wastes* **28**, 8, 963 (1956).
99. J. H. Nebiker, T. G. Sanders, and D. D. Adrian, *J. Water Poll. Control Fed.* (Res. Suppl., part 2), p. R255 (Aug. 1969).
100. J. T. Novak, "Character and Dewatering Properties of Sludges from Water Treatment," in *Water—1974. II. Municipal Wastewater Treatment*, AIChE Symp. Ser. **71** (No. 145), 235 (1975).
101. J. H. Nebiker, "Application," in "Source Control of Water Treatment Waste Solids," D. D. Adrian, P. A. Lutin, and J. H. Nebiker, eds., Civil Engineering Report EVE 7-68-1, Univ. of Mass., Amherst, 1968, Ch. IV.
102. D. D. Adrian, "Dewatering Sewage Sludge on Sand Beds", in *Water—1972*, AIChE Symp. Ser. **69** (No. 129), 188 (1973).
103. H. Blunk, "Beitrag zur Berechnung von Faulräumen," *Gesundheitzingenieur*, **48**, 4, 37 (1926).
104. W. Vater, "Die Entwasserung, Trocknung und Beseitigung vom städtischen Klarschlam," Doctoral Dissertation, Technische Hochschule, Hannover, 1910.
105. "Water Pollution Research 1965," Ministry of Technology, Her Majesty's Stationary Office, London, 1966, p. 91.
106. J. H. Nebiker and D. D. Adrian, *Filtr. Sep.* 245, (May/June 1969).
107. K. M. Lo, "Digital Computer Simulation of Water and Wastewater Sludge Dewatering on Sand Beds," Civil Engineering Report EVE.26-71-1, U. of Mass. Amherst (1971).
108. E. Lindsey, *Chem. Eng.* **60** (4), 227 (1953).

9
Dredging Operations and Waste Disposal

Lawrence K. Wang
Lenox Institute for Research Inc., Lenox, Massachusetts

I. INTRODUCTION

Dredging is an operation for the removal of silt, sand, clay, and miscellaneous materials from underwater surfaces by excavation, and the subsequent conveyance to and disposal of the material at an appropriate disposal site. It is thus doubly important environmentally—in the area of removal and in that of accumulation—besides having other, often significant side-effects. In general, there are two types of dredging operations: new-work dredging and maintenance dredging. The former comprises the improvement (i.e., deepening or widening) of a channel and/or harbor area by removing mainly stones and compacted sediments that were deposited through the geologic ages. The latter is employed mainly to remove the loose sediments that tend to fill up previously excavated channels.

The US Army Corps of Engineers (ACE) annually dredges some 380 million cubic yards (290.6 million cubic meters) of materials from the bottom of rivers and harbors. Of 380 million cubic yards of dredged materials, 80 million cubic yards (61.2 million cubic meters) are from new work dredging, some 300 million cubic yards (229.4 million cubic meters) are from maintenance dredging. Much concern has arisen over the potential environmental impact of the dredge disposal. Several recent articles [1–16] introduce the dredging operation and its pollution effect on water resources.

II. TYPES OF DREDGES

The clamshell dredge, the dipper dredge, the pipeline dredge, and the hopper dredge are four common types of dredges used for excavation by the US Army Corps of Engineers. The first two types are mechanically operated, whereas the latter two are hydraulically operated. These four types of dredges are shown in Figs. 1–4, inclusive. They range from 13.5 to 40 m in length. The main features and the operation method of each of these types of dredge are briefly described in the following sections.

A. Clamshell Dredge

A *clamshell dredge* (Fig. 1) excavates with a "clamshell bucket" suspended by cables from a forward-extending boom that can swing about the bow of the dredge. "Spuds", extending to the bottom, are used for keeping the dredge stable in the proper working position. The boom and bucket are swung around to empty the bucket into a scow or small barge positioned alongside the dredge. The clamshell dredge is designed for maintenance dredging in river channels where soft or cohesive underwater materials are to be removed periodically. This type of dredge is exceptionally useful for deep digging and for dredging in close quarters

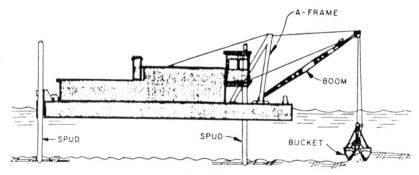

Fig. 1. *Clamshell dredge.*

alongside structures. The total cost of clamshell dredging is estimated to range from US$ 2.95–5.92/m³ (January 1983 cost data).

B. Dipper Dredge

A *dipper dredge* (Fig. 2) has a power-operated dipper stick which, by sliding through the center plane of a boom, allows the operator to control the movement of the dipper bucket in any direction. Spuds are used for maintaining a stable working position. The boom and dipper bucket are swung around to empty the bucket into a scow that is waiting alongside the dredge. The dipper dredge is a heavy-duty excavator designed for breaking up hard compacted clay and ledge-rock with its bucket. New-work dredging is usually done with this type of dredge. The total cost of dipper dredging is equal to or higher than US$ 5.92/m³ (January 1983 cost data).

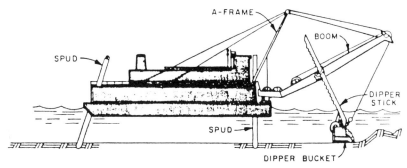

Fig. 2. *Dipper dredge.*

C. Pipeline Dredge

A *pipeline dredge* (Fig. 3) excavates with a revolving cutter surrounding the intake end of a suction pipe. The dredged materials are sucked up and transported by a pumping unit through a trailing pipeline to an appropriate disposal site. The dredge is generally equipped with two stern spuds and forward anchors to swing the hull around one of the stern spuds. Pipeline dredges are designed for excavating clayey, sandy, or silty bottoms. The total cost of pipeline dredging is about US$ 2.95/m³ (January 1983 cost data).

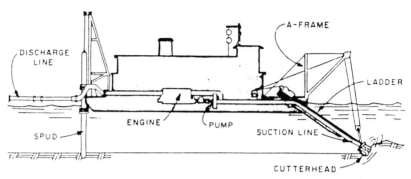

Fig. 3. *Pipeline dredge.*

D. Hopper Dredge

Hopper dredges (Fig. 4) are essentially self-propelled ships equipped with drag-heads for excavating the sediments, pumps for sucking up and transporting the dredgings, and hoppers (movable containers) for settling and storing the dredgings. When operating, the underwater sediments are excavated and eventually settled in the hoppers. Excess water with suspended 'fines' overflows at the tops of the hoppers. Dredging continues until a sufficient load of sediments has been accumulated in the hoppers. The dredge itself will then transport its load to a disposal site where the hoppers are emptied by opening the discharge doors or by pumping the dredgings out through a discharge pipe. Hopper dredges are efficient in excavating a thin layer of loose sediment covering extensive areas. About 75% of the nearly 8 million m^3 (10,000,000 yd^3) of annual maintenance dredging in the Great Lakes is done by hopper dredges. The total cost of hopper dredging (open-water disposal) is estimated to be at least US$ 8.96/m^3 (January 1983 cost data).

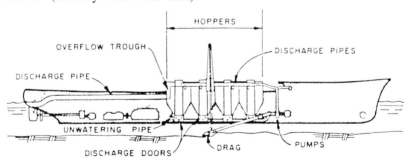

Fig. 4. *Hopper dredge.*

III. SOURCES OF POLLUTION FROM DREDGING OPERATIONS

The US Army Corps of Engineers alone dredges each year an average of approximately 290,548,000m^3 (380,000,000 yd^3) of sediments, which amounts to more than 90% of the total volume of dredging performed in the USA. A comparatively small volume of dredging is conducted by local industries to make private channels and alongside unloading piers.

The nature and characteristics of dredged materials will depend on the type of dredging, the body of water, and the location of dredging. The relative abundance of the major portions of the dredged materials from new-work dredging reflects the sedimentology of the area dredged and ranges from stone to sand, silt, or clay. The sediments excavated from industrial areas are often polluted with obnoxious or toxic contaminants that are detrimental to the land, air, and/or water environment.

The sources of pollution associated with channel sediments may be categorized as municipal and industrial discharges, storm-runoff, agricultural runoff, soil erosion, or accidental spills. Domestic sewage and industrial waste water are point-pollution sources that can be collected and treated. The extent of pollution of the water and sediments in the receiving body will depend on the degree of treatment received by such waste effluents. If they are untreated, both of these discharge sources will contribute to heavy organic loadings and toxic contaminants in the dredged deposits [13].

Storm-runoff, agricultural washes, and leachates can cause nonpoint-source pollution. The stormwaters from roofs, grounds, and streets in urban areas are collected in sewers during and immediately following storms. The pollutant load in storm waters in the areas having harbor, airport, and/or heavy industries will be high. Agricultural pollution derives from agricultural areas producing animal wastes and using fertilizers and pesticides. These pollutants are washed or leached into rivers or channels. Soil erosion will similarly contribute sediments and associated chemical pollutants. The concentration of pesticides, heavy metals, and nitrogenous and phosphatic nutrients, is generally much higher in eroded-soil sediments than when merely dissolved in runoff water. Nonpoint-source pollutants are not easily amenable to treatment before discharge into the receiving waters. Control measures to redirect or limit such waste water drainage may, however, be effective.

Spillage of oil and hazardous chemicals in rivers, lakes, and harbors from navigation accidents and other hazards or carelessness is a serious and almost random source of pollution. When the frequency and volume of spills are high, the sediments in such a location can be heavily polluted.

The pollutants that are found in dredged materials are for the most part of the same types as those found in most domestic and industrial wastes. The major differences noted as characterizing dredged materials are the high sediment content (i.e., high content of suspended solids) and the effect of pollutants on the physical behavior associated with the sediments.

Under the auspices of the US Army Corps of Engineers, Calspan Corporation, of Buffalo, NY, has conducted characterization studies of dredging spoils from Ashtabula Harbor and Fairport Harbor in Ohio. The characteristics of dredgings collected at these two harbors are shown in Table 1. For each harbor, four samples were collected from different locations. The results given in Table 1 indicate that the sediments varied widely in their chemical characteristics between harbors and even at different locations within each harbor; they also varied widely in their physical makeup. But in spite of the wide variations in the characteristics of the dredging spoils, some conclusions can be drawn from the test results. The dredged materials from Ashtabula Harbor are apt to contain higher concentrations of phosphorus, volatile solids, iron, and zinc than do the dredgings from Fairport Harbor. However, the sediments in Fairport Harbor are apt to have comparatively higher chemical oxygen demand (COD) and contents of oil and grease.

Another important characterization parameter of the dredged material is its settleability. Most of the polluting substances in the dredged materials are settleable because they were originally sediments at the bottoms of rivers or harbors. When these dredged materials are discharged on a disposal site, a muddying of the waters will occur, and oil and grease buried in the muck may be released to form a film on the surface. Initially, the sediments are suspended in the surface water of a disposal site. The resettleability of the dredged material is governed, in part, by the original composition and nature of the sediments, the nature of the dilution water (if an aquatic disposal operation is practiced), and the dredging method employed.

Settleability of the dredging samples from Ashtabula Harbor and Fairport Harbor was investigated by Calspan Corporation. The effectiveness of adequate detention-time of dredgings in a diked disposal area before discharge of the supernatant liquid back to a receiving stream or the lake is illustrated in Tables 2 and 3. From these data it is seen that removal of metals, oil and grease, and phosphorus is highly significant at

TABLE 1
Pollutant Contents as Percentage of Total Solids in Dredged Materials [11]

Parameters	Ashtabula samples				Fairport samples			
	1A	2A	3A	4A	1F	2F	3F	4F
Chemical oxygen demand	4.334	2.483	1.987	4.037	0.237	16.567	4.567	4.010
Oil and grease	0.210	0.094	0.065	0.265	TRACE	0.275	0.468	0.520
Kjeldahl nitrogen	0.138	0.043	0.048	0.152	0.016	0.065	0.167	0.020
Total phosphorus	0.004	0.001	0.008	0.030	0.007	0.009	0.006	0.003
Mercury	0.00004	0.00001	0.00002	0.00007	0.00001	0.00003	0.00004	0.00006
Iron	2.104	1.750	1.299	2.100	1.105	1.000	1.599	1.790
Chromium	0.008	0.010	0.007	0.014	0.002	0.002	<0.018	0.020
Cadmium	<0.0001	<0.0001	0.0001	<0.0001	0.001	0.0001	<0.0001	<0.0008
Arsenic	<0.0001	<0.00008	<0.0002	<0.0005	<0.00009	<0.0003	<0.0002	<0.0001
Zinc	0.018	0.015	0.013	0.017	0.006	0.006	<0.005	0.008
Lead	0.010	0.010	0.010	0.002	0.002	0.035	0.005	0.009
Copper	0.003	0.001	0.002	0.031	0.0009	<0.0009	0.002	0.003
Nickel	0.001	0.001	0.001	0.001	0.0009	0.009	0.001	0.001
Manganese	0.055	0.045	0.040	0.060	0.035	0.030	0.040	0.040
Volatile solids	4.905	4.150	18.100	—	2.400	—	—	8.499
Dissolved solids	0.093	0.082	0.116	0.693	0.053	0.139	0.188	0.589

TABLE 2
Settling-Test Data, Sample 2A from Ashtabula Harbor, Ohio [11]

Parameters	Before settling	Remaining after 1-h settling	Remaining after 4-h settling	Remaining after 18-h settling	Remaining after 40-h settling
Chemical oxygen demand, mg/L	14,900	110	108.8	81.5	77.6
Total solids, mg/L	600,000	800	—	780	872
Dissolved solids, mg/L	490	490	490	490	490
Suspended solids, mg/L	599,510	480	—	290	—
Turbidity, JTU[a]	87,500	500	—	150	39
Total phosphorus, mg/L	8.0	0.03	0.09	0.05	0.10
Surfactants, mg/L	—	0.06	—	—	—
Sediment, mg/L	—	600	—	—	—
Kjeldahl nitrogen, mg/L	259	—	29.0	25.1	25.6
Oil and grease, mg/L	565	<1.00	—	Nil < 1	Nil < 1
Iron, mg/L	10,500	41.00	—	—	—
Chromium, mg/L	60	0.50	0.045	0.020	0.020
Cadmium, mg/L	<0.6	0.04	0.055	0.020	0.022
Zinc, mg/L	90	1.1	0.170	0.095	0.075
Manganese, mg/L	270	3.10	—	0.021	0.0187
Copper, mg/L	8.0	0.12	0.055	0.040	0.020
Nickel, mg/L	<6.0	—	—	<0.015	—
Chlorides, mg/L	—	25.0	—	—	—

[a]JTU = Jackson Turbidity Unit.

TABLE 3
Settling-Test Data,[a] Sample 4F from Fairport Harbor, Ohio [11]

Parameters	Before settling	Remaining after 1-h settling	Remaining after 4-h settling	Remaining after 18-h settling	Remaining after 40-h settling
Chemical oxygen demand, mg/L	14,540	202	133	97	101
Total solids, mg/L	363,000	4240	—	2712	2460
Dissolved solids, mg/L	2140	2140	2140	2140	2140
Suspended solids, mg/L	360,860	2130	—	672	320
Turbidity, JTU[b]	155,000	780	—	195	52
Total phosphorus, mg/L	10	0.280	—	0.58	0.16
Surfactants, mg/L	—	0.06	—	—	—
Sediment, ml/L	—	700	—	—	—
Kjeldahl nitrogen, mg/L	696	56.5	—	38.2	—
Oil and grease, mg/L	1865	4	—	<10	<10
Iron, mg/L	6500	22.5	—	—	—
Chromium, mg/L	73	0.20	—	0.040	0.035
Cadmium, mg/L	<0.4	0.03	0.052	0.051	0.048
Zinc, mg/L	29	0.15	0.099	0.025	0.025
Manganese, mg/L	145	2.20	—	0.021	0.019
Copper, mg/L	11	0.07	—	0.060	0.070
Nickel, mg/L	<3.6	—	—	<0.015	—
Chlorides, mg/L	—	284	—	—	—

[a]Analyzed by the Calspan Corp., Buffalo, NY.
[b]Jackson Turbidity Unit.

only 1 h of settling time and is greater than 95% for these parameters after 18 h settling time. Further detention time does not result in significantly greater removal. It is apparent from the data that oil and grease and metals are intimately associated with sediments and are removed upon settling of solids.

The exceptionally effective reduction in the content of toxic metals and oils and greases with a few hours of settling is best appreciated when Federal Drinking Water Standards [17] are compared with the quality of the supernatant fluid after 18 h settling:

Parameters	Water quality after 18 h settling, mg/L		Proposed Federal drinking water standards, mg/L
	Sample 2A Ashtabula	Sample 4F Fairport	
Oil and grease	Nil<1	<10	Absent
Chromium	0.020	0.040	0.05
Cadmium	0.020	0.051	0.01
Zinc	0.095	0.025	5.0
Copper	0.040	0.060	1.0

Although turbidity of 1 JTU (Jackson Turbidity Unit) has been recommended for drinking water and the supernatant fluid after 40 h does not approach this in clarity, the health and environmental significance of greater JTU reductions are not clearly established. Whereas reduction of chemical oxygen demand (COD) is great after 18 h settling, further reduction to, say, 30 mg/L might be desirable. The residual COD probably results from the content of dissolved or highly colloidal organic matter. A simple treatment, such as ozonation or aeration, might be adequate for further reduction of COD and also BOD (Biochemical Oxygen Demand), although this was not ascertained in these studies.

Mechanical and hydraulic dredges excavate and transport the dredged material at different degrees of turbulence. Some would generate fine suspended solids that do not settle rapidly. Some would disturb the oil agglomerates and release part of the oil to the water's surface. Changing the chemical equilibrium through the contact of dredged materials with foreign water at the disposal site may also cause parts of the originally suspended solids to become soluble. These unsettled or difficult-to-settle substances at the disposal site will contribute as major pollutants in the overflow water that should be treated to prevent contamination of groundwater or surfacewater resources.

Additional data on the classification and engineering properties of dredged materials can be found from a Federal government report [38].

IV. DREDGE DISPOSAL AND ENVIRONMENTAL ENHANCEMENT ALTERNATIVES

A. Open Water Disposal

The approved dumping sites for the US side of the Great Lakes total approximately 102 "open water" sites. A large percentage of this total is in regions that are close to the dredging site, in relatively shallow water, and less than 3 mi (4828 m) from shore. It is interesting to note that Canada requires dredging contractors to dispose of dredging spoils in "waters no less than 15 m in depth nor within 3 mi (4828 m) of the dredging site." The current practice of disposal sites close to shore in Lake Erie thus can detract from water quality nearshore in addition to the disposal site because of the prevailing current conditions.

An example of open water disposal of dredged material is in Lake Erie near Ashtabula, Ohio. During 1975 and 1976, the ACE Waterways Experiment Station performed experimental open lake dumping to determine long-term and short-term effects. In addition to the coliform counts available in the US Army Corps of Engineers (storage retrieval) system, information available in the ACE Dredging Report [1,2] indicated that the normal background of coliform bacteria in Lake Erie varies, but usually averages 100 organisms/100 mL approximately 1.5 miles (2413.5 m) from the Ashtabula River [18]. Total coliform counts near the city water intakes averaged approximately 250 and 490 organisms/100 mL for 1970 and 1971 for the data examined [19]. East of the outfall of the city's sewage treatment plant and dredge disposal dumping site, a count of 920 organisms/100 mL was obtained in 1968. This appears to be a result of the combined effect of river discharge and disposal of dredged materials from the highly polluted shipping channel. Scarce et al. [20] showed that coliform organisms can persist in the vicinity of dumping areas for some time. A detailed discussion of the water quality of the area is discussed in Terlecky et al. [21]. Table 4 gives an analysis of the metals content of sediment contributed to the dredged materials in this area.

The disposal of dredged materials in open water almost always leads

TABLE 4
Characteristics of Fields Brook Sediment[a] [12]

Metal	Content, µg/g dry weight
Arsenic	<1.0
Barium	860
Cadmium	17
Chromium	158
Copper	50
Iron	33 000
Mercury	2.6
Nickel	59
Lead	61
Selenium	<2
Zinc	210

[a]Sediments from the mouth of Fields Brook near Ashtabula, Ohio, contributed to harbor bottom sediments; sampled on 6 September, 1973.

to a temporary loss of natural habitat for benthic invertebrates and fish, an increase in turbidity, and introduction of materials that may have a high biochemical and chemical oxygen demand (BOD and COD), possible release of toxic heavy metals as indicated above, and reduction of productivity of the biological community.

A study of US Army Corps of Engineers disposal and dredging methods was begun by the Buffalo District, ACE, in 1966. The effort was conducted to determine whether removal of even relatively small quantities of dredged material and transfer to the open lake might be harmful to the aquatic environment. The fate of this material and its effect upon water quality, as well as what alternatives were available and their costs, were considered. A part of the investigation included construction and use of alternative disposal sites. Some of the findings are summarized here [22].

1. The removal of dredged material is not harmful over the long term to water quality in the harbors where the dredging takes place.
2. The effects of dredge dumping on open lake disposal areas are not fully known. Biological tests showed that some

benthic forms that inhabit sediments were killed or forced to move. The study concluded that "in-lake disposal of heavily polluted dredgings must be considered presumptively undesirable."
3. Treatment of dredged material, even if effective, is very costly compared to open-water disposal.
4. Disposal behind enclosed dikes is also expensive, but in general is less costly than any other means of handling the dredgings except open-water disposal.
5. Diking creates other problems. Disposition of dredged material on land or in marshy areas might harm wildlife and the environment.
6. The specific "benefits" of halting open-lake disposal are not quantitatively measurable. However, the general benefits include improvements in the ecological environment of areas where the dredgings are deposited; removal of some amount of undesirable sediment material that could find its way into the ecosystem; and the esthetic advantage of reducing the local, temporary turbidity, odor, and oil slicks that appear in open-water disposal.
7. Dredging equipment and procedures can be improved to mitigate some of the adverse effects of open-lake disposal.

These items are of specific interest for discussion here because it is for these same reasons that diked disposal alternatives are being implemented in many areas.

B. Creating New Land in Diked Disposal Areas

Disposal of dredged materials in alongshore diked areas was initiated at Buffalo, New York, and Cleveland, Ohio, in 1967 to receive portions of the maintenance dredging spoil [6, 7]. An offshore diked disposal area in Green Bay, Wisconsin was first constructed in 1967 for disposal of new work dredgings from a project to deepen the outer channel, and thereafter continued to be used for disposal of maintenance dredgings from sections of the outer channel. Both alongshore and offshore diked disposal areas can be used to create new land if satisfactory fill can be provided. A partial list showing many locations of diked disposal sites and some other disposal alternatives in use or proposed for several locations in the United States is given in Table 5.

Prior to 1967, materials from maintenance dredging on the Great

TABLE 5
Partial List of Current or Proposed Diked Disposal Areas [12]

Location	Disposal method
St. Mary's River, Michigan	Diked disposal islands
Toledo Harbor, Ohio	Contained disposal facility
Buffalo, New York	Diked disposal
Sunny Pt., Southport, North Carolina	Diked disposal
Ontonagan Harbor, Michigan	Diked disposal
Ashland Harbor, Wisconsin	Diked disposal
Duluth-Superior Harbor, Minnesota	Diked disposal
Goose Creek, Somerset County, Maryland	On-land disposal, retaining dikes
Ft. Myers Beach, Florida	Beach enrichment
Gila Gravity Main Canal, Arizona	Fill
Longboat Pass, Manatee, City, Florida	On-beach—beach enrichment
Grand Haven Harbor, Michigan	Marsh fill
Willapa River and Habor, Washington	Lowland and marshland fill—dikes
Grand Lagoon, Florida	Beach enrichment
Skagit County, Washington	Landfill
White River, Arkansas	Bank disposal (alongshore)
Detroit and Rouge River, Michigan	Contained disposal
Panama City, Florida	On land
Gulf Intracoastal Waterway, Texas	Leveed disposal
Carrabelle to St. Marks River, Florida	Diked land disposal
Lewis and Clark Connecting Channel, Oregon	Flood dike reinforcement
Hemstead Harbor, New York	Diked disposal
Kainakakai Harbor, Molakai, Hawaii	On land *or* open sea
Elizabeth River Basin, New Jersey	On marshland
Ashtabula Harbor, Ashtabula, Ohio	Diked disposal
Cleveland Harbor, Cleveland, Ohio	Diked disposal
Conneaut Harbor, Conneaut, Ohio	Diked disposal
Erie Harbor, Erie, Pennsylvania	Diked disposal
Fairport Harbor, Fairport, Ohio	Diked disposal
Huron Harbor, Huron, Ohio	Diked disposal
Lorain Harbor, Lorain, Ohio	Diked disposal
Oswego Harbor, Oswego, New York	Diked disposal
Rochester Harbor, Rochester, New York	Diked disposal
Vermillion Harbor, Vermillion, Ohio	Diked disposal

Lakes were rarely deposited inside offshore or alongshore diked areas to create new land [2]. The fine-grained sediment dredged from harbors polluted with industrial and domestic wastes is not usually suitable fill because it is hygroscopic and lacks shear strength.

Maintenance dredging from the Rouge River near Detroit and Maumee River at Toledo have been deposited in diked disposal areas since 1960 and 1961, respectively. Each of the three areas at Toledo were equipped with an overflow for run-off of excess water. Rouge River spoil is deposited in the Grassy Island disposal area in the Detroit River, 4.5 mi (7240·5 m) downstream from the dredging project. This island does not discharge effluent, but holds it in a large ponding area. The Grassy Island area provides a wildlife habitat for waterfowl in the area.

An offshore diked disposal area in Green Bay, Wisconsin, which was initiated with new work dredgings in 1967, is expected to be used for disposal of maintenance dredgings from the outer channel.

C. Wildlife Enhancement in Diked Disposal Areas

The possible value of tidal marshes and wetlands both as an energy source and as a nursery ground for much of our commercial and sports fishery species has been suggested many times [23–25]. Woodhouse et al. [26] under the aegis of the US Army Corps of Engineers, reported a study of the possibility of developing new marsh to replace some of that lost to human activities. They proposed that dredge spoil, produced by maintenance dredging in our sounds and estuaries, could be a logical source of materials, and dredging operations would enable the continuation of necessary navigation. Two objectives were thus necessary—to get the materials into the area to establish a new tidal marsh, and to stabilize these materials so that they would not re-enter the dredged channels. The advantages of systems such as these are manifold. They provide breeding and nesting grounds for many species of waterfowl, and nursery grounds and a food location source for fish.

There are positive indications that a beach area near the Coast Guard Station on Lake Erie, Buffalo, New York used for disposal of maintenance dredging is also being used extensively by many species of birds and waterfowl as a feeding ground.

As indicated by Odum [27], the primary productivity of coastal waters compares favorably with that of land and fresh water. Coastal marshes are usually thought to have the most productive ecosystem of any found to date. If marshlands or artifically created marshlands are produced as a byproduct of diked disposal, nutrients released by dredging may possibly be valuable in the flow of energy through the community, as emphasized by Cronin et al. [28]. A natural result of properly planned dredging operations might be stimulation of some aspects of community well-being. In addition, controlled or partly controlled release and direction of sediment loads could be useful to stimulate specific parts of the physical and biological community. However, there may also be dangers from the possible release of harmful substances as the result of the stirring-up of bottom materials [28]. Conceivably, this can be properly directed and controlled, depending upon the nature of the substances, the effects, and local environmental and biological factors. Experiments at the Bear's Bluff Laboratories [29] in South Carolina, for example, have indicated that diked disposal areas might possibly be used for the creation of suitable rearing areas for the commercial production of shrimp. Proper salinity management and knowledge of the nature of possible industrial wastes (PCBs, heavy metals, or other chlorinated hydrocarbons) would be required. *Spartina* spp. (salt marsh grass) can be used for cattle feed; hence, diked disposal areas might be used agriculturally as sources of animal food. It is estimated that the average marsh pond can produce 250–400 lb (113–181 kg) of fish, 100 lb (45 kg) of crab, and 300–400 lb (136–181 kg) of shrimp per acre per year (1 ac = 4047 m^2). Thus, with proper management and site selection, a multi-use scheme can be envisioned for diked disposal areas.

Any study of the enhancement of the wildlife of an area (or at least the minimum adverse effects) must take into consideration available food supplies. Shore birds, for example, feed along the shore, in very shallow waters, or among grasses on insects, fish, or benthic organisms. Enhancement plans must then consider the effects of dredging alternatives on the food chain at all levels so as not to affect the source of food for a higher trophic level. Fish spawn timing, and bird nesting schedules can be obtained and taken into account when scheduling dredging operations so that a minimal disruption by turbidity, noise, or degradation of water quality will occur.

Proper dredging management should include precautions to control adverse sediment effects by taking advantage of tidal and littoral currents in order that suspended sediments might be collected [30]. Where green plants are present in a diked disposal area, for example, water clarity during the daylight hours is important for continuance of primary productivity; perhaps dredging at night might be considered. On the other hand, if faunal nighttime feeding were shown to be affected by increased turbidity and dredging, the dredging might then be confined to daylight.

Consideration of the physical factors important in the estuarine environment [31] leads to the following checklist for physical factors that must be considered in planning of dredging operations.
1. Characteristics of substrates (shear strength, water content, size distribution, other material properties).
2. Tides (magnitude and extent of influence).
3. Wave heights and prevailing direction.
4. Current strength.
5. Salinity distribution (i.e., gradients sharp or gradual).
6. Dissolved oxygen.
7. Temperature.
8. Concentration of certain ions (heavy metals).
9. Possible food sources.
10. Dikes as physical barriers.

Consideration should also be given to water circulation patterns in dredged areas and adjacent dikes. The extent of modification of currents may be a determining factor in whether disturbances will affect the food chain. If a free interchange with open water is prevented, a resupply of oxygen and the delicate salinity balance of the waterway can be disturbed. If euryhaline species (i.e., those having a wide range of tolerance with respect to salinity) are involved, this presents no problem. If, on the other hand, the fish present are stenohaline species (i.e., those having a narrow range of tolerance with respect to salinity), they may survive and reproduce only within narrow ranges of salinity even though they are osmoregulators (i.e., organisms that are capable of maintaining a relatively fixed internal salt concentration by means of various metabolic processes). A thorough knowledge of the species characteristics present in the area is of profound importance, as for example, in the appearance of chloride secreting cells in the gills of juvenile salmon [32]. Lobsters, when moving into the brackfish waters of estuaries, maintain their body fluids hypertonic (i.e., the fluid salt concentration inside their membranes is greater than that of the aquatic environment surrounding them) to the external medium and thus become osmoregulators. Normally lobsters are isosmotic with their environment. *Mytilus edulis* (bay mussel), like many species of oysters, is adapted to widely ranging salinities [33]. If the osmotic concentration of the blood of organisms follows changes in the environment, and these characteristics are known, minimal damage might be done by restricted circulation.

The degree of change in temperatures present in the dredged area or in the disposal sites must be known because temperature changes may have multiple effects upon organisms living in an environment [34]. Knowledge of major species' tolerances to temperature fluctuations should be evaluated.

Experience with diked disposal areas on Lake Erie at Buffalo Harbor, New York has shown that a series of events can be predicted on partially exposed disposal areas [6]. Invasion of these areas by annual plants and early seeding deciduous trees occurs very soon on exposed spoil surfaces. The high nutrient value of this material promotes lush growth; therefore, within a year or two, lush plant communities develop.

As plant communities in disposal areas develop, the increased cover provides a nesting and brooding area for water fowl, as well as a proximate location to the shore zone as a source of food. Insects and soil organisms usually abound and provide an additional food source in the newly created environment.

During the period when the Buffalo diked disposal area is partially filled, but with large expanses of open water, it attracts migratory water fowl. In Buffalo Harbor, mallard ducks have nested and produced young. Other ducks forage in the lagoons and swampy margins, and shore birds such as sandpipers, plovers, and killdeer spend the summer exploiting food along the shore.

Modification of lagoon design, water depth, revisions of dredging schedules, weir effluent treatment, possible supernate treatment, or a combination of the aforementioned might be considered in an investigation of wildlife enhancement alternatives.

D. Additional Disposal Alternatives

There are many dredging operation areas, particularly new-work dredging, that produce materials with bearing and shear strength properties suitable for landfill or as reinforcing materials. Many areas along our coastal waters have materials that are relatively nonpolluted; e.g., Skagit County, Washington [35]. This site will be used for industrial facilities, providing employment opportunities for the Swinomish Indian Reservation. The materials to be dredged from the Lewis and Clark Connecting Channel, Oregon, will provide reinforcement for existing flood control dikes [3].

It has been suggested that, if offshore sand dredging coupled with beach fill were employed in many areas, thorough planning could provide a synthetic ecosystem that would result in the establishment of deepwater refuges in the borrow areas offshore [30]. This approach might have the additional bonus of restoring beaches for recreation and natural community succession inshore. Deepwater refuges, such as borrow pits, furnish protection for many species of fish and shellfish. According to Thompson [30] the effectiveness of these refuges depends upon the geographical, hydrographical, and ecological requirements of the organisms involved, and coordination with commerical and sport fishing intersects.

Several beach enrichment projects are underway or planned; e.g., Grand Lagoon, Florida [4] and Fort Myers Beach, Florida [5]. The value of increased recreational opportunities must be considered an important benefit.

Abandoned strip-mine areas have often been mentioned as alternatives for land disposal operations. Disposal of dredged materials on abandoned mine sites may provide beneficial effects in land restoration as well as pollutant neutralization. Disposal of dredged material on land sites far removed from natural waterways will require increased costs and these may be justified by the environmental gains to be derived. Currently, sewage sludge has been applied to denuded areas with significant results in restoration of the productivity of the land. Research at the Pennsylvania State University has developed much of the state-of-the-art in this area.

E. Environmental Enhancement by Modifications of Dredging Schedules and Equipment

One modification which can be made quite easily is a change in the timing of dredging schedules. Studies of the wildlife in the dredging and disposal area should help to avoid nesting, mating, and spawning schedules. If turbidity is a problem in interfering with photosynthetic activity in dredging areas, perhaps dredging during darkness could replace daylight dredging.

Modifications of dredging equipment is another alternative. In the Great Lakes most hopper dredges have been modified to allow direct pump ashore or to a disposal area rather than simply opening the hoppers. Additionally, equipment modifications could be made to prevent the fine sediments from escaping from the overflow pipes in a hopper dredge. In areas where disposal sites are close at hand, the allowance of two or more hours' settling of solids in the hopper may be possible followed by even overflow of supernatant containing little suspended solids.

V. DREDGE TRANSPORT ALTERNATIVES

Dredged material is conveyed into disposal areas from dredges either hydraulically or mechanically. Hydraulic handling of dredged material is by far the more common method. It is accomplished by pumping through submerged or floating pipeline from a pipeline dredge, through direct pump-out from a hopper dredge moored at the disposal site, or by a combination of the two methods through the use of a rehandling basin that receives dredged material from the hopper dredge or from scows that is then pumped out into the permanent disposal area.

Rehandling basins serve to reduce the piping distances and number of booster pump stations necessary to pump dredged material to centrally located disposal areas. These also take advantage in confined disposal operations of the operational speed of hopper dredges and dump scows. For a portion of the Houston Ship Channel maintenance dredging, for example, dredged material was pumped from the dredge to a temporary rehandling location in a turning basin and then redredged with a larger dredge for disposal in a permanent retention facility away from the channel.

Some 44% of the hopper dredges owned by the US Army Corps of Engineers are equipped for direct pump-out of dredged material by use of mooring facilities. Hopper dredges so equipped can be used effectively in heavily trafficked ports and waterways where confined disposal is often required and where pipeline dredges hinder navigation. The Detroit District uses hopper dredges for most of its maintenance dredging work. Disposal is by direct pump-out from the hopper to the containment facility. Mooring facilities at Craney Island allow hopper dredges to unload in 1 h into the disposal area through a submerged pipeline. Direct pump-out of hopper dredges may eventually replace barge scows in the Buffalo District.

A. Long Distance Piping

Long-distance piping has been used by the US Army Corps of Engineers for dredging projects, especially along the Gulf Coast, at distances up to about 5 mi (8045 m), but is not used extensively. The Philadelphia and Charleston Districts have proposed such operations. The Jacksonville District's Fort Pierce Harbor project involves long-distance hydraulic disposal of dredged material. The Chicago District considered using an existing 40-mi (64,360 m), small-diameter coal slurry pipeline to pump material from the dredging area to disposal in an abandoned strip coal mine, but concluded that the pipe diameter was too small for the distance involved. The Craney Island replacement study in the Norfolk District may recommend a disposal site in the Dismal Swamp area. This site would require the redredging of material deposited initially in a part of the existing Craney Island facility for piping 10 mi (16,090 m) upland to a proposed permanent 5000-ac (2,023,500 m^2) disposal site.

The concept of long-distance piping of dredged material is not without problems. Objectionable dredged material, such as scrap iron and rock fragments greater than a few inches (1 in. = 2.54 cm) in diameter, must be segregated from material entering the pipeline. Thus, rehandling facilities will be required in most cases. Booster stations, pumps, power requirements, and extra personnel add appreciably to the cost of such systems.

B. Mechanical Transport Methods

Dipper dredges, bucket dredges (especially draglines), and ladder dredges are used occasionally for conveying dredged material into a disposal area, but are limited to very small dredging jobs, dredging of oversized debris, and to secondary tasks such as dike building and clearing out of rehandling basins on major jobs. Bottom dumping of scows has been used in filling a confined disposal site at Indiana Harbor in the Chicago District. There are two diked containment facilities in this harbor owned and built by the Inland Steel Company in fill extensions for its plant.

It may be possible to introduce management techniques into procedures for discharging dredged material to impoundment areas that will enable the treatment of oil and grease-contaminated supernatant to be more practical and economical. As an example, it may be found that oil and grease are partitioned in the upper level water and fine sediments of the dredge hoppers. In this case, it may be practical to discharge this portion of the hopper to sectioned parts of the disposal area having oil and grease removal equipment, whereas the remaining heavier sediments could be discharged to the main part of the disposal area. It may also be found practical to segregate dredgings known to be derived from a particularly contaminated river or harbor reach. Other management practices to enable practical segregation and treatment of oil dredging within the configuration of dredging transfer and containment technology currently employed are being investigated by Calspan Corporation and the Lenox Institute for Research Inc.

VI. DREDGE CONTAINMENT ALTERNATIVES

Confinement of dredged material is generally achieved by construction of diked disposal areas built on coastal marshes or into appropriate shore promontories or offshore islands. The major features of these areas are the dike itself and a sluice or sluices that allow excess water to be drained from the disposal area.

The shapes, heights, and composition of retaining dikes are generally dictated by containment capacity requirements, local availability of construction materials, and prevailing foundation conditions. There are many types of dikes. A large number of dikes have been built entirely with the soil and sediments borrowed from the containment area. More recently, well-designed dikes containing graded stone or slag and granular fill have been constructed, enabling higher dike construction and ensuring against dike failure. Many areas are confronted with poor foundation soil at containment facility sites. Available sites are normally marginal lands not economically suitable for private development. Foundation soils are commonly natural deposits of soft clays and silts of various organic contents. In many instances, disposal sites have been used for past unconfined disposal, and the dikes must be constructed on previously deposited unconsolidated dredged material. Dredge material often consists of fine-grained wet materials of poor engineering quality. Low shear strengths of natural and dredged materials can limit initial dike construction to heights of only a few feet. Dikes of greater heights can be attained through construction of incremental dike sections, which are normally built a short-time prior to disposal operations. As dikes are raised periodically, substantial heights can be achieved even on very weak foundations. This is possible because there is a gain in the shear strength of certain foundation soils as they drain and consolidate under the loading of dredged material during periods of inactivity. Dike raising is usually accomplished by incorporating the initial dike into new dike construction although, in some cases, interior dikes are constructed at some distance from the inside toe of the existing dike.

It is common practice to borrow materials from inside the disposal area for initial dike construction and for each dike raising, because these materials are economical to obtain. Consequently, the quality of dredged material may greatly affect ultimate dike dimensions and stability in two ways since the dredged material can be both foundation and the construction material. Because of the poor engineering quality of most dredged material, more suitable material has been borrowed in some instances from locations other than the disposal area.

A. Retaining Dike With Stone, Sand, or Carbon Filter

A new type of dike design that has been used on a small dredge disposal area in Buffalo is of special interest. In this design, as shown in Fig. 5,

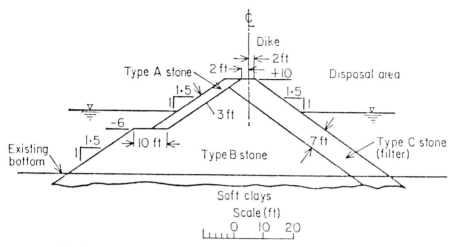

Fig. 5. Typical cross-section of the retaining dikes at the two Cleveland Harbor disposal facilities (Buffalo district) [12].

Material specifications

Type A Stone. The stone shall be graded from the coarse to the fine sizes. The maximum size shall be 1400 lb and the median size (i.e., the 50% size by wt) shall be approximately 400 lb. Stones weighing less than 90 lb each may comprise up to but may not exceed 15%, by wt, of the total mass of stones and will not be used in determining the median size. Neither the breadth nor the thickness of any stone shall be less than one-third of its length.

Type B Stone. The maximum-size stone shall be at least 50 lb. At least 50%, by wt, shall be in pieces weighing not less than 15 lb each, and not more than 15%, by wt, shall be in pieces weighing less than 5 lb each.

Type C Stone. The stone shall be reasonably well graded throughout the fall within the following limits:

Percent passing by wt	Stone size
100	125 lb
85–100	70 lb
50–70	5 lb
15–30	No. 4
10–23	No. 16
7–22	No. 40

the dike is constructed so that dredging supernatant water seeps through the dike and back to the lake. The dike contains well-graded stone on the containment area side (Type C stone filter) which filters out sediments as the water passes through. The presence of this filter makes the use of

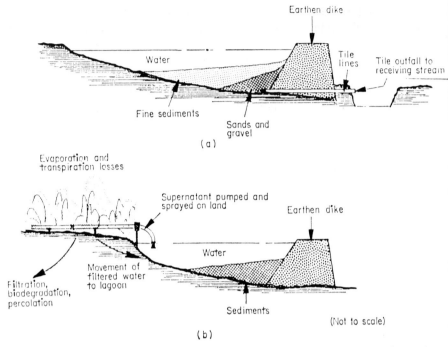

Fig. 6. Possible marshland creation schemes for hazardous materials disposal [12].

sluices unnecessary, although sluices are constructed in the event that the stone filter becomes plugged.

Sand or carbon bed filtration are other methods that should be considered during creation of diked disposal areas. Diked disposal area design and management modifications to accommodate differences in sediment characteristics must be investigated when proposing wildlife enhancement alternatives. Soil and sediments are possible filters for the removal of pollutants, as illustrated in Fig. 6 [36]. They can be considered as part of marshland creation schemes for fisheries and waterfowl enhancement alternatives, particularly if the possibility of hazardous materials must be overcome.

B. Retaining Dike with Sluices

Sluices are provided in some dredged material retention facilities to allow excess water of acceptable quality to be drained from the disposal area.

Sluice configurations vary from a simple outfall pipe placed atop or through a dike to large wood and steel-framed rectangular structures with multiple discharge pipes and stoplogs for an adjustable weir. Sluice design has received little attention in most ACE Districts. Attempts at standardizing sluice design and construction have been unsuccessful because little is known of sluice characteristics as they affect effluent quality. There is a lack of data on the effects of sluice design on discharge water quality including content of oils and greases.

The simplest sluice is a pipe placed horizontally within the dike near its crest. As the level of the slurry rises, the upper portion runs off through the pipe. No precise level control is provided by this type sluice, and thus detection time and effluent quality are not controlled. The outfall pipes may become plugged and allow enough pressure to build up on the dikes to cause a failure. They are seldom used in ACE operations and are limited to supplementary drainage through cross dikes within a large disposal area. A siphon is similar to an outfall pipe, but is equipped with a pump to start the effluent flowing through the system.

The most widely used sluice in ACE operations is the drop inlet. It consists of a rectangular wood- and metal-framed inlet or half-cylindrical corrugated metal pipe riser equipped with a gate of several stoplogs (usually 2 × 10 in. or similar sized timbers). The stoplog can be added or removed as necessary to raise or lower the level of slurry within the disposal area. A discharge pipe leads from the base of the riser through the dike to the exterior. Ideally, the discharge pipe extends beyond the exterior of the dike or into a catch basin to prevent scouring of the exterior slope. Various degrees of sophistication are achieved in this basic form by the addition of protected stilling basins or plunge pools and multiple-sided or "Y" gates.

A box sluice of flume consists of a timber structure built through the dike section so that it interrupts the dike line. The timber floor of the sluice structure forms the spillway along which the effluent escapes after topping the stoplogs in the weir gate. Box sluices or flumes are seldom used by the ACE because of the danger of failure and subsequent release of impounded materials.

It appears that the widely used drop inlet with stilling basins may be modified to enable oil and grease removal. Thus, some physical, chemical, and/or biological methods may be incorporated into these structures to remove oil. Such modifications should be investigated for feasibility efficiency and cost in the future.

C. Retaining Dike with Divider

New types of diked disposal sites, as mentioned earlier, are really filtering and percolation basins that hold the dredged materials, immobilize the pollutants, and filter water back into the environment. One suggestion that might be employed would be the use of a divider dike within the diked disposal site. This dike would divide the site into two or more portions that could be filled in alternate years. The use of divider dikes could become part of a two-phased plan to minimize wildlife disruption by allowing the filling of one side and allowing the other side to remain undisturbed. This of course has two advantages, a disruption of only one-half of the site while allowing the engineering and scheduling activities to continue, and it does not interfere with breeding or nesting activities. Another advantage would be to minimize the turbulence caused by the intermittent discharge of dredged materials into a diked disposal site, so that the settleable solids can be deposited without serious disturbance. Use of divider dikes, increased settling time through dredging schedule revision, aeration, immobilization of heavy metals or use of chemical precipitants or flocculants may all serve to improve water quality within diked disposal areas.

Other containment area management skills to promote natural dewatering of fine-grained dredged material are reported by the US Army Engineer Waterways Experiment Station [39].

VII. DREDGE TREATMENT ALTERNATIVES

Recognizing that sediment and water characteristics vary from harbor to harbor and region to region, it becomes necessary to consider these differences when evaluating dredge treatment and disposal alternatives.

The dredge disposal alternatives discussed in the last section can be classified as direct dredge disposal without treatment or with a negligible degree of treatment. Although dredge treatment is very costly compared to the simplest open-water disposal, partial treatment in conjunction with a beneficial disposal technique could result in the best solution for pollution control and environmental conservation. Accordingly, various processes available for dredge treatment are introduced here and discussed. Some of the processes may be impractical or very expensive for widespread application.

A. Processes Available for Sediment–Water Separation

The polluted dredged materials excavated from the bottom of harbors or rivers contain contaminants that can be broadly classified as suspended solids, dissolved organics, inorganic salts, heavy metals, and oil and greases. The amount of organic material in dredgings will normally be significantly less than organic content of domestic sewage. More than 90% of pollutants in the dredged materials are usually associated with settleable solids. Thus, preliminary treatment of dredged materials involving rapid and efficient sediment–water separation will be effective.

Gravity thickening (i.e., plain sedimentation), chemical coagulation, centrifugation, and flotation are common processes applicable to sludge–water separation. Because the density of the sediments in the raw dredged materials is much higher than that of conventional raw sludge encountered in water and wastewater treatment, gravity thickening is expected to be the most efficient and economical method for the desired sediment–water separation.

The sediments in polluted maintenance dredgings can be roughly classified as follows: (1) large particles down to fine sand (about 100 μm), which are generally inert and could be used as a clean, stable fill in diked areas; (2) smaller silt and clay particles containing volatile organics, oil and greases, and deflocculated inorganic solids originating from various pollution sources. This latter class of suspended particles, needing a longer settling time than that of large inert particles, can be separated by a specifically designed gravity thickening process. Thus incoming dredgings are subjected to a short-term settling to allow removal of large particulate solids. The supernatant may require further treatment.

In summation, preliminary treatment of raw dredged materials may include the following steps.

1. Separation or disposal of relatively clean dredged sand, gravel, and larger particles from raw dredgings during their discharges at the disposal or treatment site.
2. Accumulating and storing the bulk of the polluted fine solids in diked disposal areas. If stored in diked disposal area, solids will not need further treatment.
3. Treating contaminated water fractions (i.e. spoil water) of the dredgings before discharge to public waterways.

The separated sediments usually do not need any further treatment.

The sediments, however, can be stabilized by a new oxygenation–ozonation process [44] when necessary.

B. Processes Available for the Treatment of Spoil Water

The spoil water separated from the raw dredged material may contain polluted fine suspended solids as well as dissolved solids. It may or may not require treatment, depending on the degree of contamination and desired quality and use of the receiving waters.

Conventional facilities used for treating domestic wastes are generally designed to handle uniform loading rates with known and highly organic waste characteristics. Since dredging operations are characterized by periodically large inorganic loadings with highly variable pollutant contents, biological treatment systems feasible for conventional sanitary wastes might not be adaptable to treating polluted spoil waters. Treatment systems with a favorable and cost-effective combination of physical-chemical and biological waste treatment processes, nevertheless, appear necessary to solve the dredging water treatment problem.

Some of the potential chemical, physical and biological processes available for treating dredge spoil water are summarized in Tables 6 and 7. The capabilities, treatment efficiencies, and approximate operating costs of these unit processes are shown in these tables. Brief descriptions of these processes are given below.

1. Physical-Chemical Processes Available for Treatment of Spoil Water

Equalization is a process for averaging the variations in flow and composition of a liquid. Usually, a holding basin is used to provide a flow of reasonably uniform volume and composition to a wastewater treatment unit.

Screening, grinding, and comminuting are generally used as pretreatment tehniques for coarse solids handling and removal. The first two unit processes are self-explanatory. Comminuting (or comminution) is defined as a unit process of cutting and screening coarse solids contained in wastewater flow before it enters the flow pumps or other units in the treatment plant.

Centrifugation is a physical operation involving the separation of suspended matter from the mixture of water and suspended matter by centrifugal force. Centrifugation can also be used for oil–water separation in some circumstances.

TABLE 6
Efficiencies and Costs of Preliminary, Primary, and Secondary Treatments[a]

Waste constituent	Treatment processes	% Removal	Cost, ¢/1000 gal[b]
Coarse solids	Screening	90	<11.4
	Grinding and comminuting	—	<11.4
	Centrifugation	90	45–90
Suspended solids	Sedimentation	60	<11.4
	Dissolved air flotation	60–99	11–15
	Coagulation	80	11–45
	Microstraining	60	<11.4
Dissolved organics	Activated sludge	60	11–45
	Trickling filter	60	11–45
	Anaerobic contact	50	45–90
	Aerated lagoon	50	45–90
	Stabilization pond	50	45–90
	Carbon adsorption	70	45–90
	Adsorption flotation	95	>90
Refractory surfactants	Foam separation	90	11–45
Oil	Gravity separation	95	<11.4
	Dissolved air flotation	90	11–15
	Absorption	30–80	11–45
	Adsorption filtration	99	45–90
	Conventional filtration	30–90	<11.4
	Adsorption flotation	99	>90
Neutralization	Acid or base treatment	99	11–45
Bacteria, viruses	Chlorination	99	<11.4
	Irradiation	99	11–45
	Ozonation	90	<11.4

[a]January 1983 cost.
[b]1¢/1000 gal = 0.2642¢/m^3.

Sedimentation is also a physical unit process involving the separation of settleable matter from water or wastewater by gravity.

Microstraining (or microscreening) is a form of simple filtration for suspended matter removal. The minimum particle size removed depends upon screen mesh size and pattern of weave.

Chemical clarification is a process system consisting of the following steps: coagulation and flocculation, sedimentation, and filtration. Co-

TABLE 7
Efficiencies and Costs of Advanced Treatments[a]

Waste constituent	Treatment processes	% Removal	Cost, ¢/1000 gal[b]
Fine suspended solids	Sand filtration	70	2.2–11.4
	Microstraining	60	2.2–11.4
Turbidity	Dissolved air flotation	99	11–15
	Diatomaceous earth filtration	70–90	11–45
	Ultrafiltration	100	>90
Ammonia–nitrogen	Biological nitrification	90	6.3–45
	Ion exchange	90	11–45
	Ammonia stripping	85	11–45
	Breakpoint chlorination	99	11–45
Nitrate–nitrogen	Anaerobic denitrification	85	4.5–45
	Ion exchange	90	11–45
	Algae harvesting	50–80	11–45
	Carbon adsorption	20–80	45–114
Phosphorus	Chemical precipitation	95	11–45
	Ion exchange	90	11–45
	Biological uptake	30	11–45
	Dissolved air flotation	90	11–15
Taste and odor, trace dissolved organics	Carbon adsorption	95	45–90
	Dissolved air flotation	90	11–15
Soluble inorganics (heavy metals, radioactivity, salts, hardness)	Ion exchange	90	>90
	Distillation	95	>90
	Reverse osmosis	90	>90
	Chemical precipitation	20–95	11–45
	Freezing	80	>90
	Electrodialysis	90	>90
	Carbon adsorption	5–90	45–90
	Precipitate flotation	90	11–15
Color	Chemical clarification	80–90	11–45
	Carbon adsorption	95	22–90
Bacteria	Chemical oxidation	90–99	<11.4
Viruses	Reverse osmosis	90	>90

[a] January 1983 cost.
[b] 1¢/1000 gal = 0.2642 ¢/m³.

agulation is achieved by adding chemicals which, when mixed with the wastewater, produce small floc particles removable by subsequent sedimentation and filtration. Both suspended solids and colloidal solids can be removed by this technique. Alum coagulation and magnesium carbonate coagulation processes can be used for water purification.

Chemical precipitation is also a process system involving chemical feed-mixing, coagulation–flocculation, sedimentation, and filtration. The term chemical precipitation is used only if the primary concern of the process is removing dissolved impurities, not suspended solids. For instance, soluble phosphate or heavy metals can be removed by the lime precipitation process.

A new chemical clarification (or chemical precipitation) process system involving mixing, flocculation, flotation, and filtration has just been developed by the Lenox Institute for Research [41,42]. The new process system, called SANDFLOAT, is as effective as conventional chemical clarification (or chemical precipitation) process systems, but saves the capital cost by about 60% [41].

Sand filtration is a process for the removal of suspended solids from water that is passed downward through a filtering medium. The medium consists of a layer of sand, prepared anthracite coal, or other suitable material resting on a supporting bed of gravel or a porous medium. The purified filtrate is removed by an underdrainage system.

Diatomaceous earth filtration is a form of mechanical separation that uses a layer of fine filter media and diatomaceous earth built on a relatively loose septum to screen out the suspended solids. When the filter becomes clogged with deposit, it is back washed.

Granular carbon adsorption (or carbon filtration) is a unit process used to remove dissolved organics, oily substances and/or chlorine from water. A receptacle is filled with granular carbon, and the water being treated is passed through either fixed or moving beds. The exhausted granular carbons are usually regenerated and reused. The fixed bed operation is also termed adsorption filtration.

Powdered carbon adsorption is a process system used to remove mainly dissolved organics, oily substances, taste- and odor-causing substances, and suspended matters from water. The powdered carbon is first added to the treatment unit for adsorbing pollutants from water. Coagulants and coagulant aids are then added to the unit for removing the spent carbons as well as other suspended by coagulation, sedimentation, and filtration.

Foam separation is a unit process involving the selective separation of surface-active substances from water by a bubbling and foaming technique. Hydrophobic pollutants are attached on the rising gas bubbles' sur faces, and withdrawn as foam from the top of the separation unit, while the purified water is discharged from an outlet near the bottom. A surface-active agent can be added to the unit as foaming agent, if necessary. The foam separation process can be further subclassified to foam fractionation and froth flotation. The former involves the separation of dissolved pollutants from a homogeneous system. The latter involves the separation of suspended and dissolved pollutants from a heterogeneous system. The common process unit frequently used for foam separation is dispersed air flotation. When a cationic surfactant is used in foam separation process, bacteria can be killed by the added surfactant; thus no other disinfectant is needed.

Electrodialysis is a process by which charged species are separated by selective membranes that are usually made of chemically treated plastic. When water or wastewater containing mineral salt passes through the electrodialysis cell, the positive ions are attracted to a negative electrode, and the negative ions are attracted to a positive electrode. Selective separation of the charged species by the membrane thus occurs. The principle of this process, and the controlling flux, is the ionic transport of anions and cations through the membranes to opposite chambers in an electrodialysis stack. An applied voltage difference across the stack is the sole driving force.

Membrane filtration deals either with suspended or colloidal solids and separation occurs based on the physical size of the particular matter in relationship to the physical size of the pores contained with the membrane. Membrane with pore sizes down to 100 Å have been used. For this membrane process, the important aspect is that the separation represents a true filtration mechanism with water passing through the uniform pores of the membrane filter under a driving force of a modest pressure differential, about 10 $lb/in.^2$

Ultrafiltration differs from membrane filtration principally in that the material that is blocked by the membrane is now in solution. The impurities that are blocked are typically higher molecular weight organic materials (colloidal and/or dissolved) ranging in size from about 10 Å up. Thus, a tighter membrane is required wherein the pore sizes or the average diameters of the tortuous paths that exist within the plastic membrane will be substantially smaller than in membrane filtration. The driving force moving water through this membrane will also be applied pressure ranging from 10 to 600 $lb/in.^2$

Reverse osmosis is very similar to ultrafiltration except that a tighter membrane is required wherein the pore size or the average diameters of the tortuous paths that exist within the membrane are about 1 Å. Water is caused to pass from a feed solution of high concentration through the semipermeable membrane to a solution of much lower concentration (i.e., the opposite of natural osmosis). To cause this flow, a pressure of 600–1500 lb/in.^{2}, which is much higher than the osmotic pressure, must be applied to the feed solution. A reverse osmosis membrane is capable of rejecting inorganic ions as well as organic molecules above about 150 mw.

Dialysis is a technique for separating ions and low molecular weight organics in solution by diffusion through a membrane. One side of the membrane is the feed solution containing the impurities. The other side of the membrane is the dialyzate fluid. When large volumes of dialyzate fluid are used, a solute concentration gradient will be developed across this membrane. The impurities (i.e., ions and low mw organics) in the feed fluid being processed will then diffuse through the membrane and mix with the dialyzate fluid together. The feed fluid is thus purified.

Ion exchange involves the displacement of ions of given species from insoluble exchange materials by ions of different species when solutions of the latter are brought into contact with the exchange materials. The process can be used to remove ionic pollutants or impurities from water.

Aeration is a process by which air and water are brought into intimate contact with each other for the purpose of removing volatile substances or supplying oxygen.

Chemical oxidation converts the organic matter into carbon dioxide, water, oxides of nitrogen, and oxides of other elements, and thus will potentially remove the organics or kill the bacteria in water. Oxidation agents used may include molecular oxygen, ozone, hydrogen peroxide, chlorine, and potassium permanganate. When ozone and chlorine are used, the oxidation processes are termed ozonation and chlorination, respectively. Another chemical oxidation process, wet combustion, involves pumping organics-laden wastewater and air into a reactor vessel where the organic fractions undergo rapid oxidation at elevated pressure and high temperature. Recently, a new chemical oxidation process, oxygenation–ozonation, has been developed and proven to be effective [44].

Ultraviolet irradiation is a process of disinfection by which the water can be disinfected by exposure to the ultraviolet radiation emitted by a quartz-mercury vapor lamp. Such radiation destroys bacteria, bacterial spores, moulds, mould spores, viruses, and other microorganisms.

Freezing is a process for separating dissolved solids from water at freezing temperature.

Precipitate flotation involves the conversion of dissolved impurities, such as lignin and phosphate, to insoluble particulates by adding a precipitating agent, such as alum, in a flotation unit. The precipitated substances are then separated by the flotation technique.

Absorption is a process for oil–water separation or sludge–water separation. For example, a unit equipped with a hydrophobic sponge-type foam can remove oil from the water–oil mixture. If the sludge–water separation is desired, the unit equipped with hydrophilic foam should be used for absorbing water.

Ammonia stripping is a modification of the aeration process for removing gases in water. Ammonium ions in wastewater exist in equilibrium with ammonia and hydrogen ions. As pH increases to above 9 (preferably near 12) ammonia may be liberated as a gas by agitating the wastewater in the presence of air. This is usually done in a packed tower with an air blower.

Sorption is a unit process developed to remove various forms of phosphate from water or wastewater by a sorption column containing activated alumina or other synthetic sorbents.

Evaporation is a vapor–liquid transfer operation in which vapor is driven from wastewater by heat.

Carbonation and recarbonation involve the diffusion of carbon dioxide gas through a liquid to render the liquid stable with respect to precipitation or dissolution of alkaline constituents. The term "recarbonation" is used only if carbon dioxide gas is diffused through liquid to replace the carbon dioxide previously removed by the addition of lime.

Adsorption flotation involves the addition of powdered (or very fine granular) adsorbent(s) to the flotation unit where the added adsorbent(s) adsorb the dissolved and oily pollutants in the water. The spent adsorbent(s) as well as other suspended impurities in the flotation unit will be attached on to rising bubbles and eventually be removed from the overhead of the flotation unit. A flotation aid may be added to the system in an amount necessary to sustain flotation. The adsorbent(s) to be used may consist of activated carbon and/or any other insoluble adsorbents. The process has been used for industrial waste treatment, oil–water separation, and trihalomethane precursor removal and found to be very promising [43].

2. Biological Processes Available for the Treatment of Spoil Water

Biological wastewater treatment consists of those forms of wastewater treatment in which bacterial or biochemical action is intensified to stabilize, oxidize, and nitrify the unstable organic matter. Activated sludge, trickling filter, aerated lagoons, and contact stabilization processes are examples of conventional biological treatment methods. A number of biological processes, such as biological nitrification, anaerobic denitrification, algae harvesting, and so on are being used for advanced wastewater treatment. These are described briefly as follows.

Activated sludge process involves the agitation and aeration of a mixture of wastewater and activated sludge. The activated sludge are mainly bacteria, protozoa, and other microorganisms that consume the organic pollutants present in the wastewater under an aerobic environment. The sludge is subsequently separated from the treated wastewater by sedimentation. A new modified method named aerobic process modification can further reduce nitrogen and phosphorus by the proper selection and adjusting of food sources for the microorganisms.

A trickling filter consists of artificial bed of coarse material such as broken stone, clinkers, slate, brush, or plastic materials over which wastewater is distributed or applied in drops, films, or spray from troughs, drippers, moving distributors, or fixed nozzles, and through which the wastewater trickles to the underdrains, giving opportunity for the formation of zoogleal slimes that clarify and oxidize the wastewater. Generally, there are two types of trickling filters, slow-rate filters, and high-rate filters. A new type of trickling filter named FLOCOR, by Ethyl Corporation, uses a synthetic packing (made of rigid vinyl plastic) to replace conventional filter beds. The efficiency of waste treatment by this method is reported to be improved.

Contact stabilization process is a modification of the activated sludge process in which raw wastewater is aerated with a high concentration of activated sludge for a short period, usually less than 60 min, to obtain organics removal by adsorption. The solids are removed by sedimentation and transferred to a stabilization tank, where aeration is continued to further oxidize and condition them before their reintroduction to the raw sewage.

Lagooning is a process by which wastewater containing waste solids is placed in a basin, reservoir, or artificial impoundment for the purpose of storage, treatment, or disposal. There are two types of lagoons, the anaerobic lagoon and the aerated lagoon.

Anaerobic denitrification is a process used to remove nitrates from wastewaters. An organic matter such as methanol, ethanol, acetone, or organic sewage serves as a carbon source and the waste is placed in an anaerobic environment. Under these conditions, nitrates will be reduced by denitrifying bacteria to nitrogen gas and some nitrous oxide, which escapes to the atmosphere.

Algae harvesting employs the same principles as those in the biological removal of nitrogen and phosphorus compounds by conventional treatment processes. In specially designed shallow ponds, nitrogen and phosphorus compounds are converted to algal cell tissue at maximum sustainable rates.

The UNOX system, an oxygenation system developed by Union Carbide, is based on the combination of recent advances in gas–liquid contacting and fluid control systems with well-established air separation and wastewater treatment technology. The enriched oxygen is economically supplied by an air separation unit, and efficiently mixed and dissolved in the "mixed-liquor" of the activated sludge process. This direct oxygenation system has high efficiency and can reduce total treatment costs by up to 40%.

The Bio-Disc process, an Autotrol Corporation development, removes biodegradable pollutants by aerobic biological action. Closely spaced rotating polyethylene discs are partially submerged in wastewater. An aerobic biological growth develops on the surface of the discs and provides an abundant microorganisms propulation to treat the wastewater. Rotation of the discs brings the biological growth into contact with the organic impurities and aerates the wastewater. High efficiency and short retention time is claimed. The process is also called rotating biological contractor (RBC).

3. Processes Available for the Treatment of Leachate from Dredged Material in Upland Areas

The disposal of dredged material in upland areas (removed from tidal influence) is an attractive alternative to other disposal options. Because of the physical and chemical characteristics of dredged material, concern has risen about the environmental impact of leachates from such disposal areas on local water resources. The pH, orthophosphate, cadmium, copper, mercury, and soluble lead in dredged material leachates are not expected to present water quality problems. However, the levels of alkalinity, iron, manganese, lead, ammonia-nitrogen, nitrate-nitrogen, and zinc in such leachates, when exceeding selected water quality criteria, and in the absence of any dilution effects, could pose potential water quality problems. The leachate problems can be mitigated by installing artificial liners or underdrain leachate collection systems [40].

When dredged material leachates are collected, treatment may consist of recirculation, biological, and/or physical chemical treatment process systems that have been introduced in the last two sections.

VIII. QUALITY CRITERIA OF TREATED SEDIMENT AND WATER

It is desirable that polluted dredgings be stored at the removal or disposal site for some time prior to any treatment. One storage area should be provided as the surge basin (a holding basin in which variations in flow and composition are averaged) to equalize the treatment rate, and initially to separate the sediments from the dredged materials. Two approaches can be followed: (1) separation of settleable particles from the raw dredge materials regardless of their sizes, and (2) size-separation of relatively clean dredged sand, gravel, and larger particles, from the polluted finer particles. The former approach requires the total treatment of all concentrated sediments, regardless of their sizes; the latter requires treatment of the polluted finer particles to a form that will make them acceptable for use as a stable fill or for ultimate disposal.

The treated sediments or particles should be free from oil and grease, from toxic heavy metals, and from any substances that are putrescent, deleterious, or odorous. They should not cause any organic decomposition or groundwater contamination. More detailed criteria or standards defining the quality of the treated sediments will depend on geological characteristics of the ultimate disposal site.

To establish the degree of treatment required for the treated-water discharge, effluent quality standards must be defined. Based on effluent discharge standards recommended by the *Water Pollution Board (WPB), Department of Health, Columbus, Ohio,* and by the *International Joint Commission (IJC) for Lake Erie, Lake Ontario, the International Section of the St. Lawrence River, and the Connecting Channels of the Great Lakes,* tentative water-quality criteria of receiving waters have been compiled. Table 8 indicates the specific water-quality criteria recommended by the above agencies for application to Ashtabula River, Grand River, and Lake Erie. Five general guidelines applicable to the discharge of water to Lake Erie and its tributaries have also been suggested by the WPB and the IJC, as follows [11, 37]:

1. Freedom from substances attributable to municipal, industrial, or other discharges, and from agricultural practices, that will settle to form putrescent or otherwise objectionable sludge deposits.
2. Freedom from floating debris, oil, scum, and other floating materials attributable to municipal, industrial, or other discharges, or to agricultural practices, in amounts sufficient to be unsightly or deleterious.

TABLE 8
Specific Water-Quality Criteria for Ashtabula River, Grand River, and Lake Erie [11]

Parameters	Water quality criteria of the receiving waters	
	Ashtabula River and Grand River	Lake Erie
pH	6.0–8.5	6.7–8.5
Toxic substances	Not to exceed one-tenth of the 48-h median tolerance limit, except that other limiting concentrations may be used	
Iron		<0.3 mg/L
Dissolved oxygen	>5 mg/L calendar day; >4 mg/L at any time	>6 mg/L in the connecting channels
Microbiology (Coliform group)		<1000/100 mL Total coliforms and <200/100 mL fecal coliforms for the geometric mean of not fewer than five samples taken over not more than a 30-d period; free from other bacteria, fungi, and viruses
Taste and odor		Phenols not to exceed a monthly average of 1.0 µg/L; other taste- and odor-producing substances absent
Dissolved solids	<750 mg/L as a monthly average value; <1000 mg/L at any time	<200 mg/L
Phosphorus		Loading is limited to 0.39 g/m^2/yr
Radioactivity		Gross beta activity <1,000 pCi/L; radium-226 <3 pCi/L; strontium-90 <10 pCi/L

(continued)

TABLE 8 (*continued*)

Parameters	Water quality criteria of the receiving waters	
	Ashtabula River and Grand River	Lake Erie
Temperature	<35°C at any time; max. temp. rise at any time or place above natural temp. <15°C; max. limits of the water temp. are indicated in the notes[a]	No change that would adversely affect any local or general use of these waters

[a] Maximum temperature in °C during month:

Jan.	Feb.	Mar.	Apr.	May	June	July	Aug.	Sept.	Oct.	Nov.	Dec.
10.0	10.0	15.6	21.1	26.7	32.2	32.2	32.2	32.2	25.6	21.1	13.9

3. Freedom from materials attributable to municipal, industrial, or other discharges, and from agricultural practices, producing color, odor, or other conditions, in such degree as to create a nuisance.
4. Freedom from substances attributable to municipal, industrial, or other discharges, or to agricultural practices, in concentrations or combinations which are toxic or harmful to human, animal, plant, or aquatic life.
5. Freedom from nutrients derived from municipal, industrial, and agricultural, sources in concentrations that create nuisance growths of aquatic weeds and/or algae.

It is proposed that an acceptable treated-water quality standard be established. The recommended standard will take into account all major pollutant components that are characterized in typical dredged materials that may be expected in a wide range of dredging operations. Such a standard can then be applied to the assessment of technological and economic feasibilities of the various candidate treatment processes.

IX. ECONOMIC ASPECTS OF DREDGING TREATMENT

It has been estimated by the US Army Corps of Engineers that, in January 1983, the open-lake dredging disposal costs were about \$2.35/yd^3 (1 yd^3 = 0.7646 m^3), while the addition of the least costly treatment would probably add $10.50/yd^3 to the disposal costs at large harbors and $23.51/yd^3 at small ones. In this report, general aspects of dredge treatment and disposal are introduced from engineering and management points of view, but there is no intention to recommend an optimum dredge treatment/disposal system for all dredging operations. A desirable goal of a continuous program is to develop one or more technically satisfactory and environmentally compatible dredging and disposal alternatives that are economically acceptable.

Many waste treatment methods described here are potential candidates for treating dredged materials. The combination of various candidate unit processes will conceivably result in many technically feasible treatment systems. However, only a few of these systems may provide economically feasible solutions. For instance, incineration is a well-established process for the disposal of wet sludge. This process could be impractical and costly for the highly inorganic nature of most dredge sediments. According to Calspan Corporation studies [8, 9, 11], plain sedimentation in diked disposal area will usually provide adequate primary treatment. Additional secondary and tertiary treatment will certainly improve the quality of spoil water. The question is whether or not such additional treatment is necessary for preserving or enhancing environmental quality. If the answer is positive, then, to what degree must the spoil water be treated? This is a typical manager's problem. Knowing the capability of each candidate unit process, a project manager should be able to select several technically feasible treatment systems based on the receiving water quality standards and other government publications. Knowing the cost of selected feasible treatment systems, the manager can then choose the most cost-effective system for them.

Somewhat simplified cost estimations for the various candidate processes were listed previously in Tables 6 and 7. More extensive data will be needed to provide the cost-effectiveness information of candidate unit processes and to conduct the conceptual design of one or more promising dredging treatment and disposal systems.

X. CONCLUSIONS AND SUMMARY

Dredging operations are often needed for the improvement of a channel and/or harbor area. Unavoidably the water resources close to the dredging site will be contaminated to some degree. This chapter presents the main features and operational methods of new-work dredging and of maintenance dredging. The former comprises the improvement (i.e., deepening or widening) of a channel and/or harbor area mainly by removing stones and compacted sediments that were deposited through the geologic ages. The latter is employed mainly to remove the loose sediments that tend to fill up previously excavated channels and harbors. The clamshell dredge, the dipper dredge, the pipeline dredge, and the hopper dredge, are four common types of dredges that are used for river and harbor excavation by the US Army Corps of Engineers. Their functions, operations, and cost data are indicated.

In general, only the sediments excavated from maintenance dredging operations are polluted with obnoxious or toxic contaminants that are detrimental to the land, air, and water environment. The sources of pollution that are associated with the channel and harbor sediments may be categorized as being from municipal and industrial discharge, storm-runoff, agricultural runoff, soil erosion, and accidental spills.

Typical dredged materials from Ashtabula and Fairport Harbors in Ohio were collected and analyzed for investigation. It was found that the sediments varied widely in their physical and chemical characteristics between harbors and even at different locations within each harbor. Nevertheless, the pollutants in the dredged materials of a specific harbor (or channel) can be generally characterized. Settleability of the dredging samples was investigated in detail. It was observed that oil and grease and heavy metals were intimately associated with sediments and could be removed with settlement of solids.

The current dredging practices in the USA includes the disposal of dredged materials in open sea, open lake, along-shore diked areas, offshore diked areas, abandoned strip-mine areas, and other disposal sites. Because of the excellent settleability of the dredged materials, it is desirable that the polluted dredgings be stored on a disposal site for some time prior to any treatment. A disposal area can be used to create new land if satisfactory fill is produced. The satisfactory fill can be the nonpolluted dredged materials from the new-work dredging operations, or well-treated sediments from maintenance dredging operations. The treated sediments or particles should be free from oil and grease, from toxic heavy metals, and from any substances that are putrescent, deleterious, or odorous. They should not cause any organic decomposition or groundwater contamination.

The dredged materials can be separated into the sediment and the supernatant water (i.e., spoil water) by settling. From the dredging-settling data presented in this chapter, it appears that sufficient settling of dredging in a diked disposal area, followed by discharge of supernatant to Great Lakes water, will not contravene the proposed Great Lakes Water Standards, and so the goal of water resources conservation may be achieved.

XI. PRACTICAL EXAMPLES

Example 1

Estimate the evaporation of a diked disposal area for a day during which the following averages are obtained: water temperature = 15.56°C; air temperature = 26.67°C; relative humidity = 40%; wind velocity = 12.87 km/h; and barometric pressure = 736.60 mm Hg. It has been known that the following formula illustrates the dependence of evaporation on the factors just cited:

$$E = 0.0497 (1 - 0.52 \times 10^{-3} P_a) \times (1 + 0.167W)(V - v) \quad (1)$$

where E is evaporation in cm/d; P_a is the barometric pressure in mm Hg; W is the wind velocity in km/h; V is the vapor pressure at water temperature in mm Hg; and v is the vapor pressure at dew-point temperature of the atmosphere in mm Hg. The vapor pressure is a function of temperature that can be expressed by:

$$V \text{ or } v = 4.571512 + 0.352142T + 0.007386T^2 + 0.000371T^3 \quad (2)$$

where T is the temperature in °C.

Solution

$$V = 4.571512 + 0.352142 (15.56)$$
$$+ 0.007386 (15.56)^2 + 0.000371 (15.56)^3 \quad (2a)$$
$$V = [4.571512 + 0.352142 (26.67)$$
$$+ 0.007386 (26.67)^2 + 0.000371 (26.67)^3]$$
$$\times \text{(relative humidity)}$$
$$= [26.254617] \times (0.40)$$
$$= 10.502 \text{ mm Hg}$$
$$E = 0.0497 (1 - 0.52 \times 10^{-3} \times 736.60)$$
$$\times (1 + 0.167 \times 12.87)(13.237 - 10.502) \quad (1)$$
$$= 0.264 \text{ cm/d}$$

Example 2

Wang and Elmore [45] developed Eqs. (3), (4), (5), and (2b) for determination of the saturation concentration of dissolved oxygen (DO) in either fresh water or saline water:

$$DO_{sfn} = 14.53475 - 0.4024407T$$
$$+ 0.834117 \times 10^{-2}T^2 - 0.1096844 \times 10^{-3}T^3$$
$$+ 0.6373492 \times 10^{-6}T^4 \quad (3)$$

$$DO_{ssn} = DO_{sfn} + CL(-0.1591768 + 0.5374137 \times 10^{-2}T$$
$$- 0.1152163 \times 10^{-3}T^2 + 0.1516847 \times 10^{-5}T^3$$
$$- 0.8862202 \times 10^{-8}T^4) \quad (4)$$

$$DO_{ssp} = DO_{ssn}(P - V)/(760 - V) \quad (5)$$

$$V = 4.581148 + 0.3058575T$$
$$+ 0.1954036 \times 10^{-1}T^2 - 0.7095922 \times 10^{-3}T^3$$
$$+ 0.3928136 \times 10^{-4}T^4 - 0.5021040 \times 10^{-6}T^5 \quad (2b)$$

where:

DO_{sfn} = saturation concentration of DO in fresh water (negligible concentration of chloride) at normal barometric pressure (760 mm Hg), and any water temperature, mg/L

DO_{ssn} = saturation concentration of DO in either saline or fresh water at normal barometric pressure (760 mm Hg), any water temperature and any chloride concentration, mg/L

DO_{ssp} = saturation concentration of DO in either saline or fresh water at any barometric pressure, any water temperature and any chloride concentration, mg/L

T = water temperature, °C
CL = chloride concentration in water, g/L
P = barometric pressure, mm Hg
V = pressure of saturated water vapor at the water temperature, mm Hg

After a dredging operation, the dredged material was settled for 18 h at a disposal site for separation of heavy sediments. The water fraction (i.e., spoil water) of the dredgings was treated by dissolved air flotation (DAF) before discharge to a fresh water river. It was found that the DAF effluent would cause a DO drop of 1.2 mg/L at the river's critical location where water temperature was 29°C under normal barometric pressure (760 mm Hg). The receiving water's DO standard was greater than 6 mg/L at all times. Could the DO standard be met?

Solution:

DO_{sfn} = 14.53475 − 0.4024407 (29)
   + 0.834117 × 10⁻² (29)² − 0.1096844 × 10⁻³ (29)³
   + 0.6373492 × 10⁻⁶ (29)⁴

DO_{sfn} = 7.66 mg/L
DO drop = 1.2 mg/L
DO remaining in water = 7.66 − 1.2 = 6.46 mg/L

Yes, the receiving water standard on DO (>6 mg/L) could be met if the saturation concentration of DO_{sfn} was attainable.

Example 3

Suppose the dissolved air flotation (DAF) effluent in Example 2 was discharged into a saline tidal river under the following conditions: DO drop at critical location = 1.2 mg/L; water temperature = 29°C; chloride concentration = 5 g/L; barometric pressure = 700 mm Hg; and DO standard of river = 6 mg/L or higher. Could the DO standard of this tidal river be met?

Solution:

DO_{sfn} = 7.66 mg/L at 29°C
CL = 5 g/L
DO_{ssn} = 7.654594 + 5[−0.1591768
   + 0.5374137 × 10^{-2}(29) − 0.1152163 × 10^{-3}(29)^2
   + 0.1516847 × 10^{-5}(29)^3 − 0.8862202 × 10^{-8}(29)^4]
DO_{ssn} = 7.30 mg/L
V = 4.571512 + 0.352142 (29)
   + 0.007386 (29)^2 + 0.00371 (29)^3 (2a)
V = 30.04 mm Hg
P = 700 mm Hg
DO_{ssp} = DO_{ssn} (P − V)/(760 − V)   (5)
   = 7.30 (700 − 30.04)/(760 − 30.04)
DO_{ssp} = 6.70 mg/L
DO drop = 1.2 mg/L

DO remaining in water = 6.70 − 1.20 = 5.5 mg/L

No, the receiving water standard on DO (>6 mg/L) could not be met. It should be noted that V could also be determined with Eq. (2b), which is slightly more accurate, but applicable to computer analysis only.

REFERENCES

1. US Army Corps of Engineers (ACE), "Preliminary Report on Effects of Spoil Disposal at Ashtabula, Ohio," 1968.
2. US Army Corps of Engineers, "Dredging and Water Quality Problems in the Great Lakes," vols. 1 and 4, 1969.
3. US Army Corps of Engineers, Portland District, "Lewis and Clark Connecting Channel, Oregon, Draft Environmental Impact Statement," 1971, 11 pp.
4. US Army Corps of Engineers, Mobile District, "Draft Environmental Impact Statement-Grand Lagoon, Florida," 1971, 7 pp.

5. US Army Corps of Engineers, Jacksonville District, "Ft. Myers Beach Channel, Florida—Final Environmental Impact Statement," 1971, 37 pp.
6. US Army Corps of Engineers, Buffalo District, "Final Environmental Impact Statement, Diked Disposal Area Number 2," Buffalo Outer Harbor, Buffalo, NY, 1972.
7. US Army Corps of Engineers, Buffalo District, "Final Environmental Impact Statement, Diked Disposal Site No. 12," Cleveland Harbor, Cuyahoga County, Ohio, 1973, 26 pp.
8. R. P. Leonard, "Assessment of the Environmental Effects Accompanying Upland Disposal of Pollued Harbor Dredgings, Ashtabula Harbor, Ohio." Calspan Tech. Rept. CAL No. NC-5191-M-2, 1972, 31 pp.
9. R. P. Leonard, "Assessment of the Environmental Effects Accompanying Upland Disposal of Polluted Harbor Dredgings, Fairport Harbor, Ohio." Calspan Corporation, Buffalo, New York, Tech. Rept. CAL No. NC-5191-M-1, 1972, 51 pp.
10. W. Weber and J. C. Posner, "Release of Chemical Pollutants From Dredge Spoils." Technical paper presented at the Association of Environmental Engineering Professors Workshop on Environmental Impact and Linkages, Charleston, SC, 1974.
11. L. K. Wang and R. P. Leonard, *Env. Conserv.* **3,** 123 (1976).
12. L. K. Wang, and P. M. Terlecky, and R. P. Leonard, *J. Env. Mgmt.* **5,** 181 (1977).
13. J. E. Drilmeyer, and W. E. Odum, *Env. Conserv.* **2** (1), 39, (1975).
14. L. J. Paulson, T. G. Miller, and J. R. Baker, "Influence of Dredging and High Discharge on the Ecology of Black Canyon," Lake Mead Limnological Research Center, Las Vegas, Tech. Report No. 1, 1980, 59 pp.
15. US Army Corps of Engineers, "Gowanus Creek Channel; Brooklyn, New York; Review of Reports; Main Report and Environmental Impact Statement," Tech. Report, 1981, 81 p.
16. L. Tent, *Wasserwirtschaft* **72** (No. 2), 60 (Feb. 1982).
17. US Environmental Protection Agency, "Proposed EPA Regulations on Interim Primary Drinking Water Standards," Washington, DC, Technical Report 40 FR 11990, 1975, 9 pp.
18. US Department of Health, Education and Welfare, "Report on Pollution of Lake Erie and Its Tributaries; Part I, Lake Erie," 1965, 50 pp.
19. US Environmental Protection Agency (EPA), "Lake Erie, Ohio, Pennsylvania, New York Intake Water Quality Summary, 1971," 1972, 501 pp.
20. L. E. Scarce, "Survival of Indicator Bacteria in Receiving Waters Under Various Conditions," in *Proc. 7th Conf. on Great Lakes Research,* University of Michigan, Ann Arbor, Publ. No. 11, 1964.
21. P. M. Terlecky, Jr., J. G. Michalovic, and S. L. Pek, "Water Pollution Investigation: Ashtabula Area," US Environmental Protection Agency Report No. EPA-905/9-74-008, 1975, 145 pp.
22. R. S. Hansen, *Limnos* **4** (1), 3 (1971).
23. E. P. Odum, *The Conservationist* **15,** 12 (1961).
24. A. W. Copper, "Salt Marshes," in *Coastal Ecological Systems of the United States, A Source Book for Estuarine Planning,* Vol. 1, Odum, H. T., Copeland, B. J. and McMahon, E. A., eds. pp. 567–611. Institute of Marine Sciences, University of North Carolina, 1969.
25. R. B. Williams and M. B. Murdock, "The Potential Importance of Spartina Alterniflora in Conveying Zinc, Manganese, and Iron Into Estuarine Food Chains. In Proc. of the Second Nat. Symp. on Radio-ecology. USAEC, CONF-670503, 1969.

26. W. W. Woodhouse, Jr., E. D. Seneca, and S. W. Broome, "Marsh Building with Dredge Spoil in North Carolina." Coastal Engineering Research Center, US Army, CERC R-2-72, 1972, 28 pp.
27. E. P. Odum, *Fundamentals of Ecology,* 2nd Ed., Philadelphia, Saunders, 546 pp., 1959.
28. L. E. Cronin, G. Gunter, S. H. Hopkins, "Effects of Engineering Activities on Coastal Ecology," Interim Report to the Office of the Chief of Engineers, US Army, Mimeo. Rept., 1969, 40 pp.
29. J. Teal and M. Teal, *Life and Death of the Salt Marsh,* Little, Brown and Co., Boston, 1969.
30. J. R. Thompson, "Ecological Effects of Offshore Dredging and Beach Nourishment: A Review," US Army Corps of Engineers, Coastal Engineering Research Center, Misc. Paper No. 1-73, 1973.
31. D. S. McLusky, *Ecology of Estuaries,* Heinemann, London, 1971, 144 pp.
32. C. E. Warren, *Biology and Water Pollution Control,* Saunders, Philadelphia, 1971, 434 pp.
33. C. L. Prosser and F. A. Brown, *Comparative Animal Physiology,* 2nd ed., Saunders, Philadelphia, 1961, 688 pp.
34. C. L. Prosser, "Temperature: Metabolic Aspects and Perception." in *Comparative Animal Physiology,* C. L. Prosser, ed., Saunders, Philadelphia, 1950, pp. 341–380.
35. Economic Development Administration, "Industrial Park Project for the Swinomish Indian Reservation Skagit County, Washington, Final Environmental Impact Statement." Washington, DC, 1971, 58 pp.
36. P. M. Terlecky, Jr. and A. H. Kunzer, "Dredge Spoil Disposal Alternatives and Environmental Enhancement in the Great Lakes". Technical paper presented at Geological Society of America Annual Meeting, Dallas, Texas, 11-14 November, 1973.
37. US Environmental Protection Agency, "Proposed Criteria for Water Quality (Public Water Supply Intake)," Washington, DC, Vol. I, 1974, 425 pp.
38. US Army Engr. Waterways Exp. Sta. "Classification and Engineering Properties of Dredged Material" Technical Report D-77-18, 1977, 119 pp.
39. US Army Engr. Waterways Exp. Sta. "Containment Area Management to Promote Natural Dewatering of Fine-Grained Dredged Material" Technical Report D-77-19, 1977, 86 pp.
40. US Army Engr. Waterways Exp. Sta. "A Study of Leachate from Dredged Material in Upland Areas and/or in Productive Uses" Technical Report D-78-20, 1978, 428 pp.
41. M. Krofta and L. K. Wang, (1984). ASPE *J. Eng. Plumbing,* **0, 1,** 1.
42. M. Krofta and L. K. Wang, Civil Eng. for Practicing & Design Engrs. (1984). **3,** 253 Pergamon, NY.
43. M. Krofta, L. K. Wang, and M. Boutroy, "Development of a New Treatment System Consisting of Adsorption Flotation and Filtration" Lenox Institute for Research, Technical Report. LIR/10-84/7, 1981. 28 pp.
44. L. K.Wang, "Principles and Kinetics of Oxygenation-Ozonation Waste Treatment System" US Dept. of Commerce, National Tech. & Inform. Ser., Springfield, VA. Technical Repot, PB 83-127744, 1982, 139 pp.
45. L. K. Wang and D. C. Elmore, "Computer-Aided Modeling of Water Vapor Pressure, Gas Adsorption Coefficient, and Oxygen Solubility" US Dept. of Commerce, National Tech. & Inform. Ser., Springfield, VA Technical Rept., PB82–118787, 1981, 137 pp.

Subject Index

A

Aquifer, 183, 230, 241
  protection of, 241–252
  yield of, 183, 189
Assimilation ratio, 63
Atmospheric reaeration, 21

B

Benthic respirometer, 20
Biochemical oxygen demand, 11–18, 78
  carbonaceous, 11–15
  nitrogenous, 11, 15–18
Blaney-Criddle method, 162
BOD, *see* Biochemical oxygen demand
Bowen ratio, 159

C

Chemical control,
  of nuisance organisms, 270–274
  of organisms pathogenic to man, 263, 264
  of organisms pathogenic to plants, 268–270
  pollution from, 274–277
  of vector organisms, 264–268
Clamshell dredge, 448
Coliform die-off, 21–24
  rate, 23
Cooling ponds, 108–122
  design, 109–122
  design chart, 115
  equilibrium temperature, 112
  heat dissipation evaluation, 109–113, 116–119
  heat dissipation mechanism, 108
  surface heat transfer coefficient, 110
  surface temperature prediction, 113–116
Cooling towers, 123–136
  heat dissipation mechanism, 123
  problems associated with, 134
  types, 126–136

D

Deep-well disposal, 215–260, 349
  confinement conditions, 223

evaluation of well site, 222–229, 234
potential hazards, 229–234
receptor zones, 224
well designs, 217–222
Dewatering, 374–386
theory, 423–432
Diked disposal areas, 459–464
Dipper dredge, 449
Dissolved oxygen
deficit, 25, 72, 78
sag point, 26–27
theoretical, 19
Dredges, types, 448–450
Dredging, 447–490
containment, 467–472
disposal, 457–465
economics, 485–486
pollution from, 451–457
quality criteria, 483–485
transport, 465–467
treatment, 472–482
Dry cooling tower, 133
Dryers, 388–397
Drying, 373–446
curves, 404–410
lagoons, 385–386
rate, 402–404
theory, 400–432
thermal, 386–397
time, 413–417

E

Estuaries
mathematical models, 40–44, 46–47, 67–83
water quality, 40–44, 46–47, 61–106
Estuarine dispersion coefficient, 65
Estuarine number, 64
Estuary analysis, 40–44, 46–47
Eutrophication, 44, 193
Evaporation, 146, 157–161, 397–400
determining amount of, 158
theory, 432–440
Evaporators, 397–400, 436
Evapotranspiration, 147, 161–163

F

Flash dryers, 390
Flood routing, 172
Forced-draft tower, 130

G

Groundwater, 182–193, 226–229
direction of flow, 184

H

Hopper dredge, 450
Hydrograph, see stormwater runoff
Hydrologic cycle, 146
Hydrology, 146–157

I

Induced-draft tower, 131
Infiltration, 163
Injection wells, see Deep-well disposal
International Groundwater Modeling Center, 193
Isohyetal method, 154

K

Kjeldahl nitrogen, 16

L

Lake, also see Reservoirs
impact of pollution, 193–200
water quality analysis, 44–45, 48–51

M

Mechanical forced-draft tower, 130
Mechanical induced-draft tower, 131
Meyer's equation, 160
Moment method, 78
Multiple hearth furnaces, 395

SUBJECT INDEX 495

N

Natural draft atmospheric tower, 126
Natural draft wet hyperbolic tower, 128
Nitrification, 15–18
Nuclear fuel cycle, 296, 298–314

O

Oxygen consumption, 11–21
Oxygen demand,
    biochemical, 11–18
    sediment, 19–21
Oxygen sag equation, 24, 26

P

Percolation, 163
Pests, 261–282, *also see* Chemical control
Photosynthesis, 18–19
    rate of, 18
Pipeline dredge, 449
Precipitation, 148–156
    causes of, 149
    forms of, 148
    geographic variation, 152
    measurement of, 154

R

Radiation, maximum permissible levels, 284
Radioactive wastes
    disposal, 341–351
    management, 332–351
    sources of, 298–324
    surveillance, 352–356
    transport mechanisms, 324–332
    treatment, 336–341
Radioactivity, effects of, 285–290
Reaeration, 21
    coefficient, 63
Reservoirs, 139–214
    flood routing through, 172
    storage in, 173

Respiration rate, 18
Respirometer, benthic, 20
Retaining dike, 468
Rippl method, 180
River analysis, 24–40
Rotary dryers, 389, 417
Runoff, 164–182

S

Sandbeds, 375–385
Sediment oxygen demand, 19–21
Soil characteristics, 185
Spray dryers
Stream self-purification coefficient, 21, 26
Stormwater runoff,
    estimating, 168
    hydrograph, 167
    time of concentration, 165
    unit hydrograph, 169
Streeter-Phelps equation, 25, 28, 29
Subsurface hydrodynamics, 189, 226–229
Surface transfer coefficient, 63
Surface water quality analysis, 1–59
Surface water quality criteria, 56–57

T

Temperature correction factor, 15
    nitrification, 17
    sediment oxygen demand, 20
Theis method, 189, 226
    Jacob modification of, 192
Thermal discharges
    cooling of, 107–138
    impact of, 200–207
Thermal drying, 397
Thiessen method, 153
Tidal rivers, *see* Estuaries
Toroidal dryers, 394
Transmissibility, coefficient of, 190, 226
Transpiration, 161
Tray dryers, 388

U

Unit hydrograph, 169

V

Vectors, *see* Pests

W

Water quality criteria, 2, 56–59

Water quality model
   estuary analysis, 40–44, 46–47
   formulation, 3–6
   lake analysis, 44–45, 48–51
   river analysis, 24–40
   summary, 37–38, 46–47, 50–51
   two-dimensional, 36
Well injection, *see* Deep-well disposal
Wells, 187

HETERICK MEMORIAL LIBRARY
363.73947 W324   onuu
/Water resources and natural control pro

3 5111 00163 0726